COMMUNITIES AND MUSEUMS IN THE 21ST CENTURY

Communities and Museums in the 21st Century brings together innovative, multidisciplinary perspectives on contemporary museology and participatory museum practice that contribute to wider debates on museum communities, heritage and sustainability.

Set within the context of globalisation and decolonisation, this book draws upon bi-regional research that will enrich our understanding of the complex relationships between Europe, Latin America and the Caribbean through museum studies and practice. Chapters reflect upon the role of museums in defining community identities; the importance of young people's participation and intergenerational work for sustainability; the role of museums in local development; and community-based museums and climate change. Contributors examine these issues through the lens of museum partnerships and practices, as well as testing the continued relevance of the notion of 'integral museum' and its relatives in the form of ecomuseums. With its focus on regional museums in Latin America and Caribbean, this book highlights how the case studies promote greater intercultural dialogue, global understanding and social cohesion. It also demonstrates how the methodology can be adapted to other communities who are facing the perils of climate change and unsustainable forms of development.

Communities and Museums in the 21st Century proposes creative and sustainable strategies relevant to a globalised future. With its focus on global societal challenges, this book will appeal to museologists and museum practitioners, as well as those working in heritage studies, cultural studies, memory studies, art history, gender studies and sustainable development.

Karen Brown is Professor of Art History at the University of St Andrews, UK.

Alissandra Cummins is Director of the Barbados Museum & Historical Society, and lecturer and coordinator in museum and heritage studies at the University of the West Indies, Cave Hill.

Ana S. González Rueda is a researcher focused on critical approaches to learning in contemporary art curation.

ICOM Advances in Museum Research

This research series, developed by the International Council of Museums (ICOM), draws on the expertise of ICOM's worldwide network of museum professionals, representing a range of museum- and heritage-related disciplines.

Bridging theory and practice, the series addresses diverse issues of broad interest to the museum field and is of relevance for institutions around the world, featuring contributions by representatives of a range of cultures. Focusing on different types of museums and diverse fields of activity within the museum, the titles in the series will provide useful and thought-provoking insights for today's museum professionals. Its multi-perspective approach ensures its relevancy for academics, researchers and students of museology. The behind-the-scenes glimpses offered into the state of the field will also appeal to the general museum-going public.

The following list includes only the most recent titles to publish within the series. A list of the full catalogue of titles is available at: https://www.routledge.com/ICOM-Advances-in-Museum-Research/book-series/ICOMAMR.

The Future of Natural History Museums
Edited by Eric Dorfman

Communities and Museums in the 21st Century
Shared Histories and Climate Action
Edited by Karen Brown, Alissandra Cummins, and Ana S. González Rueda

ICOM international council of museums

COMMUNITIES AND MUSEUMS IN THE 21ST CENTURY

Shared Histories and Climate Action

Edited by
Karen Brown, Alissandra Cummins, and Ana S. González Rueda

Routledge
Taylor & Francis Group
LONDON AND NEW YORK

Designed cover image: Heritage Identity Workshop at the Túcume Site Museum. © Luis Repetto Málaga

First published 2024
by Routledge
4 Park Square, Milton Park, Abingdon, Oxon OX14 4RN

and by Routledge
605 Third Avenue, New York, NY 10158

Routledge is an imprint of the Taylor & Francis Group, an informa business

Funded by International Council of Museums (ICOM).

British Library Cataloguing-in-Publication Data
A catalogue record for this book is available from the British Library

ISBN: 978-1-032-29127-7 (hbk)
ISBN: 978-1-032-28841-3 (pbk)
ISBN: 978-1-003-28813-8 (ebk)

DOI: 10.4324/9781003288138

Typeset in Times New Roman
by codeMantra

CONTENTS

FIGURES AND TABLES

Figures

Tables

CONTRIBUTORS

Anne Bancroft is the Head of Conservation & Collection Care at the Barbados Museum & Historical Society. Anne has worked as a conservator in Barbados, India, Italy and the UK for international, national and community museums, libraries and archives. She has a focus in collection care in tropical environments. Her main area of research is on the conservation of sacred objects. She has been a guest lecturer on conservation and heritage programmes in the Caribbean, the UK and India, where she runs conservation/preservation workshops in different regions as a consultant with a focus on capacity building. She has authored/coauthored a number of articles and post prints including 'Worth a hundred Milibands'; Conservation's role in Embracing Cultural Identity at the V&A', ICOM-CC 16th Triennial Conference, 2011; "Minus 20 Degrees in the sun" in *Integrated Pest Management for Collections. Proceedings of the 2011: A Pest Odyssey, 10 Years Later,* English Heritage; and 'Hanging Sacred Cloth: The Practice of Displaying Thangkas', *Orientations magazine* 4/4, Sept. 2020.

Jamie Allan Brown is a research fellow at the School of Art History at the University of St Andrews. His experiences include working and supporting multidisciplinary projects across the Global South, his research interests include community heritage, sustainable development and youth participation in community-based museums. He previously led the bi-regional youth exchange between Latin America and Europe for the EU-LAC Museums project (EC Horizon 2020, 2016–2021), was Co-Investigator for the Community Crafts and Culture project (GCRF, 2019–2021) and will coordinate the research-led youth exchange between Scotland and the Caribbean for the Shared Island Stories project (UKRI 2022–2027).

Karen Brown is Professor of Art History at the University of St Andrews. She specialises in the role of community museums and heritage for addressing global issues, including social inclusion, well-being and climate action. Recent publications include the issue 'Museums and Local Development' of *Museum International* (2019) and the co-edited volume *On Community and Sustainable Museums* (2019). She has coordinated multiple national and international research projects including Shared Island Stories between Scotland and the Caribbean: Past, Present, Future (2022–2027) and EU-LAC Museums (2016–2021), and she is winner of the 2021 Europa Nostra Ilucidare Award for Heritage-led International Relations.

Catherine Anne Cassidy holds a bachelor's degree in Anthropology from the University of Central Florida, a MLitt (dist.) in Museum and Gallery Studies and a PhD in Computer Science from the University of St Andrews. She led the virtual museum design work in the EU-LAC Museums project and developed workflows for community-led digitisation efforts in Scotland, Europe, Latin America and the Caribbean, creating virtual tours, 3D objects and virtual museums. She is continuing this work through Northern Periphery and Arctic Programme–funded projects CINE: Connected Culture and Natural Heritage in a Northern Environment, PHIVE and HIVE, as well as Interreg North Sea Region Programme project CUPIDO. Catherine Anne brings an interdisciplinary approach to the research group Open Virtual Worlds, which develops emergent technologies for cultural and natural heritage organisations. Her doctoral research included developing strategies to 3D digitisation that allows the value of digital heritage to be recognised while strengthening connections between heritage, its community and the museum through emergent technologies and their democratisation.

Marie Claverie holds a *Maîtrise* in Art History from the Panthéon-Sorbonne University, Paris, and a Master of Engineering from the Ecole Centrale Marseille. She worked at ICOM as a project manager and coordinator (2012–2020). In the Programs and Partnerships Department, she coordinated partnerships, fundraising activities and digital projects aiming at supporting the fight against illicit traffic of cultural goods. From 2018, she contributed to the development of the newly created Museums and Society Department. Her work there aimed to support and promote the social and environmental role of museums around three thematic pillars: sustainability, peace and human rights, and cultural democracy and inclusion. More recently, she has worked as a cultural heritage officer for the French local government at Conseil Départemental de la Haute-Garonne (2021–2022).

Tracy Commock holds a Bachelor of Science in Botany from the University of the West Indies and a Master of Philosophy in Plant Systematics and Conservation from the Universities of Reading and Birmingham. She is currently Director of the Natural History Museum of Jamaica (NHMJ). She also served as the Natural Science Specialist for Jamaica's delegation at UNESCO's World Heritage Committee (2014–2017) and is a member of Jamaica's Scientific Authority for the Convention on the Trade of Endangered Species of Wild Fauna and Flora (CITES). She conducts botanical research and is currently a PhD candidate at the University of the West Indies, where she is researching a genus of plants endemic to Jamaica. She led the Museum Connect Project, coordinating the overall administration of the project in Jamaica.

Alissandra Cummins is Director of the Barbados Museum & Historical Society and lecturer and coordinator in museum and heritage studies at the University of the West Indies, Cave Hill. She specialises in Caribbean heritage, art and museums. She has served in a number of prestigious positions, including Founder President, Museums Association of the Caribbean; President, International Council of Museums; Chairperson, Executive Board of UNESCO; and Editor-in-Chief, *International Journal of Intangible Heritage*. She is currently President of the International Coalition of Sites of Conscience. Key publications include the co-edited and co-authored volumes *Curating in the Caribbean* (2012), *Plantation to Nation: Caribbean Museums and National Identity* (2013) and *Exhibiting Migration and Gender: Reflections, Response, Resilience-Companion Reader* (2020).

Bárbara Elmúdesi Krögh holds a bachelor's degree in History from the Pontificia Universidad Católica de Chile and a master's degree in Public Humanities from Brown University. She has led initiatives linked to the enhancement of cultural heritage as well as formal and informal education with public schools, libraries, museums, non-profit organisations and communities both in Chile and abroad, providing a vision that allows people and their communities to develop in a comprehensive and contextualised way in a fair and collaborative manner.

Javiera Errázuriz Contreras is a Journalist. She graduated from Universidad de Santiago and holds a master's degree in Rural Development from Universidad Austral. She specialises in researching, writing and editing content on culture, heritage, tourism and local development.

Laura Fúquene Giraldo holds a degree in Industrial Design from the Universidad Nacional de Colombia and a masters in Human Scale Development and Ecological Economics from the Universidad Austral de Chile, as well as a diploma in Cultural Management. She is a Research Assistant for the Chilean team of the EU-funded project 'EU-LAC Museums and Community'. She is currently a Project Formulation Coordinator at the Office of Relations and Cooperation of the Mayor's Office of Santiago de Cali, and Advisor to the Office of the Superintendence of Family Subsidies for the Colombian Pacific region.

Kaye Hall holds a Master of Education (MEd) in Social Context and Education Policy from the University of the West Indies (UWI) as well as a teaching certificate in Heritage for Human Resource Management and Training from the University of Florence. She currently works as the Education and Community Outreach Officer at the Barbados Museum and Historical Society (BMHS) where she manages its public education programming.

Sherilyne Jones is a doctoral student at the University of South Florida in Tampa, Florida. She earned her master's degree in Museum Studies from the University of Leicester, UK. As a museum professional from Belize, she has over 18 years of experience in archaeology and museum management. She was the former Director of the Museum of Belize, and her academic interests include multiculturalism and cultural diversity, museum anthropology, critical museum and heritage studies, and national narratives. Her current research explores issues and practices in heritage and its intersection with museums.

Kate Keohane is a Leverhulme Early Career Fellow in the History of Art at the University of Oxford. She completed her PhD at the University of St Andrews in 2020 as part of the Horizon 2020 project EU-LAC Museums. Her research centres on the interplay between text and image and the ways in which art can offer alternative models for being-in-the-world. She has published with Wasafiri and Afterimage, and has written for Tate, Art History and the International Curators Forum. She is currently working on two book projects. The first, 'Some Otherwhere: Edouard Glissant and the Caribbean in Contemporary Art', tests the limits of Glissant's relevance to art making, writing and curation. The second, 'Locating Common-Places: Artistic Practices for Existing Differently in a Damaged World', focuses on collaborative, site-specific artistic strategies of resistance against toxic and colonial forces.

Natalie McGuire holds a bachelor's degree in History of Art at the University of Leicester and a master's degree in Museums and Cultural Heritage at the University of Auckland. She is currently a PhD candidate in Cultural Studies at the University of the West Indies Cave Hill with a research focus on Anglophone Caribbean museums and community agency. She is the Curator – Social History and Engagement at the Barbados Museum & Historical Society and serves on several committees, including as Public Relations Officer for ICOM Barbados, and is a member of the board for the Barbados National Art Gallery.

Alan Miller is a lecturer in Digital Heritage and Communications in the School of Computer Science at the University of St Andrews. He holds a bachelor's degree in Politics from the University of York, a master's degree in Software Systems and a PhD in Computer Networking from the University of Glasgow. His research focuses on the use of emergent 3D and immersive technologies for the preservation of heritage and promotion of sustainable development with an emphasis on exploring climate futures through Virtual and Augmented Reality. Alan has extensive experience in applying immersive technologies to create heritage mobile applications, VR exhibit frameworks and virtual museum infrastructures. He has worked with museums and galleries in developing award-winning digital visitor-facing exhibits. Alan oversaw digital development of the virtual museum within the EU-LAC MUSEUMS project and the Northern Periphery and Arctic Programme–funded project CINE: Connected Culture and Natural Heritage in a Northern Environment, and he continues to do so for the Northern Heritage Network, a result of clustering digital heritage project outputs in the PHIVE project. He is part of the University of St Andrews research group Open Virtual Worlds and a founding member of the company Smart History.

Teresa Morales Lersch is a research professor in the National Institute of Anthropology and History of Mexico. From 1985 to the present, she has helped establish 27 community museums in the state of Oaxaca, together with her husband, Cuauhtémoc Camarena. They also helped create the grassroots networks, the Union of Community Museums of Oaxaca in 1991 and the National Union of Community Museums of Mexico in 1994. Since 2000, they have fostered the development of the Network of Community Museums of America.

David A.J. Murrieta Flores is a postdoctoral fellow at the Department of Art of the Universidad Iberoamericana (Mexico City, Mexico), under the supervision of Dr Ana Torres Arroyo. He holds a PhD in Art History & Theory from the University of Essex (UK), a master's degree in Art History & Theory from the same institution and a bachelor's degree in History from the National Autonomous University of Mexico (Mexico). As a postdoctoral researcher, he works on the collectives articulated around the 'little magazines' of Crononauta (Mexico), Rebel Worker and the American Situationist International (US).

Dionne Newell is Senior Research Officer of Entomology at the Natural History Museum of Jamaica (NHMJ). She holds a bachelor in Zoology and Botany and a masters in Zoology with a focus on Pest and Pesticides Management from the University of the West Indies. She is currently a PhD candidate in the Department of Life Sciences in the same university. She joined the staff of the NHMJ in August 1999. She was appointed Jamaica's National Focal Point for the Inter-American Biodiversity Information Network (IABIN) in 2005 and served as the Caribbean representative on the IABIN Executive Committee. She currently represents the Institute of Jamaica on several committees, including the Biodiversity and Game Bird Committee and the local Scientific Authority for the Convention on International Trade in Endangered Species of Wild Fauna and Flora (CITES). Under the Museums Connect project, she served as one of the local scientists training students in biodiversity assessments and species identification.

Cuauhtémoc Camarena Ocampo holds a degree in Social Anthropology from the Escuela Nacional de Antropología e Historia (ENAH) and has been a research professor at the Instituto Nacional de Antropología e Historia (INAH) since 1981. He was Director of the Museo Regional de Oaxaca from 1989 to 1992. In 1985, he began to collaborate with indigenous people for the creation of community museums in Oaxaca, along with his wife, Teresa Morales Lersch. They promoted the exchange of experiences and mutual support among communities, which led to the founding of the Unión de Museos Comunitarios de Oaxaca, A.C. in 1991. Over the course of 34 years, they have advised the creation of 27 community museums in the State of Oaxaca.

Luis Repetto Málaga (1953–2020) was a Peruvian museologist. He served as Director of the National Institute of Culture and of the Museum of Popular Arts and Traditions of the Riva-Agüero Institute. He also served as President of ICOM-Peru from 2010 to 2016. He was an ardent supporter of the role museums can have for sustainable community development and was one of the pillars of the EU-LAC Museums project.

Ana S. González Rueda is a researcher focused on critical approaches to learning in contemporary art curation. Recent publications include the co-edited volume *Decolonising the Curriculum* (2022), 'Disorienting the Gaze: Ngozi Onwurah's Early Films' (2021), and 'Possessing Nature: The Mexican Pavilion as a Site of Critical Analysis' in the *Journal of Curatorial Studies* (2020). González Rueda worked as a research assistant for the SFC GCRF Community Crafts and Cultures (2021) and the EC Horizon 2020 EU-LAC Museums projects. She currently teaches at the American College of Greece in Athens.

Allison Thompson PhD is an art historian, writer and curator living in Barbados. She has been a lecturer in the Division of Fine Arts at the Barbados Community College specialising in modern and contemporary art of the Caribbean, Africa and the African diaspora. She is co-director of PUNCH Creative Arena, an initiative for creative action and has worked with a number of cultural organisations in the Caribbean including the Barbados National Art Gallery, ICOM Barbados, and is the founding president of AICA Southern Caribbean, a regional chapter of the International Art Critics Association.

Natalie Urquhart is the Director and Chief Curator of the National Gallery of the Cayman Islands and the Cayman National Cultural Foundation. She holds a bachelor's degree in Art History, a Master of Arts in Arts Policy and Management, and specialises in arts sector strategic planning and development, programming and exhibition making. She is the author of the 'Art of the Cayman Islands' (Scala, 2016) and writes regularly about Caymanian art for national and regional journals. She is a former core committee member of the Tilting Axis Caribbean contemporary arts alliance (2016–2020) and board member of the Museums Association of the Caribbean (2016 – 2022), serving as MAC's president from 2017 to 2020. She currently heads the Visual Arts and Creative Industries Committee for the Cayman Islands National Culture and Heritage Policy Project and is the Creative Director of Cayman Art Week, an initiative she founded in 2021 in response to the impact of the pandemic on the creative sector.

Karin Weil is an anthropologist. She holds a master's degree from Universidad Austral de Chile and a diploma in Curatorship from the UAI with training in Climate Change and risk assessment for cultural heritage. She is currently coordinating internationalisation strategy at Prorrectoría and is in charge of cultural and natural heritage at the Río Cruces Wetlands Centre of the Universidad Austral de Chile. She has extensive experience in the management and coordination of projects related to the heritage of southern Chile. During her professional career and as an adjunct academic at the Universidad Austral de Chile, she has led interdisciplinary research projects, heritage management, community museography, curatorship and others, addressing various situated dimensions of the culture and communities of the southern south of Chile.

PREFACE BY ALISSANDRA CUMMINS

In his conception of a modern museum of the Americas as a place of continuous dialogue, mixing tangible with intangible, capable of conserving both diversity and creolisation, Édouard Glissant's idea of the museum as itinerant (living), rather than fixed and permanent, was conceived on the model not of a continent but of an 'archipelago as a network of interrelations between traditions and research, opening onto the unknown, locating itself at a sort of focal point of this universe…' and was intended to counteract the forces of homogenisation generated by globalisation. Fernando Ortiz Fernández's seminal thinking, particularly his profoundly influential concept of 'transculturation', embraced simultaneity, conflict, inherent contradiction and hybridity to grasp the complex transformation of cultures brought together in the crucible of colonial and imperial histories. For Kamau Brathwaite, the shared histories, languages and structures of feeling of the Caribbean islands formed the fabric of a submarine unity premised on creolisation and cultural diversity. Other Caribbean intellectuals have conceptualised the archipelago as an analytical framework for disrupting the notion of insularity and for thinking beyond linear narratives of historical, national and cultural development.

This volume directly addresses this deficit in global understanding and the significant gap in the available literature which results. It explores some of the complex issues arising from recent approaches to comprehending what the term 'shared history' truly means to collaboration between museums and their communities. Recognising and respecting the multiple perspectives of community participants is one thing, but how can museums incorporate this successfully into exhibitionary and educative practice? What are the strategies that were explored and deployed in the various crossdisciplinary approaches described by the museum practitioners in South and Central America, the Caribbean and Europe who, as both experienced and emerging professionals, have grown in their learning and have come to value their experiences of working closely with the audiences they serve? How to position themselves vis-à-vis the cultural specificities of diasporic societies, such as those of the Caribbean, has been a challenge acutely faced by remote and island museums and galleries, large and small, and this has been a driving question in the development of memory institutions and their relations to the communities and cultures that surround and encompass them.

The aim of this book is to gather experiences from small museums and remote communities, often silenced by academia and their authenticity and credibility ignored because they lie outside what much of the museum and art world persist in defining as the mainstream. In order to explore and explain in an equitable manner their perspectives with respect to working with communities, this book provides first-hand evidence of how both institutions and individuals work together to achieve legitimacy and sustainability through public education and audience engagement, through conservation of traditional culture, and cyber museology of cultural communities and diasporas seeking to re-connect. It also explores the new museum pedagogies they have begun to articulate to achieve such goals.

The November 2018 *Itinerant Identities* event, an international museum conference, co-hosted by the University of the West Indies and the Museums Association of the Caribbean with the funding support of the EU-LAC Museums project under Horizon 2020, provided the crucial underpinnings for the development of this publication. It aimed to provide a timely and interactive international platform to meet, discuss and debate museologies and intersecting disciplines through a range of engaging discursive and experimental gatherings. This book has drawn on the wide-ranging, interdisciplinary experiences and expertise presented at this gathering where museologists and museum administrators, art historians and archivists, curators and community leaders, exhibit designers and educators, critics and cultural theorists, came together in Barbados to interrogate the museum conditionalities of the past, with a view to informing the present and the multidisciplinary debate on the new frontiers of museums and community engagement. In this context, the gathering has inspired the future direction of these institutions in addressing the changing needs of their communities.

Nevertheless, the preceding decade since ICOM had last co-hosted the annual conference of the Museums Association of the Caribbean (MAC) with the support of the International Curators Forum (ICF) also provides an important frame of reference within which this book's emergence should be understood. That conference held in Barbados in 2009 produced two major results. It forged the beginnings of the bi-regional curatorial partnership entitled Black Diaspora Visual Arts (BDVA), which ultimately resulted in the generation and development of the *Arrivants* art exhibition discussed later in this book. The other legacy was the publication of some of that conference's papers within the first book on Caribbean museums – *Plantation to Nation: Caribbean Museums and National Identity*.

However, while that book focused primarily on revealing the histories and genealogies of the region's institutions, this volume has quite a different trajectory. The chapters in this book are based on contemporary museum practice, particularly through engagement with local and diasporic communities. The experiences outlined by institutional and academic specialists from multiple fields focus on their work's intersection with issues of museums and memory/history and heritage and in effect become 'a network of interrelations between traditions and research, opening onto the unknown'. The contents are therefore expected to stimulate debate, promote advocacy and provoke action, based on cutting-edge presentations, informed by intense dialogue and interaction.

A book like *Communities and Museums in the 21st Century: Shared Histories and Climate Action* could not have been possible without the cooperation and collaboration of several individuals and institutions, many of whom were instrumental in the writing of its content. The editors are particularly gratified that the technical, academic and funding support provided under the project EU-LAC Museums, which received funding from the European Union's Horizon 2020 programme under Grant Agreement number 693669, has resulted in this critically important resource, garnering significant recognition in 2021 for the project with the award of the ILUCIDARE Special Prize for Excellence in Heritage-led International Relations. Continued research has been supported by the UKRI Engineering and Physical Sciences Research Council grant number EP/X023036/1-Shared Island Stories Between Scotland and the Caribbean: Past, Present, Future (2022–2027). The Government of Barbados, most particularly Prime Minister the Hon. Mia Amor Mottley and then Minister the Hon. John King, with responsibility for culture, afforded the project a generous and sustainable environment which was essential for the fulfilment of many of its objectives. The enduring partnership between the University of the West Indies at Cave Hill, its Vice Chancellor, Principal and its Office of Research, and the Barbados Museum and Historical Society, its management and staff, as well as the emerging collaboration with the University of St Andrews in the UK, provided the fundamental framework for the conceptualisation and dissemination of new museological approaches included in this book. The Museums Association of the Caribbean (MAC) with the generous support of the Smithsonian's Museum of African American History and Culture provided crucial opportunities, particularly for Caribbean colleagues, to participate in the dialogues and discourses which have seen results within the pages of this book. The Barbados Community College's academic and student corps formed a crucial support team for the *Itinerant Identities* conference and the *Arrivants* exhibition, which provide a key backdrop to this book. Finally, our grateful thanks go to the International Council of Museums (ICOM) for supporting this publication, especially the commissioning editors, editor and indexer – Aedín MacDevitt and Antonia Ivo, Sashivadana Ambikadas and Averill Buchanan – for their patient assistance in its production. This preface has provided an opportunity to acknowledge all these parties for their contributions with grateful thanks.

FUNDING ACKNOWLEDGEMENT

This edited book has been supported by two main research grants: the European Union's Horizon 2020 programme under Grant Agreement number 693669; and the UKRI Engineering and Physical Sciences Research Council grant number EP/X023036/1.

We also wish to thank ICOM and Routledge for their generous editorial and production expertise.

INTRODUCTION

Museum communities/community museums

Karen Brown

This book explores case studies of community museums, ecomuseums, grassroots heritage organisations and their networks from Europe, Latin America and the Caribbean. It builds upon and complements the growing literature on the broad topic of 'museums and community' through its specific focus on museums that have been created from community action, respond to local challenges and are reliant upon local systems of governance. The volume is birthed out of the international conference *Itinerant Identities: museum communities/community museums* held at the University of the West Indies, Cave Hill, in November 2018. This was a major academic meeting for the European Commission Horizon 2020 research project European Union and Latin American and Caribbean (EU-LAC) Museums that brought together museum and university partners from the Caribbean, Chile, Costa Rica, Peru, Portugal, Scotland (project co-ordinator), Spain and International Council of Museums (ICOM) (http://www.eulacmuseums.net).[1] On this occasion, we partnered with the Museums Association of the Caribbean and launched the innovative exhibition of contemporary art entitled *Arrivants: Art and Migration in the Anglophone Caribbean World* at The Barbados Museum and Historical Society. These transatlantic events and the compilation of this volume are part of an ongoing process seeking to address imbalances in the discourses of museology and art curation that have to date occluded the Caribbean and Central American regions, and where possible we endeavour to give greater agency and voice to the communities represented. The conference title *Itinerant Identities* reflects Alissandra Cummins' point that the Caribbean is in essence a region where (virtually) everyone came from (virtually) everywhere else, whether voluntarily or by force, and the subtitle 'museum communities/community museums' seeks to reflect the problematics of terminology in the field while also reinstating the possibility of characterising a 'community museum' informed by transatlantic research.

DOI: 10.4324/9781003288138-1

In the past, scholars, including Hooper-Greenhill (2000), Weil (2002), Witcomb (2003), Watson (2007), Crooke (2007) and Davis (2011), have tended to refute the idea of a set definition for community in relation to museums. As Peter Davis concluded, the term 'museum community' is 'an almost meaningless expression' from a sociological perspective (Davis 2011, p. 36). Even so, in recent decades, a great many volumes on the topic have appeared by these authors and more including Karp, Lavine and Kreamer (1992), Golding and Modest (2013), Kadoyama (2018) and Allison (2020). The complexity of the field is largely owing to the multifaceted working relationships between museums as institutions and communities on the ground. For example, Sheila Watson notes the misunderstandings and assumptions that frequently arise between museums and their stakeholders when they attempt to work with and for communities (Watson 2007, pp. 8–12), while Elizabeth Crooke recognises the political side of museums, entwining her arguments with concepts of place, belonging and memory when she asserts that 'collectively we form a myriad of sometimes shifting communities [...] Nevertheless, we need communities in order to build our experiences and forge our identities' (2007, pp. 71–2). Almost a decade later, she reflects,

> the sustained interest in the concept of community has had a major impact on museum practice [...] it is not just a case of museums representing or symbolizing community; now it is museums forging community identity, altering community experiences, and improving community life.
>
> *(Crooke 2015, pp. 481, 486)*

At the same time, some recent scholarship has come to recognise that museums are not neutral spaces and is advocating for them to acknowledge the contestations surrounding their histories and current uses, calling on them to wake up from a state of 'sleep walking' to become more 'active', 'ethical' and 'mindful', especially in response to the climate crisis and its attendant issues for South-North relations (Sandell 2007; Newell and Wehner 2017; Janes and Sandell 2019).

This new volume will explore how community museums are gaining in recognition within this movement through the presentation of case studies from remote areas of Europe, Latin America and the Caribbean. The local actions presented are all, in their own ways, speaking to major societal shifts and global issues, such as climate justice, in a manner often under-recognised by museum support organisations and governments. For example, Chapter 6 on archaeology museums in northern Peru and sustainable development focuses on community-level responses to the impact of the cyclical El Niño weather phenomenon, which prompted museums to build community cohesion alongside sustainability of cultural heritage assets. Despite their contemporary relevance, remote museums that engage meaningfully with communities are nevertheless often systemically under-resourced and under-represented on the national and international stages, often precisely because of their small scale, lack of visibility, clear definition, constitution, conformity, museology research or inscription in national accreditation systems. This is one of the major gaps in knowledge that this book seeks to address. By seeking to *characterise* community museums in all their diversity, it brings to light their potentialities vis-à-vis global challenges, informing a new wave of what I shall refer to as 'ecological community museology' for the 21st century. This concept aligns with scholars such as Cameron (2014) and Wehner (2016) who have argued for the need to 'ecologise' museum work at large because of the ecological crisis facing the world (climate change, food and water crises, overpopulation, loss of biodiversity, species extinction and more), with profound impacts on humanity (Rockström et al. 2009; Jeffrey 2019). The potential for museums to be significant collaborators in local climate action is evident in several activities presented in this book, such as community-based exhibitions, education activities and citizen science in Jamaica (Chapter 7) and at the National Gallery of the Cayman Islands (Chapter 8).

What is a community museum, and why do they matter?

While this book advocates for the societal and environmental relevance of community museums, to define a 'community museum' in a universal way as part of this discussion is likely to prove an unhelpful endeavour. This is because discussions in this field can quickly become complicated – and contested – when scholars attempt to define 'museum' and 'community' and even more so when these words come together as 'community museum'. To begin by breaking down the term 'community museum' into 'community' and 'museum', one notes that the concept of 'community' itself is a vast topic of discussion in social science – at least 94 definitions have been identified, the only common characteristic being that they all involve people (Hillery 1955, cited in Barton and Goldsmith 2016, p. 25). When it comes to the term 'museum' itself, there are numerous definitions offered by dictionaries and museum support organisations, although in recent years, definition-making has proved a contentious and divisive task. From 2013, in ICOM debates on the revision of its Statutes, and through the formal process of the reimagining of the ICOM definition of a museum that has taken place since 2016, voices have clashed in a world structured and fractured by histories of inequality (ICOM 2019; ICOM Code of Ethics 2021). This friction is all too easily pitted as conservative versus inclusive museology, or Global North versus Global South. However, the process has also highlighted the fact that the world has firmly entered an age of museum activism, one in which a new generation is calling for a more diverse system involving community action and social participation to supplant what it sees as an outdated museum model inherited from the West (Cummins, Farmer and Russell 2013; Mairesse 2017; Brown, Brulon Soares and Nazor 2018; Brown and Mairesse 2018; Sandahl 2019; Brulon Soares 2020a). In Prague, on 24 August 2022, the Extraordinary General Assembly of ICOM finally approved the proposal for the new museum definition (ICOM, 2022). Calls for museums to become facilitators of community action and decolonisation are also gathering apace, led by museum support organisations, such as the UK Museums Association, who define decolonisation for museums as 'a long-term process that seeks to recognise the integral role of empire in museums – from their creation to the present day' (UK Museums Association 2021), and ICOM's International Committee for Museology (ICOFOM), which is investigating ways in which museology has been predominantly shaped by Western thought. Drawing on the thinking of Anibal Quijano, Walter Mignolo and others concerning the coloniality of power, ICOFOM Chair, Bruno Brulon Soares, has propounded that 'decolonising museology' is an active process of reclamation by subaltern groups whose museums were previously defined by experts and 'the hegemonic discourse of nation states' (Brulon Soares 2020b, 51; see also Mignolo and Walsh 2018; Quijano 2020; Brown, Brulon Soares and González Rueda 2022).[2] This line of thought is aligned with Chapter 1 in this book, whose authors, Cuauhtémoc Camarena Ocampo and Teresa Morales, based in Oaxaca, Mexico, have led the Red de Museos Comunitarios de América and created and promoted *museos comunitarios* in Latin America at grassroots level for decades. These museums usually do not receive State support, and their personnel are often volunteers with limited access to professional training, making these institutions at once vulnerable in economic terms, but potentially strong in terms of community-led sustainable development. This is because at their best, community museums are created from community need, curated from traditional knowledge and managed using accepted forms of local governance.

Attempts have also been made to provide practical guidance on the creation of community-led museums, including the American Alliance of Museums' *Museums and Community Toolkit* (2002) aimed at planning successful museum–community dialogues, and the *Manual para la creación y desarrollo de museos comunitarios* (Camarena Ocampo and Morales Lersch 2014), discussed in more depth later. *On Community and Sustainable Museums* (Brown, Davis and Raposo 2019) – another key output from the EU-LAC Museums project – contains landmark instructive chapters by Hugues de Varine, Morales and Davis, and a selection of case studies of community museums from countries partnering in the project. An initiative to map community museums online also originated from the EU-LAC Museums project, with researchers uploading a suite of short videos on community museums to a YouTube channel called 'Museos Comunitarios' and proposing an Observatory of Community Museums to policy makers for future funding by the European Commission. A recent book and online resource researched by Csilla Ariese-Vandemeulebroucke (2018) has correspondingly broken new ground in mapping community museum organisations in the Caribbean with an emphasis on the grassroots, and a new project, Shared Island Stories Between Scotland and the Caribbean: Past, Present, Future, is augmenting these investigations into ecological community museology from the grassroots and continues to advocate for community museums around the world.[3] A number of the museums featured in the present volume likewise grew out of collective action at community level, often as a way of rescuing heritage at risk of disappearance. For example, the Museum of Neltume in Chile discussed in Chapter 3, which began with one family making a collection concerning woodlands and subsequently engaging in collective memory work by narrating stories of dictatorship and its impact on their local population.

However, while community museums are gaining enhanced recognition in the 21st century, it is important to remember that they have been born from community need in different formats and in different parts of the world since the 19th and early 20th centuries. They include small local examples formed in the UK and USA, the Heimatmuseen in Germany, Open Air museums in Sweden and initiatives in Africa and Mexico (Chaumier, in Mairesse and Desvaillés 2011; Davis 2011, pp. 50–68). For example, in promoting sensitivity to local natural habitats and their value for people to study and enjoy, the English otologist and founder of Wimbledon Village Club, Joseph Toynbee's (1815–1866) thesis was that museums need not collect and display rare or remarkable objects, but rather 'the common objects of Nature' in the neighbourhood of the museum – in this case specimens found within a five-mile radius of a parish church of Wimbledon (Toynbee 1863). Arguing that what he called the 'New Museum' be first and foremost useful for society, the Director of Newark Public Library from 1902 to 1929, John Cotton Dana (1856–1929), created the Newark Museum in 1909 at a small, local scale because the 'museum of the old type [...] has hardened into a cake of ancient and outgrown customs' (Dana 1917, in Peniston 1999, p. 35).[4] William Noland Berkeley (1867–1945) similarly explained why 'small-community museums' are both feasible and very desirable, for their 'helpful service to every class of citizens' in small cities, towns and villages (Berkeley 1932, pp. 7–8). Such community-based principles arguably paved the way for the better-known Anacostia Neighborhood Museum, opened in Washington, DC in 1967. Described as 'probably the first really communitarian museum in the world' by de Varine, Anacostia was created as an African-American museum commissioned by the Smithsonian Institution where the founder, the Methodist Episcopal Zion preacher John Kinard, worked alongside the local community to create a museum focused on education for future generations, producing displays on issues facing the local residents such as life in prison (Kinard and Nighbert 1972; de Varine 2017a, p. 20).

Other initiatives, cited by de Varine, and also by Serge Chaumier in the *Dictionnaire encyclopédique de muséologie* (Mairesse and Desvaillés 2011), are first from Niger, where the Niamey 'modules of living culture' (de Varine 2017a, p. 19) saw seven principal ethnic and cultural groups living on five hectares of land, charged with the maintenance and interpretation (in the colonial language of French) of their own cultures, and second Mexico's Casa del Museo created in the 1970s. Overseen by the Director of the Mexican Museum of Anthropology, Mario Vázquez Ruvalcaba (1923–2000), the Casa del Museo was an experiment in decentralisation. Located in a peripheral area of the country, this museum sought to be more embedded in the community than traditional museums and to butt against the often elitist and rigid tropes of traditional museums. Museologist François Mairesse notes how it became a place of exchange and discussion of consciousness raising (as was the case with Anacostia) by bringing together awareness of social issues with ancient Mexican culture (Mairesse 2000, pp. 43–4). The 1960s and 1970s then saw the growth of community museology and ecomuseums as a movement heavily influenced by political, cultural and social forces, including environmentalism (Davis 2008; 2011, pp. 50–68; de Varine 2017a, pp. 24–5, 34–9, 55; 2017b).[5] It is no accident that just as ideas of sustainability and decolonisation are assuming increased urgency today in the face of climate change and calls for global social justice, so community museums are – 50 years after Stockholm's UN Conference on the Human Environment and the 1972 Round Table of Santiago de Chile – due to come into their own in addressing major societal and environmental issues for the 21st century (ICOM Resolution No. 5, 2019).[6]

Museos comunitarios *in Latin America*

Museos comunitarios thrived from the 1980s, often as a form of resistance against dominant regimes. Mexico saw the Declaration of Oaxtepec promoted by International Movement for a New Museology (MINOM) (see Davis 2008), and the Union of Community Museums was created in the Mexican state of Oaxaca in 1991. In 1993, the National Program for Community Museums was created, defining a community museum as one that is born in, created, run and managed by the community. The network Red de Museos Comunitarios de América was founded in 2000, and at its first meeting, it resolved to strengthen the museums located around Latin America (Camarena Ocampo and Morales Lersch 2016). Directed by Camarena Ocampo and Morales, this network maintains a carefully crafted and fixed definition of the *museo comunitario*, referring specifically to ideas of collective self-determination and memory that are crucial for Indigenous and ethnic contexts:

> A community museum is created by the community itself: it is a museum "of" the community, not generated outside "for" the community.
>
> A community museum is a tool for the community to affirm the physical and symbolic possession of its heritage, through its own forms of organisation.
>
> A community museum is a space where community members build a collective self-knowledge, fostering reflection, criticism, and creativity. It strengthens identity because it legitimises history and their own values, protecting the community's way of life inwards and outwards. It strengthens the memory that feeds their aspirations for the future. (What is a community museum?) (www.museocomunitario.org)[7]

Their practical *Manual para la creación y desarrollo de museos comunitarios* (2014) calls for the whole community to be involved in the decision-making processes of museum creation, in the gathering of the museum's collections and in the selection of the topics that are to be told. Through such involvement, it is the community's vision that is projected by the objects on display, and the authors argue in compelling ways, here and elsewhere, that the distinctiveness of *museos comunitarios* arises from their focus on 'telling a story, building a future' to bring about community self-determination (Camarena Ocampo and Morales Lersch 2019, pp. 38–53).

Museos comunitarios are important in Latin America for several reasons. Often these museums tell a different story from mainstream museums, being born from the grassroots and curated by local people using local systems of governance, especially in Indigenous territories. They are examples of Brulon Soares's reflections on the power of the subaltern in contemporary discussions about institutional power and control, and they offer a model for self-determination of Indigenous, ethnic and marginalised groups in the realm of tangible and intangible cultural heritage, which speaks powerfully to contemporary debates in decolonisation. They are represented in Chapters 1 and 2 of this volume through examples of museums and community empowerment from Mexico, Colombia and Costa Rica.

Towards a transatlantic community museology

The primary aim of this volume is therefore to advocate for the contemporary relevance of a global community museology rooted in the past but mindful of global issues, such as social justice, decolonisation and climate change, when considered through a transatlantic lens. The secondary aim is to expand the museum studies corpus on community museums, ecomuseums, critical museology or sociomuseology, in English language. In his landmark monograph *L'Ecomusée singulier et plurial* (2017a), de Varine observes that 'ecomuseology' does not exist as a discipline in academic research or university teaching and adds that 'the New Museology as a world movement and as a discipline different to traditional museology has been little studied' (p. 67).[8] His observation could be challenged by citing the teaching of museology in institutions, such as the Sorbonne in Paris, the University of Lisbon, the Reinwardt Academy in Amsterdam, the University of Newcastle upon Tyne and the Institute for Experimental Museology in the Federal University of Rio de Janeiro.[9] However, it may be fair to state that although the New Museology – including ecomuseology, sociomuseology and community museology more generally – has been explained by a number of scholars, such as Van Mensch (1992) and Peter Davis, who summarises it as a 'radical reassessment of the roles of museums within society' (2011, p. 62), these topics have so far been under-represented in the UK and Anglophone museum studies programmes.[10] This gap is also prevalent in the Caribbean region; an issue addressed in our closing Chapter 15 where Alissandra Cummins and Anne Bancroft present the history of museum studies teaching in the Anglophone Caribbean and suggest innovative ways forward, including prioritising museum conservation training for tropical climates, thereby combating systemic histories of dependency between the region, North America and the UK, in particular.

Museum communities/community museums: case studies

The chapters that follow are organised into two thematic sections. In **PART 1, 'Community Museums: Nurturing identities and resilience'**, thematic strands include: the role of community museums in the struggle for self-determination; the question of the role of museums in defining community identities; the importance of young people's participation and intergenerational work for sustainability; the role of museums in local development, reconciliation and healing; and community-based museums and climate change.

The opening **Chapter 1**, 'Community museums and decolonisation: reflections from the Network of Community Museums of America' by Camarena Ocampo and Morales Lersch, highlights the colonial context for asserting self-determination in community museum contexts in their networked museums from eight countries in Latin America. Drawing on the key thinkers and writers of the Latin American network of modernity and coloniality, including Quijano and Michael (2000), Mignolo and Walsh (2018), Castro-Gómez and Grosfoguel (2007), the authors frame the situation for how museum communities view themselves in the context of a colonial order imposed by a European cultural imaginary. Linking with the ideas of Linda Tuhiwai Smith of New Zealand (2012) in relation to Māori people, they also tackle the epistemic problem of incorporating subaltern knowledge into the processes of knowledge production to enhance self-determination, well-being and self-worth. Emphasising their own network's practices in creating community museums since 1985, the authors outline how they engage Indigenous communities in a collective methodology in the construction of knowledge through existing community assemblies' frameworks. The chapter is instructive in demonstrating methods used by different communities, including community workshops for shaping display narratives, intergenerational transmission of knowledge through oral history and research questions, as well as disentangling their stories from colonial narratives and enriching self-worth that resists a dominant imaginary and enables new ways of seeing self through their collective processes. In **Chapter 2**, 'International collaboration between ecomuseums and community museums: the experience of the EU-LAC Museums Bi-Regional Youth Exchange in fostering identity, building community sustainability and resilience', Jamie Brown and Karen Brown explore some of these ideas through a specific case study of a transnational youth exchange that involved community museums in the network discussed in Chapter 1, as well as the ecomuseum of the Isle of Skye in Scotland, and several others in the northern Porto region of Portugal. This case study draws attention to the essential value in investing in young people, with one of the core methodologies used being intergenerational transmission of knowledge, and another being community mapping, in which the young people recorded significant sites, resources and other places of importance for local identity and tradition. Through a detailed account of the process of running and disseminating this youth exchange, the authors suggest the value of such cultural exchange between youth not only to the local communities involved but also as a model for future projects.

Chapter 3, 'Passion as a mobilising tool for community-based museums: case studies from Southern Chile' by Karin Weil, Bárbara Elmúdesi, Laura Fúquene and Javiera Errázuriz also takes a case study approach to community-based museums, this time in the context of the historic Round Table of Santiago de Chile of 1972, and focusing on the social role of museums. Somewhat like Camarena and Morales, their findings highlight the role of community-based museums in telling uncensored stories and communicating memories from within the community itself, ultimately creating a sense of belonging. The chapter draws significant attention to the setting for the Santiago Round Table, which took place over ten days, convened by the government of President Salvador Allende. For the authors of this chapter, the conditions for community-based museums go beyond the sharing of attributes, to be conditional on the strength of connection between members of the community, who are 'doing something' in an active way together. Case studies include the Museo Comunitario Despierta Hermano de Malalhue, founded to tackle discrimination against Mapuche children, the Museo Escolar de la Aguada, created in response to environmental conflict, and the Centro Cultural and Museo y Memoria de Neltume, developed in the context of human rights violations. These museums are presented for their roles as 'activist museums' in offering safe spaces for reflection against dominant powers and hegemonic institutions, for processes of overcoming trauma enacted at community level, to 'perfect the art of living, not that of progress'. Similar issues are at stake in **Chapter 4**, 'Museums and Community Engagement in Belize: case studies for protection and active participation of knowledge' by Sherilyne Jones. Focused on a country located culturally in the nexus between Central America and the Anglophone Caribbean, Jones explores the role of museums as an enabling tool for the exploration and expression of identity and collective histories in an independent nation (since 1981) that includes a variety of ethnic roots. She traces the development of the Belizian network of Houses of Culture – small, local museums that are cultural spaces in district towns aimed at empowering communities through shared authority to preserve, transmit and promote their culture, such as Garifuna drumming. This expansion of the definition of community museums is then developed in **Chapter 5**, 'The EU-LAC Museums project and community-based museums', by Karen Brown, Marie Claverie and Karin Weil. The chapter reflects on the terms of the EU-LAC Museums project and presents the results of the project's international survey entitled 'What is a community museum in your region?'. By 19 April 2021, this survey had gathered 528 responses from 70 countries written in English, Spanish, Italian, Portuguese and French and yielded some fascinating insights about community museums, their governance and value for local communities within and beyond the EU-LAC regions. From these results, the emphasis is found to be on 'people', 'place', 'space', 'culture' and 'future' as holding the key characteristics of the community museum.

Furthering the discussion around museums, community and sustainability is **Chapter 6**, 'Museums as tools for sustainable community development: a study of four archaeological museums in northern Peru' by Luis Repetto Málaga and Karen Brown, which transitions the book towards an ecological community museology. This chapter communicates the processes and outcomes of the Peru case study of the EU-LAC Museums project, which worked with four community-centred archaeological museums on Peru's northern coast. The chapter also grounds itself in the 1972 Round Table and emphasises a principle of decentralising museum focus to the peripheries but moves on to discuss ways in which museums can become a resource for local cultural, educational and economic development, through territorial management and international tourism. The former was most significant for the north-coast populations, when the severe El Niño flooding that hit in February 2017 drew attention to the need for museums to get involved in territorial management, supported by the University of València in Spain. One of the ways to do this was by reviving popular traditions, such as *chicha de jora* making (a traditional alcoholic drink made from maize). Herein, sustainability is framed as necessitating involvement of community members and support for them, while also highlighting areas where there has been a disconnect between the museums and certain communities. A similar focus on engaging local populations with heritage organisations for tackling climate change issues is found in **Chapter 7**, 'Connecting museums through citizen science: Jamaica/US partnership in environmental preservation'. Herein, Tracy Commock and Dionne Newell present the case study of the project 'Citizen-Led Urban Environmental Restoration', which saw young citizen scientists (aged 14–18) in Jamaica and the USA work closely with scientists from the museums to restore two environmentally degraded urban sites. The chapter reflects on the benefits and effectiveness of education outreach outside of traditional museum walls. It also offers possible solutions and methods to improve the efforts of museum professionals in natural history and engage citizens in environmental restoration in urban spaces.

Another case study from the Caribbean closes this section on museums and climate action: **Chapter 8** by Natalie Urquhart, 'Evoking wonder to inspire action around climate change – a collaborative exhibition project in the Cayman Islands', transitions the volume towards a focus on participatory curatorial practice. Taking the stance that museums are among the most trusted institutions, the chapter presents a compelling case for their role in bringing about positive change in climate action, especially in the context of islands, which are among the most vulnerable places in the face of changing climate effects. An effective way in which this can be achieved, argues Urquhart, is by inspiring wonder through art installation, using waste as materials. By creating visitor experiences that trigger not despondency, but positive reinforcement and action, museums can make a difference. The example provided is the National Gallery of the Cayman Islands' multidisciplinary collaborative exhibition *Coral Encounters*, which drew on macro photography to consider the wonders of underwater nature and coral health in the reefs, together with science-fiction inspired colours and grids to create impact and engage a wide range of audiences, including schools and families.

The second section of this book, **PART 2 'Connecting Regions: Communities and museums co-curating heritage and memory'**, examines similar issues through the lens of contemporary museum partnerships and practices, as well as testing the continued relevance of the notion of the 'integral museum' and its relatives in the form of ecomuseums. Herein, a series of case studies present diverse opportunities and approaches that offer insights into communities partnering with museums through co-creation and co-curation strategies to enable bi-regional action. This section focuses largely on exhibition development, both from Caribbean and Latin American perspectives and for Caribbean and Diasporic audiences: studying the potential of exhibitions by examining the contribution of multivocal, co-curatorial methodologies to the development of a distinctively Caribbean approach to exhibition-making; or reflecting on the design and architecture of the Virtual Museum of Caribbean Migration and Memory (VMCMM).

Opening the section, Natalie McGuire's **Chapter 9**, 'The case for a rhizomatic research approach in Caribbean museology' focuses on community inclusiveness in museum theory and practice by putting community voice front and centre. Basing her position in the theories of Martiniquan writer and philosopher Édouard Glissant's *Poetics of relation* (1990), McGuire sees the simile of the rhizome as paradigmatic for processual community-focused museology in the Caribbean region, a region with multifaceted identities. The author advocates for knowledge-*sharing* (rather than knowledge-*collecting*) as a counteraction to top-down exhibition education practices perpetuating colonial legacies and occluding histories and voices of the local people, and particularly the Afro-Caribbean experience. Through a rhizomatic approach, she argues, multivocality can 'de-linearise' authority within meaning-making. The final section of the chapter then reflects on the value of regional networking, especially through digital tools for the contemporary era, while **Chapter 10**, 'Co-curating memory: deconstructing the silences around Caribbean migration to Britain' then illustrates some of these ideas through a case study authored by McGuire and Kaye Hall. Herein, they document how, as part of the EU-LAC Museums project, they collaborated to facilitate a community-led composite history of post-World War II Caribbean migratory experience to Britain, and its role in multi-regional exchanges. It traces the development of a VMCMM and panel exhibition titled 'The Enigma of Arrival: The Politics and Poetics of Caribbean Migration to Britain', both funded by the EU-LAC Museums project, as ways of mitigating silences or gaps in telling migration histories by Caribbean migrants to Britain. Participatory methodologies in exhibition-making are framed within recent discourse led by Nina Simon whose work influenced exhibition practice by encouraging a non-hierarchical approach to learning and exhibition-making in the curatorial framework. In addition to targeted exhibition research, an open call in the Caribbean region and its diaspora asked people to contribute their memories, stories and unique perspectives on the Windrush migration. This process of co-creation aimed to create meaning through engagement with relevant communities; a process enabled by the VMCMM and discussed in the next chapter. **Chapter 11**,

'A case study of community virtual museums in the age of crisis designing a virtual museum of Caribbean migration and memory' by Catherine Cassidy, Alan Miller and Alissandra Cummins, explains in more depth the technical development of the VMCMM discussed in Chapter 10. The framework developed, using Omeka open source software, brings together 3D models, 360-degree tours and migration stories relating to the Windrush story. The chapter argues for the value of telling stories, such as Windrush scandal and survivals in digital format, in response to the growing global trend of Internet connectivity and usage. This resource was tailored to cater to a wide range of digital capacity and literacy within resource restrictions and includes an upload facility for users to share their stories. The efficacy of the tool is further underlined through the impact of the COVID-19 pandemic on museums and heritage sites. Considered together, Chapters 7–9 contribute to this book's overall aim to inform the shift in museology towards cultural decolonisation by supporting community participatory initiatives to reclaim their histories outside of narrow, national narratives, and their underlying colonialist, imperialist assumptions.

In **Chapter 12**, 'Ecomuseology in artistic practice: post-colonial strategies of collective return in Latin America and the Caribbean', Kate Keohane then creates bridges between ecomuseology and art curation by centring her theorising in the work of Glissant, and drawing synergies with recent discussions in ecomuseology notably by Pappalardo. In so doing, she interprets ecomuseums as embedded in a landscape in a way capable of re-activating memory and difficult narratives in the realm of contemporary visual arts. Her case studies are Fresh Milk residency (Barbados) and Semillero Caribe (Mexico and Cali, Colombia), and the BetaLocal collective (Puerto Rico). Keohane's chapter connects back to McGuire's, through its focus on imagined networks and alternative forms of community-making in the region. By focusing specifically on what she calls artistic ecomuseological practices, Keohane seeks to draw art history and museology closer together in the realm of participatory practices relating to landscapes 'damaged by the effects of colonisation'. For example, through Annalee Davis's art practice themed on pre-colonial seeds in Barbados, or the group experiences of the Semillero programming designed to avoid colonially implicated strategies of knowledge dissemination in the context of landscape and the diaspora, or BetaLocal's initiatives around 'un-learning' outside formal education spaces with a view to confronting difficult heritage. Expanding on this discussion around contemporary art and its display but looking specifically at the context of Mexico is **Chapter 13**, 'Exhibition-making as storytelling: the 14th Fomento Económico Mexicano S.A.B. de C.V. (FEMSA) Biennial in Michoacán Mexico' by Ana S. González Rueda and David A.J. Murrieta Flores. Investigating the roles of Mexico's modern and contemporary art in national history, and the stories that challenge and unsettle established narratives, it focuses on the ways in which *Inestimable azar* (Inestimable chance), the 14th FEMSA biennial (February 2020–February 2021), based in the Mexican state of Michoacán, decentralised established curatorial positions. The analysis is situated in relation to Mexican muralism of the 20th century, challenging official discourse through storytelling in the context of the biennial as a space located outside Western modernities and the dominant neoliberal order. Identity and homogenising nation building had been propagated among the early 20th century muralists by myths and images illustrating key periods in the nation's history leading to the eventual liberation of Indigenous peoples. In the context of the biennale, the authors present close readings of selected artists' works to challenge and revise this system, by drawing attention to the agency of Indigenous groups in contemporary mural and art-making processes that craft counter-stories in response to their erasure.

Chapter 14, 'Centring the Caribbean in the Global: Exhibiting Caribbean Art from a Caribbean Perspective' by Allison Thompson then tackles the geopolitics of art curation, presenting the exhibition *Arrivants: Art and Migration in the Anglophone Caribbean World* (2018) as a case study that moves beyond the familiar trajectory of exhibiting Caribbean art in Europe or North America. Drawing attention to the role of curating for a Caribbean audience, this exhibition is highlighted as paradigmatic for curating regional art from within, and of appeal to both local and international displays and audiences, while also informing new discourse on contemporary Caribbean visual practice. Closing our edited volume is **Chapter 15**, 'The politics of change: new pedagogical approaches to Caribbean museology, conservation and curatorship' by Alissandra Cummins and Anne Bancroft, which maps a seminal history of museum studies in the Anglophone Caribbean region since the 1990s in the frame of the decolonisation of museum practice and theory, with particular focus on museum conservation and preservation as it relates to resource-limited collections housed in tropical climates. This fascinating trajectory is accounted for within the ecosystem of Caribbean heritage support organisations and periodic recommendations and actions, including the first Artifacts, Museums and Archives course initiated in Jamaica in 1992. However, the chapter makes the point that the courses provided to the present day do not include specialist conservation training to equip professional collections care management. This is a major shortcoming in addressing the specific professionalisation needs of the region that has only recently begun to be addressed, notably through the formation of The University of the West Indies's Caribbean Heritage Network, as well as online instructional training. This training, while useful and timely in the present day, is limited in its ability to train professionals for object intervention because it is not based on experiential learning alongside experts in the field. Moreover, in-situ intervention must be invested in for the region in order to dismantle dependencies on outside countries where the expertise and laboratories are located, and to enable linkages between the training and personal experiences in locations increasingly affected by seasonal hurricanes and growing climate crises/crisis events.

The aims and uses of this book

This book makes island and remote communities the focus of an international museological book for the first time. It focuses on partnership in co-creation as integral to the development of museums as generative rather than recipient knowledge centres. Far from being exhaustive, it will have achieved these aims if it becomes a catalyst for further research and discussion within this interwoven field of investigation and informs museum studies pedagogy. It is anticipated that the contents will appeal to museologists and museum practitioners interested in a broad range of critical issues facing heritage, museums and galleries today, including migration, the role of new technologies, sustainability and social inclusion. With its focus on global societal challenges, this book will also appeal to scholars of heritage studies, cultural studies, memory studies, art history, gender studies and sustainable development among other disciplines, as well as museum studies students as the next critically engaged and potentially activist generation of museums professionals and academics. For those employees and volunteers, students and researchers associated with museums in remote and island communities, it draws on the knowledge and experiences of communities often marginalised from the mainstream by virtue of the realities of their geographies, climates and resources. The book will also have resonance through its focus on new research issuing from the Caribbean and Central America, in particular, which have been virtually ignored as regions in the current academic literature. It will also demonstrate ways in which research questions affecting these regions are applicable to other territories, notably other Small Island Developing States and remote communities, particularly those facing the perils of climate change and unsustainable forms of development, for example, by defining, developing and disseminating new museum ideas and models of co-curatorship/co-partnership to support disaster, health emergency and climate change resilient communities in the contingent conditions of the 21st century. This volume as a whole takes a people-centred approach to heritage interpretation, memory and conservation, away from the tradition of object-centred institutions. The book is evidence that no matter how academics may define their terms, it is the perceptions of the people on the ground that matter for characterising community museums for the 21st century.

Acknowledgements

Research for this introduction has been informed by a number of projects co-ordinated from the University of St Andrews. I wish to thank all of our funders and collaborators for the field work experiences, cultural encounters, friendships and networks they enabled in the process of research. They are: European Commission Horizon 2020 grant number 693669-EU-LAC Museums (2016–2021); Scottish Funding Council Global Challenges Research Fund – Community Crafts and Cultures (2018–2021); and Royal Society of Edinburgh – Scottish Community Heritage (2019–2022). Research has most recently been supported by the United Kingdom Research and Innovation programme (UKRI) Engineering and Physical Sciences Research Council grant number EP/X023036/1 Shared Island Stories Between Scotland and the Caribbean: Past, Present, Future (2022–2027). I am especially grateful to Jamie Brown for his meticulous project management skills and endless good humour. I also thank Alissandra Cummins, Peter Davis, Ana González Rueda and François Mairesse for providing their comments on drafts of this introduction. Finally, this book would not have been possible without the support of Aedín Mac Devitt and the editorial team at ICOM. Thank you for your professionalism and patience throughout.

Notes

1 The EU-LAC Museums project (2016–2021) was funded by the European Commission Horizon 2020 Programme under Grant Agreement Number 693669. https://eulacmuseums.net/index.php (Accessed 14 September 2022).
2 Brulon Soares (2021) has rightly noted that the role of community experiences is central to the ICOM Definition of a Museum process.
3 The Shared Island Stories project (2022–2027) was selected for funding by the ERC Consolidator Grant scheme and is now funded by the UKRI EPSRC under Grant Agreement Number EP/X023036/1. See: http://sharedislandstories.wp.st-andrews.ac.uk/ (Accessed 14 September 2022).
4 That said, some of Dana's recommendations are not so utopian or egalitarian; for example, 'Centralize authority. A museum cannot be well managed by a board of directors. No business can' (p. 42).
5 Definitions of ecomuseums and distinctions between them and 'community museums' have evolved since 1970 to the present day. While this discussion is beyond the scope of this introduction, Davis usefully defines ecomuseums as, 'community-driven museums or heritage projects that aid sustainable development' (Davis 2007, p. 199).
6 This Resolution was one of the major outcomes of the collaborative EC Horizon2020 project EU-LAC-Museums. The Resolution was submitted under the auspices of ICOM Europe and ICOM LAC who also supported the project Steering Committee. This introduction has also been informed by our project online survey 'What is a Community Museum in your Region?' found here: https://standrews.eu.qualtrics.com/jfe/form/SV_5oRFHE4ScQEOdNz.
7 The original Spanish reads: 'Un museo comunitario es creado por la misma comunidad: es un museo "de" la comunidad, no elaborado a su exterior "para"' la comunidad. Un museo comunitario es una herramienta para que la comunidad afirme la posesión física y simbólica de su patrimonio, a través de sus propias formas de organización. Un museo comunitario es un espacio donde los integrantes de la comunidad construyen un autoconocimiento colectivo, propiciando la reflexión, la crítica y la creatividad. Fortalece la identidad, porque legitima la historia y los valores propios, proyectando la forma de vida de la comunidad hacia adentro y hacia fuera de ella. Fortalece la memoria que alimenta sus aspiraciones de futuro' ('Que es un museo comunitario?').
8 Translation the author's. De Varine later names André Desvallées (editor of the anthology of new museology, *Vagues*), as the best theoretician of ecomuseums in France. See A. Desvallées, 'Introduction, *Ecomusée: rêve ou réalité*', Special Issue, *Publics et musées* 17–18 (2000): 11–31.
9 The new museology is described in *Vagues* by André Desvallées (1992 and 1994). Established in 1985, the international Movement for the New Museology (MINOM) has also produced a series of edited volumes on 'Sociomuseology' in Portuguese, French, Spanish and English. In Spanish, key reference texts include de Carli (2006), and Camarena Ocampo and Morales Lersch (2016).
10 In addition to the work of Davis, see that of other Anglophone scholars, including Boylan (1992), Corsane, Davis and Murtas (2008), Crooke (2015) of the UK (and

Italy – Murtas) and Sutter et al. (2016) based in Canada. The shortcoming partly arises from barriers of language and access: de Varine's monograph is published in French and translated into Spanish, and most other literature in the field is published in Latin languages outside mainstream peer-reviewed journals, including early publications in French through ICOM (de Varine 2017a, 66–7).

References

Allison, D. B. (2020). *Engaging Communities in Museums. Sharing Vision, Creation and Development.* Oxon and New York: Routledge Focus.

American Alliance of Museums. (2002). *A Museums and Community Toolkit.* Washington DC: Rowman & Littlefield Publishers.

Ariese-Vandemeulebroucke, C. (2018). *The Social Museum in the Caribbean. Grassroots Heritage Initiatives and Community Engagement.* Leiden: Sidestone Press.

Barton, B., and M. Goldsmith. (2016). 'Community and Sharing'. In *Sharing the Costs and Benefits of Energy and Resource Activity: Legal Change and Impact on Communities*, edited by L. Barrera-Hernández, B. Barton, L. Godden, A. Lucas, and A. Rønne, 317–23. Oxford: Oxford University Press.

Berkeley, W. N. (1932). *The Small-Community Museum: Why it is Entirely Feasible; Why it is Extremely Desirable.* Lynchberg: Bell Company.

Boylan, P. (1992). 'Ecomuseums and the New Museology: Some Definitions'. *Museums Journal*, 92(4): 29.

Brown, K., B. Brulon Soares, and A. González Rueda, eds. (2022). *Decolonising the Curriculum in Museum and Heritage Studies.* Paris, ICOM: ICOFOM Study Series.

Brown, K., P. Davis, and L. Raposo, eds. (2019). *On Community and Sustainable Museums.* Translated by A. González Rueda. EU-LAC-MUSEUMS, Lisbon. https://doi.org/10.5281/zenodo.2646479.

Brown, K. E., B. Brulon Soares, and O. Nazor, eds. (2018). *Definir los Museos del Siglo XXI: Experiencias Plurals.* Paris: ICOFOM.

Brown, K. E., and F. Mairesse. (2018). 'The Definition of the Museum through its Social Role'. *Curator: The Museums Journal*, 61 (3–4): 525–39. doi:10.1111/cura.12276.

Brulon Soares, B., ed. (2020a). *Decolonising Museology: Museums, Community Action, and Decolonization.* Vol. 1. Paris, ICOM: ICOFOM Monograph Series.

Brulon Soares, B., ed. (2020b). *Defining the Museum: Challenges and Compromises of the 21st Century.* Vol. 48 (2). Paris, ICOM: ICOFOM Study Series.

Brulon Soares, B. (2021). 'Decolonising the Museum? Community Experiences in the Periphery of the ICOM Definition'. *Curator: The Museum Journal*, 64(3): 439–55.

Camarena Ocampo, C., and T. Morales Lersch. (2014). *Manual para la creacion y desarrollo de museos comunitarios.* https://mediacionartistica.files.wordpress.com/2014/02/manual-para-la-creacion-y-desarrollo-de-museos-comunitarios.pdf (Accessed 14 September 2022).

Camarena Ocampo, C., and T. Morales Lersch. (2016). *Memoria: red de museos comunitarios de América: experiencias de museos comunitarios y redes nacionales.* https://www.museoscomunitarios.org (Accessed 3 April 2021).

Camarena Ocampo, C., and T. Morales Lersch. (2019). 'Community Museums: Telling a Story, Building a Future'. In *On Community and Sustainable Museums*, edited by K. Brown, P. Davis and L. Raposo, 38–53. Lisbon: EU-LAC-MUSEUMS. https://doi.org/10.5281/zenodo.2646479.

Camarena Ocampo, C., and T. Morales Lersch. (n.d.). 'Que es un museo comunitario?' https://www.museoscomunitarios.org (Accessed 3 April 2021).

Cameron, F. (2014). 'Ecologizing Experimentation: A Method and Manifesto for Composing a Post-Humanist Museum'. In *Climate Change and Museum Futures*, edited by Fiona Cameron and Brett Neilson, 16–33. New York: Routledge.

de Carli, G. (2006). *Un museo sostenible: museo y comunidad en la preservacion activa de su patrimonio*. Costa Rica: UNESCO.

Castro-Gómez, S., and R. Grosfoguel, eds. (2007). *El giro decolonial: Reflexiones para una diversidad epistémica más allá del capitalismo global.* Bogotá: Siglo del Hombre Editores.

Chaumier, S. (2011). 'Société'. In *Dictionnaire encyclopédique du musée*, edited by F. Mairesse and A. Desvaillés, 543–57. Paris: Arman Colin.

Corsanne, G., P. Davis, and D. Murtas. (2009). 'Place, Local Distinctiveness and Local Identity in Europe and Asia'. In *Heritage and Identity: Engagement and Demission in the Contemporary World*, edited by Marta Anico and Elsa Peralt, 47–62. New York: Routledge.

Crooke, E. (2007). *Museums and Community: Ideas, Issues and Challenges*. London and New York: Routledge.

Crooke, E. (2015). 'The "Active Museum": How Concern with Community Transformed the Museum'. In *The International Handbooks of Museum Studies*, general eds. S. Macdonald and H. R. Leahy. Vol. 2: Museum Practice, edited by Conal McCarthy. Chichester, West Sussex: John Wiley & Sons, Ltd.

Cummins, A., K. Farmer, and R. Russell, eds. (2013). *Plantation to Nation: Caribbean Museums and National Identity*. Champaign, IL: Common Ground.

Dana, J. C. (1999). 'A Few Fundamental Notes'. In *The New Museum: Selected Writings of John Cotton Dana*, edited by W. A. Peniston, 42–3. Newark, NJ: Newark Museum Association; American Association of Museums.

Davis, P. (2007). 'Ecomuseums and sustainability in Italy, Japan and China: adaptation through implementation'. In *Museum Revolutions: How Museums Change and Are Changed. Proceedings of The Museum: A World Forum*, edited by S. J. Knell, S. MacLeod, and S. Watson. ER. London: Routledge.

Davis, P. (2008). 'New Museologies and the Ecomuseum'. In *The Ashgate Research Companion to Heritage and Identity*, edited by B. Graham and P. Howard, 397–414. Aldershot: Ashgate.

Davis, P. (2011). *Ecomuseums: A Sense of Place*, 2nd edn. London: Continuum.

Do Nascimento, José de, Jnr., A. Trampe, and P. A. Dos Santos, eds. (2012). *Mesa redonda sobre la importancia y el Desarrollo de los museos en el mundo contemporáneo.* Vols. 1 and 2. Brasilia: IBRAM.

Egelston, Anne E. (2013). 'From Stockholm to Our Common Future'. In *Sustainable Development: A History*, edited by A. E. Egelston, 59–88. Dordrecht: Springer.

EU-LAC Museums. Museums and Community: Concepts, Experiences, and Sustainability in Europe, Latin America and the Caribbean (September 2016–January 2021). This project received funding from the Horizon 2020 Research and Innovation Programme under Grant number 693669. Available at: http://www.eulacmuseums.net (Accessed 14 September 2022).

Glissant, E. (1990). *Poétique de la relation* (Poétique III). Paris: Gallimard; *Poetics of Relation*, trans. Betsy Wing (1997). University of Michigan Press.

Golding, V., and W. Modest, eds. ([2013] 2015). *Museums and Communities: Curators, Collections and Collaboration.* London and New York: Bloomsbury.

Guido, H. F. (1973). Report from the Director of the 'Round Table on the Development and the Role of Museum in the Contemporary World', Santiago de Chile, Chile, 20–31 May. SHC-72/CONF.28/4, Paris, 29 January. Translated from the Spanish.

Hooper-Greenhill, Eilean. (2000). *Museums and the Interpretation of Visual Culture*. London and New York: Routledge.

ICOM. (2022). Announcement on the ratified Museum Definition. https://icom.museum/en/resources/standards-guidelines/museum-definition/ (Accessed 8 June 2023).

ICOM. (2019). Announcement on the New Museum Definition. https://icom.museum/en/news/icom-announces-the-alternative-museum-definition-that-will-be-subject-to-a-vote/ (Accessed 14 September 2022).

ICOM Code of Ethics. (2021). https://icom.museum/en/resources/standards-guidelines/code-of-ethics/ (Accessed 14 September 2022).

ICOM Resolution No. 5: 'Museums and Environment' (Grenoble, 1971). https://icom.museum/wp-content/uploads/2018/07/ICOMs-Resolutions_1971_Eng.pdf (Accessed 14 September 2022).

ICOM Resolution No. 5: 'Museums, Community Sustainability' (Kyoto, 2019). https://icom.museum/wp-content/uploads/2019/09/Resolutions_2019_EN.pdf (Accessed 14 September 2022).

Janes, R. R., and R. Sandell, eds. (2019). *Museum Activism*. London and New York: Routledge.

Jeffrey, T. (2019). 'Decolonial Museums in a Time of Social-Ecological Crisis: Cultural Landscape and the Revitalisation of Museological Theory'. *South African Museums Association Bulletin*, 41: 29–36.

Kadoyama, M. (2018). *Museums Involving Communities: Authentic Connections*. London and New York: Routledge.

Karp, Ivan, D. Steven Lavine, and Christine Mullen Kreamer, eds. (1992). *Museums and Communities: The Politics of Public Culture*. Washington, DC: Smithsonian Institution Press.

Kinard, J. R., and E. Nighbert. (1972). 'The Anacostia Neighbourhood Museum, Smithsonian Institution, Washington, D.C.' *Museum International*, 24(2): 103–9.

Mairesse, F. (2000). « La belle histoire: aux origines de la nouvelle muséologie. » *Publics & musées*, 17–18: 33–55.

Mairesse, F., ed. (2017). *Définir le musée du XXIe siècle: matériaux pour une discussion*. Paris, ICOM: ICOFOM Study Series.

Mairesse, F., and A. Desvaillés, eds. (2011). *Dictionnaire encyclopédique du musée*. Paris: Arman Colin.

Mignolo W., and C. Walsh. (2018). *On Decoloniality: Concepts, Analytics, Praxis*. Durham, NC: Duke University Press.

Museums Association (n.d.) https://www.museumsassociation.org/campaigns/decolonising-museums/# (Accessed: 24 March 2021)

Newell, Robin, and Wehner, eds. (2017). *Curating the Future: Museums, Communities and Climate Change*. London and New York: Routledge.

Quijano, A., and Michael E. (2000). "Coloniality of Power, Eurocentrism, and Latin America." *Nepantla: Views from South* 1(3): 533–80.

Rockström, J., Steffen, W., Noone, K. *et al.* (2009). 'A safe operating space for humanity'. *Nature* 461: 472–475.

Sandahl, J., ed. (2019). *The Backbone of Museums*. Special Issue of *Museum International*, 71: 1–20.

Sandell, Richard. (2007). *Museums, Prejudice and the Reframing of Difference*. London and New York: Routledge.

Sutter, G. S., T. Sperlich, D. Worts, R. Rivard, and L. Teather. (2016). 'Fostering Cultures of Sustainability through Community-Engaged Museums: The History and Re-Emergence of Ecomuseums in Canada and the USA'. *Sustainability*, 8(12): 1310.

Toynbee, J. (1863). *Hints on the Formation of Local Museums: By the Treasurer of the Wimbledon Museum Committee*. London: Robert Hardwicke.

Tuhiwai Smith, L. (2012). *Decolonizing Methodologies: Research and Indigenous Peoples.* London: Zed Books Ltd.

UK Museums Association. (2021). Available at: https://www.museumsassociation.org/campaigns/decolonising-museums/#:~:text=Decolonisation%20is%20not%20simply%20the,creation%20to%20the%20present%20day. Accessed 8 June 2023.

Van Mensch, P. (1992). *Towards a Methodology of Museology*. PhD diss., University of Zagreb. http://emuzeum.cz/admin/files/Peter-van-Mensch-disertace.pdf (Accessed 14 September 2022).

de Varine, H. (2017a). *L'Ecomusée singulier et pluriel: un témoinage sur cinquante ans de muséologie communitaire dans le monde.* Paris: L'Harmattan.

de Varine, H. (2017b). Interview with Karen Brown, Collège des Irlandais, Paris.

Watson, Sheila, ed. (2007). *Museums and Their Communities*. London and New York: Routledge.

Wehner, K. (2016). 'Towards an Ecological Museology: Responding to the Animal-Objects of the Australian Institute of Anatomy Collection'. In *Curating the Future: Museums, Communities, and Climate Change*, edited by Jennifer Newell, Libby Robin, and Kirsten Wehner, 85–100. London and New York: Routledge.

Weil, S. (2002). *Making Museums Matter*. Washington, DC: Smithsonian Institution Press.

Witcomb, A. (2003). *Re-imagining the Museum: Beyond the Mausoleum*. London and New York: Routledge.

PART I

Community museums

Nurturing identities and resilience

1

COMMUNITY MUSEUMS AND DECOLONISATION

Reflections from the Network of Community Museums of America

Teresa Morales Lersch and Cuauhtémoc Camarena Ocampo

The importance of community museums as tools for the affirmation of collective rights and the struggle for self-determination cannot be understood without considering the context of colonialism. As Linda Tuhiwai Smith has argued (2012, p. 8), colonialism is realised in the multiple representations and ideological constructions of the Other, which contain underlying rules and narratives regarding their identity. Community museums follow alternative rules and create different narratives. They contribute to the processes of decolonisation at several levels, questioning the logic of the construction of colonial, Eurocentric knowledge and transforming the coloniality of self by creating sites where subaltern communities represent themselves, drawing on their internal well-springs of historical struggle, communal practices and ancestral memory.

The determinant nature of colonial relationships in the development of museums was recognised as early as 1979 by Hugues de Varine (1987, p. 34):

> Since the beginning of the nineteenth century, the development of museums in the rest of the world (the non-western world) is a purely colonial phenomenon. European countries have imposed on non-European nations their method of analysis of the phenomenon and cultural heritage; they have forced the elite of these countries and the people themselves to see their own culture with European eyes.

The extent to which colonial relationships have continued to permeate the social and political context in countries such as Mexico was underlined in 1965 by Pablo González Casanova, who developed the notion of internal colonialism (Maldonado, 2011, p. 32). He characterised the manner in which the dominant classes subjugated indigenous communities to relationships of exploitation (combining a variety

DOI: 10.4324/9781003288138-3

of economic relationships of a feudal, pre-capitalist and capitalist nature) as well as discrimination (manifest in political, social and cultural institutions), which had all the characteristics of colonial domination: 'the indigenous community is a colony within our national limits' (Maldonado, 2011, p. 33). The concept of internal colonialism was also developed by Rodolfo Stavenhagen in 1965 and Guillermo Bonfil Batalla in 1972 (Garzón, 2013, p. 4).

More recently (1996–2007), the Latin/Latin American Network of Modernity/Coloniality proposed a series of concepts and analyses that bring to the forefront a new understanding of colonial relationships within the globalised society in the 21st century.[1] Aníbal Quijano, Ramón Grosfoguel, Walter Mignolo, Enrique Dussel, Santiago Castro-Gómez and others have concluded that the international division of labour between centres of power and peripheries, as well as ethnic and racial hierarchies within the population, developed during several centuries of European colonial expansion, and did not change significantly with the formal political independence of nation-states on the periphery. Rather, there has been a transition from modern colonialism to global coloniality. The forms of domination of modernity have undergone transformations, but the fundamental structure of the relationship between centres of power and the periphery has endured. The subordination of countries on the periphery to the centres of power of Europe and North America remains constant. Power structures of long duration, entrenched since the 16th and 17th centuries, continue to play an important role in the present. Contemporary global capitalism has re-signified, but not transformed, relationships of exclusion, which result from persistent hierarchies of an epistemic, spiritual, racial or ethnic, and gendered or sexual nature (Castro-Gómez and Grosfoguel, 2007, pp. 13–14).

The approach of the Network of Modernity/Coloniality draws a distinction between colonialism and coloniality. Colonialism refers to the political, economic and military domination and exploitation of one nation or people by another. It is expressed in specific historical periods and places of imperial domination. Coloniality, on the other hand, denotes the grammar and internal logic of colonial domination. Garzón López (2013, p. 4) sums it up as follows: 'Colonialism is the territorial occupation by coercive means (military, political, economic), while coloniality is the imposition of the European cultural imaginary'.

Moreover, coloniality is institutionalised and normalised in social practices and the functioning of the state. It is present in the mass media, the educational system and everyday language. Its matrix operates at three levels: first, as coloniality of power (political and economic); second, as coloniality of knowledge (epistemic, philosophical and scientific); and third, as coloniality of being (subjectivity, individual and collective identity) (Garzón, 2013, p. 5).

This last level of domination has an impact of enormous consequences. A self-image of inferiority, of incapacity to overcome obstacles or to struggle against the bonds of oppression, nullifies resistance before it has even begun. The power of this dynamic of subordination springs from the fact that it is exercised by the colonised subjects themselves.

The effects of the imposition of a colonial imaginary on Mesoamerican peoples were described by Laurent Aubague as follows:

> This displacement of defeat in the imaginary dimension seals it as a practically permanent inferiority complex, and clearly shows how power operates at the deep levels of symbolism and self-representation. The power conflict at the level of the imaginary is transformed into an aggression against the consciousness of self. The imaginary of power then becomes the will to annihilate the identity of the Other [...]. [T]he Western military victory has been prolonged to become the conquest of one imaginary by another, the condensation of one identity instead of another, the characterization as inferior of a whole system to understand the universe and man's place in it.
>
> *(Maldonado, 2011, p. 44)*

It is important to recognise that museums have been, and for the most part still are, institutions that manifest all three levels of coloniality. They came into being as repositories for valuable objects and knowledge in the hands of the dominant classes of Europe. Large museums were built with colonial economic and political power, amassing collections through war, theft, expeditions and traffic in cultural objects (coloniality of power). As vehicles to possess and exhibit this heritage, they became instruments of power that have reflected, legitimised and reproduced a Eurocentric and colonial world view, presenting its discourse as scientific and universal truths (coloniality of knowledge). Through museums, colonial powers have represented the native peoples of their colonies as inherently inferior and savage. The elite groups of these colonies, through processes of internal colonialism, later used museums to repeat these same narratives, influencing how native peoples perceive themselves (coloniality of being).

Community museums, vehicles to contest coloniality

Within the community museum movement, and particularly within the Red de Museos Comunitarios de América (Network of Community Museums of America), we maintain that it is possible for communities to appropriate and transform museums, contesting the manifestations of coloniality at these three levels.[2] Non-hegemonic communities are creating their own museums as tools of resistance and decolonisation, reversing their role as instruments of domination and colonial power. Instead of repeating official histories, which interpret diverse communities and peoples as objects, community museums are a vehicle for communities to construct knowledge as subjects of their own history and way of life. Rather than presenting stories about the Other, community museums represent their own communities, telling stories that have been suppressed and denied, and speaking from their own collective experience.

At the level of the coloniality of power, community museums strengthen networks and promote a collective voice to defend cultural rights and community ways of life. They are able to develop methods for new forms of knowledge to emerge, resisting the coloniality of knowledge by sustaining a process of collective construction of community history and culture. They are also instruments for community members to develop initiatives regarding their own identity, through which they may resist the coloniality of being. Community museums are vehicles for community members to see themselves with their own eyes, through their own categories, historical experiences and stories.

Developing a critique of the coloniality of knowledge is a fundamental component of the approach of the Modernity/Coloniality Network.[3] Quijano and Dussel argue that the superiority assigned to European knowledge in many areas of life was key to the development of coloniality throughout the world. Subaltern knowledge was excluded, silenced and ignored: since the Enlightenment, it was typified as a mythical, inferior and pre-scientific stage of human knowledge. As Castro-Gómez and Grosfoguel (2007, p. 20) ask, how can knowledge be produced that does not repeat and reproduce the assumptions of a Eurocentric vision? The Modernity/Coloniality Network recognises that their analysis must take into account the practical knowledge of workers, women, racialised and colonised subjects, LGBTQ groups and social movements that counter the dominant world system. This is because all possible knowledge is embodied in subjects, linked to concrete struggles, interwoven with social contradictions and rooted in specific perspectives. However, Castro-Gómez and Grosfoguel (2007, p. 21) recognise that, in their view, contemporary social sciences have not yet found an effective way to incorporate subaltern knowledge into the dominant processes of knowledge production.

Tuhiwai Smith (2012, p. 21) has explored how research methodologies can engage in processes of decolonisation. She argues that, on one level, decolonisation implies a critical understanding of the underlying assumptions, motivations and values that inform research practices. Speaking specifically of the Maori people, she states:

> Research is implicated in the production of Western knowledge, in the nature of academic work, in the production of theories that have dehumanized Maori and in practices that have continued to privilege Western ways of knowing, while denying the validity for Maori of Maori knowledge, language and culture.
>
> *(Tuhiwai Smith, 2012, p. 185)*

She describes the Kaupapa Maori approach to research, which has provided a focus 'through which Maori people, as communities of the researched and as new communities of the researchers, have been able to engage in a dialogue about setting new directions for the priorities, policies and practices of research for, by and with Maori' (p. 185). She details how Graham Smith characterises Kaupapa Maori research, which

1 is related to 'being Maori';
2 is connected to Maori philosophy and principles;

3 takes for granted the validity and legitimacy of Maori, the importance of Maori language and culture and
4 is concerned with 'the struggle for autonomy over our own cultural well-being' (*Tuhiwai Smith, 2012, p. 187*).

She further refers to a list of priorities that need to be discussed to set strategic directions of Maori research:

- determining as Maori our own research needs and priorities;
- defining the ways research should proceed;
- training of Maori researchers;
- discussion of culturally appropriate ethics;
- ongoing development of culturally sympathetic methods;
- continued collaboration with our own diverse *iwi* and communities of interest;
- development and dissemination of literature by Maori on research;
- continued reflection, evaluation and critique of ourselves as a community of Maori researchers;
- extending the boundaries for Maori (and for other indigenous peoples) of our own fields and disciplines;
- education of the wider research community, including scientific, academic and policy communities, and
- accountabilities to and outcomes for Maori (*Tuhiwai Smith, 2012, pp. 194–195*).

The integrity of the approach Tuhiwai Smith describes – consistently positioning the Maori world view as a valid philosophical foundation for research, insisting on culturally respectful practices and aligning research with the interests of Maori self-determination and well-being – make it an important example for other indigenous peoples and communities. However, it does not seem to question the basic separation between researchers and community, or the researcher and researched. It stresses the training of Maori researchers, but if they are trained to consider their own people as objects of study, the process of knowledge production is still based on the separation of object and subject, both conceptually and methodologically. The researcher is the subject with agency, the authority who controls the questions asked and the methods employed, while the community remains a passive object whose fundamental role is to provide information. Even if there are measures to ensure culturally appropriate ethics, collaborations and accountability of the researchers, the basic logic of the production of knowledge remains the same.

Communities constructing collective knowledge

To decolonise the process of knowledge production, it must be possible for all participants to be active subjects in a collective effort to construct knowledge. The role of researcher must be transformed into that of a facilitator who provides methods and guides the steps to be taken. The selection of the themes, the formulation of questions to be asked, the gathering of material and documentation, as well as interpretation, then all become steps in a collective process that is undertaken by community decision-making bodies and groups.

This approach has been developed in the community museum movement that emerged in 1985 through the initiatives of indigenous communities in the state of Oaxaca in the south of Mexico. In this most ethnically diverse state of Mexico, over more than 30 years, indigenous and *campesino* communities have been creating community museums as sites to safeguard, understand and represent certain elements of their community heritage and memory. They have been mobilised by the concern that heritage objects, traditions and historical experiences will otherwise be lost or forgotten.

In these indigenous communities, a system of local governance known as *usos y costumbres* (usages and customs) has been practised for decades. Developed through a process of resistance and recreation of community life within the context of colonial imposition, in 1995, it was officially recognised as the system to elect municipal authorities through an amendment to the State Constitution. According to *usos y costumbres*, the local community assembly is the main decision-making body. Community members (initially all men but recently women as well) all participate and voice their opinions in the community assembly. In addition, members must perform community service without pay throughout their adult lives in a variety of different committees, civil posts and ceremonial roles. The community assembly discusses and makes decisions regarding collective projects and priorities, as well as electing all the committees and officials who are held accountable for carrying these initiatives forward.

The initiatives to create community museums came from different sectors, sometimes community representatives, sometimes teachers or young people. However, they had to be approved by the community assemblies if they were to succeed as community projects. At times, the community assemblies would decide that the community museum was not a priority for the moment. At others, the community assembly approved and gave its support. In this way, proposals to create community museums became collective projects, recognised as efforts that would respond to the collective interests and concerns of the community.

Community assemblies, sometimes in dialogue with councils of elders, or in coordination with assemblies of *barrios*, determined the themes to be researched and represented in the museum. They also decided on the buildings to be occupied or constructed for this purpose. Soon, the community assemblies also elected committees to coordinate the process to create the museums, as well as to direct and manage them. These community museum committees then became part of the local system of governance and renewed periodically with newly elected community members who serve without pay.

When the authors of this chapter, as research professors responding to the request of these communities for the support of the National Institute of Anthropology and

History, began to collaborate with these efforts in 1985, it was clear to us that our role was to contribute to their collective nature, respecting traditional community practices and decisions.

Together with community members, we explored ways to consult the themes to be researched. We asked different questions, such as: what do you think the museum should talk about; what stories do you want to tell in the museum; what issues or problems should be discussed and what stories of your community do you want your children to know, to remember? Each community developed a somewhat different procedure. In Santiago Suchilquitongo (1988), meetings were organised in each one of the nine *barrios* and *agencias* to discuss and select significant themes, and the three most recurrent ones were chosen. In Santiago Matatlán (2004), the community assembly of 140 people discussed themes to be researched in groups of ten and wrote their proposals on cards that were then carefully tallied to determine which were of greatest interest. In San Juan Guelavía (2010), the proposals that emerged from a community meeting of women were included in a survey answered by hundreds of community members in order to arrive at a decision. In Santo Domingo Yanhuitlán (2013), the community museum committee organised a process in which elementary school children talked to their parents and grandparents about important themes to include, and the community assembly later voted on the proposals. Sometimes the same assembly that was called to discuss the possibility of creating a community museum would approve the project and continue with the decision on which themes to study.

The community groups organised by the community museum committees to carry out the research have been equally diverse. In San Martín Huamelulpan (1989), each of the smaller settlements constituting the community chose representatives to participate in the research group. In San Francisco Cajonos (2012), the council of elders guided the study of the territory, documented by elementary school children and younger adults, while a group of more than 60 women recorded traditional healing practices, and a group of teachers developed a timeline of local history.

Working with these community groups, we have developed a series of methods to carry out the inquiry, including ways to define the fundamental questions used to guide the research, or methods of exploring the parts of the story to be told. Historians, educators and community members have all participated over the years in developing different tools and methods. Participatory oral history methods and ways to arrive at consensus through brainstorming and categorising the ideas of the group have been very important tools. For example, research groups brainstorm and organise the questions to ask in interviews or community dialogues and reiterate the fundamental importance of respectful listening. Sometimes the community dialogues or conversations are held with one individual at a time; at other times, they take the form of collective exchanges. The observation and documentation of communal practices by groups of young people and children, as well as detailed observation and documentation of communal sites and territories, have been extremely valuable.

The aforementioned experiences of selecting and documenting themes have distinct characteristics, but they all respond to the same objective: to build a collective vision of community stories and memory through a process determined by community decisions and carried out by the creative work of community members. This collective vision is a body of shared knowledge built throughout its different phases through a process controlled by the community. The community, that is, determines the subject of study, develops the focus and questions to be asked, organises who collaborates in the research and determines how it will be presented.

Eleazar García Ortega (2015), activist, educator and a member of the community authorities when the Community Museum of San Juan Guelavía was created, described the process as follows:

> The themes that will be presented in the museum are also defined in a collective process, through consultations and written surveys, as well as interviews with elders.
>
> *(p. 16)*

> In the case of the Community Museum of San Juan Guelavía we chose two themes: 'the harvest of salt from the earth,' which led us to walk through our history, and the theme of the 'planting native corn in humid soil.' We organised research teams, we developed the script of questions, we identified the people who were most knowledgeable about our local culture. We began to work, to locate sites within our local geography, where real and mythical events occurred; we drew sketches, maps, we made a timeline.
>
> *(pp. 17–18)*

> That is to say, the collective subject that observes itself becomes the protagonist, participating in its own construction, interacting with other individuals who share the same identity. And this subject begins to create a museum that is not oriented to folklore, to the extravagant expectations of tourists, but a museum that is a resource for the development of other community members in a process of endoculturation. The positions of 'subject' and 'object' begin to fluctuate.
>
> *(p. 19)*

Another case that illustrates this process is the 'Määtsk Mëjy Nëë' Community Museum of San Juan Bosco Chuxnabán, a Mixe community of 120 families located in the northern mountains of the state of Oaxaca. When community members discussed the themes for their museum in their general assembly in 2008, they agreed to speak about a recent archaeological discovery, and the 'agrarian problem'.

The community museum committee organised the process to carry out the research on these themes with the participation of secondary school students. Many elders helped the students to create family trees for the founding families of Chuxnabán. Civil and agrarian authorities were also very active working with the students in documenting the agrarian problem. In the process of constructing the questions to ask about this issue, it appeared that the story involved Chuxnabán's struggle to end decades of violent conflict with neighbouring communities over territorial boundaries. In the 1990s, the community representatives had begun a process of negotiating directly with their neighbours without intervention from the government. As secondary school students and community representatives documented their conversations, a different title to the exhibition emerged: 'The importance of the agrarian struggle and solutions through independent dialogue'.

When young people and adults came together in the workshop to develop the design of the exhibition, they split into teams to propose how each part of the story should be represented. One of the teams had the task of visualising the process of dialogue. They proposed the recreation of a scene in which five people were sitting at a table talking. These people represented one delegate from Chuxnabán and one from each of the four communities with which they had had successful negotiations. When the team presented their proposal to the other participants of the design workshop, they were asked when this scene had taken place. Their answer was that it had never actually taken place. The negotiations had developed with each community separately. The team explained that this scenario was a symbol of how the representatives of Chuxnabán were able to sit down together at the negotiating table and peacefully come to agreements, with all of these villages.

When the students were listening to the testimonies, they commented that they had heard references to these stories, but their vague notions were gradually clarified as the process developed. The adults commented that the decision taken in the assembly had been a good one, because it was essential to pass on knowledge to their youth of how they had weathered a conflict that has cost many lives. It was a sensitive issue, one that could only be authoritatively addressed from their own experience. The facts were important, but so, too, were the lessons encompassed in this story about the value of communities talking directly amongst themselves without governmental support for any of the parties involved. This was evident in the scene community members recreated – no officials were sitting at the table during the symbolic dialogue they presented.

Since the museum opened to the public in 2011, several communities have visited it in the context of developing dialogues with their neighbours over land disputes. In this way, Chuxnabán has been able to tell a story that has value for both its own members and other communities. However, it would not have been possible if community members had not themselves constructed this vision in a collective process. Throughout the research and the design process, what emerged were their needs, their understanding of what had occurred and the symbolic truths embedded in their experience.

As García Ortega states, the interaction between community members of different generations and experience generates a collective learning process, and a shared body of knowledge, in which the subject and the object of study are not divorced but rather engage in a conversation. It is an internal process, in which participants are conscious of learning something about themselves and their collective identity. They are aware of being part of an initiative to represent themselves.

> In general terms we can say that the community museum is a participatory, collective-communal process of construction, which allows us to [become] conscious of our role and situation in the world. Furthermore, most importantly, this process of construction must be an endogenous process, an introspective journey. As we intentionally develop the initiative to look within ourselves, we recover, gradually, the self-esteem that has been denied us as subaltern or subordinate cultures. To accomplish this valuation of ourselves is to experience a process of decolonisation.
>
> *(García Ortega, 2015, pp. 20–21)*

When community groups come together to discuss and select themes to represent themselves, elements of their unique experiences and traditions appear. They turn towards their internal well-spring of tradition, communal practices and meaning and discover sources of strength that have been obscured by the imaginary of dominant culture. For example, when community members of Santa Ana del Valle discussed the themes to be included in their museum, they decided to portray their experience of the Mexican Revolution and how they opposed the federal forces of Carranza from 1915 to 1920. When the museum opened, some academics were critical of this choice. Why did Santa Ana want to profile the counter-revolutionary movement that proclaimed Oaxaca to be a sovereign state? The community answered that this was an important experience for them. During the years covered, their village was burned to the ground. The federal troops destroyed their homes and pillaged their crops and cattle. The community took refuge in the foothills and aligned with the guerrilla forces in the northern mountains which headed the movement to declare sovereignty. They defended the pass that runs through their lands and stopped the advance of the federal troops that were attacking the *soberanistas*. Later, they rebuilt Santa Ana from the ashes. Although it occurred in the context of national movements, the story they wanted to tell was about their experience facing aggression and their capacity for resistance.

Some community museums have been founded precisely to denounce experiences of aggression and violence. The Community Museum of Historical Memory of Rabinal in Guatemala defines its objective as: 'to recover and disseminate the historical memory of the Maya Achí people, through a site of reflection, critical analysis and consciousness concerning the grave violations of human rights and genocide implemented by the military governments from 1980 to 1984' (Museo Comunitario de la Memoria Histórica de Rabinal, n.d.). This community museum, the first of its kind in Guatemala, tells stories of extreme violence that had not been fully acknowledged by the state when it opened its first exhibition in 1999. It has been an instrument to demand justice and increase public awareness of the atrocities committed, telling a story that could no longer be denied.

The process of defining and telling community stories also implies processes that reveal internal contradictions. Recognising these stories may contradict how community members have adopted a self-image defined by dominant culture. García Ortega (2015, pp. 26–7) shares an experience of what occurred when the research group found a chest full of historical documents that had been forgotten:

> As we were cleaning a storage space for the museum we found an old chest. We gave notice to the Alcalde Único Constitucional: in his presence we opened the chest and found documents that were hundreds of years old. We were amazed. There was information about how San Juan Guelavía was a República de Indios until 1820; that before, all its land had been communal; that some lands had been rented to neighbouring villages and individuals to harvest salt; that the community and the church owned communal cattle. They also documented how everything was administrated by a 'Gobernador'. There was a young secretary of the municipality who was compiling the inventory. She couldn't accept what she saw. She said, 'They were Indians before, but we aren't anymore! Now everything is private property, everyone has their own property, communal things are old!'
>
> This young woman couldn't accept this information. Immediately she put up a barrier to understanding a fact of this magnitude, that neither her monocultural education nor the mass media had provided. Here we enter into the other side of this issue: decolonisation implies dismantling values, attitudes and knowledge. In this case we had conclusive evidence. That is why the museum's value is in its context. It isn't the distant hero from a textbook, but the people from our community, who walked and suffered here, like us, but at a different moment.

The experience of examining their own history in a collective process offers community members the opportunity to make evident internal contradictions within their attitudes and beliefs. Community members are confronted by how they have accepted or been complicit in the denigration of their own culture by dominant cultural norms. As they become more aware of the historical conditions that have propitiated the imposition of these norms, they are better able to detach from them. The experience of learning about their own struggles, and considering in greater depth the meaning of their communal practices, strengthens their connection to their communal identity while also enriching their sense of self-worth.

> As we refuse to become homogenized beings, all cut in the same pattern, we begin a slow process of decolonisation. As we affirm our right to be who we are, we find the certainty to seek our own destiny. For this reason, even the most remote villages now desire this re-encounter. The community museum is a tool that helps us develop this exercise fully, so that our cultures can assert our right to resist and survive, in a world where inclusive diversity prevails.
>
> *(García Ortega, 2015, pp. 27–28)*

Contesting the coloniality of self and the coloniality of power

The community museum is a tool to dismantle the coloniality of self because it contributes to the creation of a self-image based on the autonomous source of collective memory. It brings forth strength by recognising the lived experience of resistance. It materialises images of concrete experiences, common experiences, which unite community members. Importantly, it expresses the community's own narratives and sense of meaning: it depends on neither the approval of another nor that of dominant society. Finally, the community museum provides a centre around which community members can stand together, generate awareness and build a collective alternative.

Telling their own stories empowers communities to open new perspectives without the epistemic assumptions of colonial narratives. Colonial discourse no longer determines narratives about the past or what is possible for the future. Recovering stories and constructing a critical view of the myriad processes of colonial subjugation allows communities to build strength from within, identifying their own centre and foundations of tradition, experience and meaning.

The Network of Community Museums of America brings together community museums in eight countries that embarked on this process, in a joint commitment to defend community rights in the context of the increasing destruction of their natural resources and ways of life. Their collective reflection on the characteristics of colonial domination, the importance of community memory and the relationship between memory and self-determination has strengthened their vision for and practice in the community museum. During the ninth meeting of the network, held in Mulaló, Colombia, in 2018, the participants discussed and created the following manifesto:

> We make the following public statement because we understand our community museums as key elements to articulate the voice and perspective of our communities as they face current global processes. They are a collective tool for our communities to defend our rights as peoples. They are sites of identity, denouncement, reflection, construction and a place of encounter to strengthen our own structures of community organisation.
>
> Our community museums fortify the exercise of our own cultural practices that sustain the identity and memory underlying the power of our community and generating impact regarding public policies that affect us. They provide a favourable environment to generate life plans according to our own models of development. They strengthen life projects born within our community, opposing cultural impositions of the dominant power system. The community museum creates consciousness of collective memory as a source of resistance and survival of our peoples. Memory becomes a fundamental resource to build our own pathway, and unites our communities in the struggle to recover and dignify our identity and the living conditions in our territories.
>
> *(Red de Museos Comunitarios de América, 2018)*

The network is not based on vertical hierarchies and does not depend on the recognition of any institution external to the communities. It is a nexus of mutual support and solidarity, which helps transform relationships of subordination and disempowerment by constructing horizontal bonds and autonomous projects. In addition, the network helps project the capacity for community self-governance to higher levels, expanding the reach of organised community action.

Networks of community museums oppose the coloniality of power, contributing to the development of new forms of power from the grassroots, linking diverse communities in a common purpose to overcome the injustices of economic and political domination. As part of their daily practices, community museums empower community members to be active subjects in building knowledge through a process driven by community decision-making bodies and groups; in this way, they contest the coloniality of knowledge. Community museums also enable communities to contest the coloniality of self, resisting the imposition of the dominant imaginary and offering a way for community members to see themselves through their own eyes.

Notes

1 The Latin/Latin American Network of Modernity/Coloniality is a research group that developed an influential body of work regarding concepts such as 'decoloniality' and 'coloniality of power'. This transnational and transdisciplinary group includes professors from Duke University, the University of North Carolina, Universidad Javeriana of Bogotá, Universidad Andina Simón Bolívar of Quito, the University of California (Berkeley) and the State University of New York. It was formed in 1998 and by 2006 had organised seven international meetings and numerous publications. The term Modernity/Coloniality Network is used to reference this group in the present article (Castro-Gómez and Grosfoguel, 2007, pp. 7–14).

2 The history and focus of the Network of Community Museums of America has been documented by Camarena Ocampo and Morales Lersch (2016). Current information regarding the network can be found on its website: https://www.museoscomunitarios.org/somos (Accessed 24 March 2021).

3 One of the first, and most influential, publications of the Modernity/Coloniality Network is *La colonialidad del saber: eurocentrismo y ciencias sociales. Perspectivas latinoamericanas*, edited by Lander (2000).

References

Camarena Ocampo, C. and Morales Lersch, T. (eds.) (2016). *Memoria: Red de Museos Comunitarios de América. Experiencias de museos comunitarios y redes nacionales.* Oaxaca: Red de Museos Comunitarios de América.

Castro-Gómez, S. and Grosfoguel, R. (2007). 'Prólogo. Giro decolonial, teoría crítica y pensamiento heterárquico', in Castro-Gómez, S. and Grosfoguel, R. (eds.) *El giro decolonial: Reflexiones para una diversidad epistémica más allá del capitalismo global.* Bogotá: Siglo del Hombre Editores, pp. 9–23.

García Ortega, E. (2015). *El museo como recurso descolonizante. La experiencia del Museo Comunitario de San Juan Guelavía.* Mexico: Lxs Desechables Editorxs.

Garzón López, P. (2013). 'Pueblos indígenas y decolonialidad: sobre la colonización epistemológica occidental', *Andamios*, 10 (22), pp. 305–31.

Lander, E. (ed.) (2000). *La colonialidad del saber: eurocentrismo y ciencias sociales. Perspectivas latinoamericanas*. Buenos Aires: CLACSO.

Maldonado Alvarado, B. (2011). *Comunidad, comunalidad y colonialismo en Oaxaca: la nueva educación comunitaria y su contexto*. Oaxaca: CSEIIO.

Castro-Gómez, S. and Grosfoguel, R. (2007). 'Prólogo. Giro decolonial, teoría crítica y pensamiento heterárquico', in Castro-Gómez, S. and Grosfoguel, R. (eds.) *El giro decolonial: Reflexiones para una diversidad epistémica más allá del capitalismo global*. Bogotá: Siglo del Hombre Editores, pp. 9–23.

Marco, J. et al. (1979). 'Entrevista con Hugues de Varine-Bohan', in Marco, J. et. al (eds.) *Los museos en el mundo*. Barcelona: Ed. Salvat, pp. 9–14.

Museo Comunitario de la Memoria Histórica de Rabinal (n.d.). Available at: https://www.adivima.org.gt/museo-comunitario/ (Accessed 23 February 2020).

Red de Museos Comunitarios de América (2018). *Proclama de la Red de Museos Comunitarios de América*. [Online]. Available at: https://museoscomunitarios.org/noticias/321-proclama-de-la-red-de-museos-comunitarios-de-america (Accessed 23 February 2020).

Tuhiwai Smith, L. (2012). *Decolonizing methodologies: research and indigenous peoples*. London: Zed Books.

2

INTERNATIONAL COLLABORATION BETWEEN ECOMUSEUMS AND COMMUNITY MUSEUMS

The experience of the EU-LAC Museums Bi-Regional Youth Exchange in fostering identity, building community sustainability and resilience

Karen Brown and Jamie Allan Brown

Ecomuseums and community museums: nature and culture

Understanding relationships between culture and nature has seldom been more urgent for museum and heritage professionals. In recent years, cultural heritage sites and their communities around the world have been acutely affected by natural disasters, conflict, lack of security, youth unemployment and related societal challenges.

In Europe, Latin America and the Caribbean, young people participate in their respective societies under unequal circumstances and expectations. At the most recent European-Community of Latin American and Caribbean States (EU-CELAC) Youth Day held in Quito, Ecuador (2015), it was reported that youth unemployment is at 22.8 per cent in Europe, and 18.7 per cent in Latin America (European Youth Forum, 2015). More recent reports produced by Eurostat demonstrate that 28.8 per cent of young people in the EU face poverty and are at risk of social exclusion, with young women at slightly higher risk than young men, while the Economic Commission for Latin America and the Caribbean reports that 41 per cent in Latin America face the same circumstances (European Parliamentary Research Service, 2018; Eurostat, n.d.; Youth Policy Labs, 2007; CEPALSTAT, 2020).

The situation is exacerbated in low- to middle-income countries where maintaining equilibrium, well-being and community resilience is an urgent necessity in the face of global imbalances and rapid change. Social exclusion for young people as they transition into adulthood has the potential to have ongoing consequences for both individuals and society, as it often affects all aspects of young people's lives (UN DESA, 2016). Increased social exclusion and poverty can contribute to an increased risk of inequality in terms of young people's well-being, lifestyle,

DOI: 10.4324/9781003288138-4

access to culture, education and employment opportunities (Eurostat, 2021). For communities in remote rural and island locations, the challenges of globalisation are also intensified by socio-political and environmental instability, lack of access to resources, depopulation and unethical development. Viewed in this context, museums and heritage organisations bear a considerable responsibility for the communities they serve, and they need to invest in their young people through intergenerational dialogue to strengthen their roots because, in the words of one Scottish ecomuseum director, 'young people are us tomorrow' (Ghilleasbuig, 2019).

Studies have shown that museums are among the most trusted public institutions around the world, and as such, they have an ethical obligation to support social cohesion and development, as well as to maintain traditional standards in collections care and management.[1] Discussions about relationships between nature and culture are also gathering pace, as museums—and especially ecomuseums—become increasingly appreciated for the work they do in attaining a wide range of Sustainable Development Goals. Heritage Studies has always been preoccupied with landscapes, but in recent years, the international museum world has also begun to acknowledge its responsibilities towards cultural landscapes as a fundamental resource for a sustainable future (Mac Devitt, 2017; Riva, 2017). Increased participation and access to museums and culture can facilitate a 'sense of place' (Davis, 2011) and belonging to a community, promoting social inclusion and lifelong learning. In addition, museums are increasingly addressing the climate emergency (Museums Association, 2020; Janes and Sandell, 2019). Within this emerging corpus, we focus on the integral role of youth in maintaining sustainable museums through the example of the Horizon 2020 project's Work Package 4, 'Museum Education for Social Inclusion and Cohesion' Bi-Regional Youth Exchange.

EU-LAC Museums project background

> Museums are important because they serve to remind us of who we are and what our place is in the world. [...] Museum professionals, with reference to their visitors, frequently use the expression 'museum community', but can this be defined? We also need to discover how museums interact with their community, and the community with its museums, and place this in historical perspective.
>
> *(Davis, 2007, p. 53)*

The EU-LAC Museums project at large seeks to carry out a comparative analysis of small- and medium-sized rural museums and their communities in Europe, Latin America and the Caribbean, and to develop an associated history and theory. The basis of the project is that community museums allow under-represented communities to stake a place in history, as well as to contribute to environmental sustainability and community empowerment. Funded by Horizon 2020, the

European Union's most extensive research and innovation programme to date, the project places emphasis on sustainable economic growth and industrial leadership while tackling societal challenges.[2] We have eight international partners from Scotland (Coordinator), Portugal, Spain, France, Peru, Chile, Costa Rica and the Anglophone Caribbean (Barbados, Jamaica, and Trinidad and Tobago) (see EU-LAC Museums, 2021b). Dr Karen Brown is the Project Coordinator; Jamie Allan Brown is the Project Administrator and dedicated Youth Programme Worker. A project Advisory Board and Steering Committee consists of world-leading experts in EU-LAC relations and selected for their distinct areas of expertise. They include prominent ecomuseum and community museum specialists, Hugues de Varine, Peter Davis and Teresa Morales. Luís Raposo, the former President of International Council of Museums (ICOM)-Europe, and Samuel Franco Arce, former President of ICOM-LAC and an expert in disaster management in cultural heritage, were instrumental in helping us to plan the project, set goals and measure impact (see EU-LAC Museums, 2021a). The project is rooted in a belief in the potential for youth to transform society. It aims to help those young people become tomorrow's leaders with an awareness of their heritage and identity and how these are changing, an understanding of the challenges they face, and how these are perceived within a global context. EU-LAC Museums encourages mutual understanding between the regions to build on existing and new partnerships and aims to overcome challenges for mutual sustainability and continuous dialogue within our museum communities.

To reach the Horizon 2020 goal of 'fostering inclusive, innovative and reflective societies', the EU-LAC Museums project sought to research state-of-the-art initiatives in museums and community empowerment and move beyond those initiatives to implement actions in each partner country. The project also produced a number of academic and scientific outputs, notably an extensive bibliography dedicated to ecomuseology and community museology, and a new collection of essays, *On Community and Sustainable Museums* (2019), in which many of the project's guiding principles are explained by our project advisers (Brown et al., 2019; EU-LAC Museums, 2019b).

Recruiting young people in Europe, Latin America and the Caribbean

Participants in the EU-LAC Museums Bi-Regional Youth Exchange were Druim Nan Linntean (the Isle of Skye Ecomuseum), in Scotland; Ecomuseo de la Cerámica Chorotega de San Vicente de Nicoya (San Vicente de Nicoya Ecomuseum), Museo Comunitario Indígena de Boruca (the Community Museum of Boruca) and Museo Comunitario Yimba Cajc de Rey Curré (the Community Museum of Yimba Cajc de Rey Curré), all in Costa Rica; and Museu da Chapelaria (The Hat Museum) in São João da Madeira, Museu de Olaria (the Museum of Pottery) in Barcelos, and the Museu Municipal in Penafiel, Portugal. These locations and entities were selected precisely for their remote locations, their precarious socio-economic sustainability and the significant work they are carrying out to maintain tradition and foster resilience for the future.

The Youth Exchange project involved young people aged between 15 and 18 from geographically rural communities and different socio-economic, ethnic, cultural and religious backgrounds linked to museum communities in Costa Rica, Scotland and Portugal. Through a tailored recruitment process, the young people were brought on board according to their proven commitment and contribution to volunteering, local heritage and community museum projects. During the programme the focus was on raising awareness of their culture and their identity, how these are changing, and how they are understood in the context of the wider world. In Scotland, a rigorous application process was developed which required each young person to produce a video, take part in assessed group work, complete a comprehensive written application and attend an interview with a member of our team, a Scottish Gaelic language schoolteacher and a local ecomuseum representative. The programme's local community partner, the Staffin Community Trust, and Ecomuseum Druim Nan Linntean in Skye advised that the project should involve young people learning or speaking Scottish Gaelic to ensure an understanding of the local community and heritage. This process was an opportunity for young people to display their skills and personality outside their formal school education, and to demonstrate a commitment to their local community and the Scottish Gaelic language.

In the end, the project involved over 100 young people: 90 from Costa Rica (11 selected for travel), 6 from Portugal and 6 from Scotland. Our project goals were

1 to empower each young person to learn more about their own community, language, identity, heritage and culture, and to locate similarities with the other communities involved;
2 to foster confidence in each young person to take an active role within their individual communities and
3 to encourage each young person to reflect on and document their journey as they took part in the Youth Exchange.

Discovering nature and culture through the landscape and intergenerational dialogue

Within the programme, the young people engaged in a range of activities that fostered a 'sense of place', advocated by Peter Davis (2011) as the main role of ecomuseums, through a greater understanding of nature and culture located around the ecomuseum or community museum. The rural setting of the Isle of Skye in the north west of Scotland was selected because of its unique and protected landscape, its investment in the Scottish Gaelic language and its rural way of life. It is also home to Scotland's first ecomuseum, founded as Ceumannan (translated from the Gaelic as 'Footprints' and currently being re-branded as Druim Nan Linntean, 'Ridge of Ages').

In Costa Rica, La Red de Museos Comunitarios, the country's network of community museums, was identified as a strong partner because of its history of community participation and empowerment. The Costa Rican Youth Exchange Programme was led by the National Museum of Costa Rica and operated within the indigenous village communities of Boruca and Rey Curré in the south, and San Vicente de Nicoya Ecomuseum in the north. Monthly workshops were facilitated in partnership with members of the local community, trained by La Red de Museos Comunitarios de América (the Network of Community Museums of America) using their programme first developed by Teresa Morales and Cuauhtémoc Camarena in Oaxaca, Mexico, called 'Our Vision of Change' (Museos Comunitarios, n.d.).

Similarly, in Portugal, the project recruited young people from the rural communities of Barcelos, Penafiel and São João da Madeira within the Porto region to ensure that they had much in common with their fellow travellers; in this case, the National Museum of Archaeology worked with a consultant from the University of Porto.[3] Within Scotland, the development of the activities and workshop programme, jointly led by the University of St Andrews, the Staffin Community Trust and the Comunn na Gàidhlig (Gaelic language and culture society), fostered an opportunity not solely for the young people but also for the professionals and organisations working at grassroots level in the Isle of Skye. The ecomuseum further consolidated its role as a hub of local knowledge, showcasing, safeguarding and respecting the landscape, the way of life, the Scottish Gaelic language and knowledge of traditional crofting (Scottish subsistence farming).

The monthly workshops in each country involved a number of activities that encouraged young people to engage with their landscapes, heritage and identities, and to think critically about the issues affecting them, such as over-tourism, depopulation, globalisation and access to resources. During the exchange programme, they took part in 'Community Walkabouts', in which each host conducted a guided walk through the landscape of their community, identifying places of significance such as sacred sites, buildings, natural features and animals. The young people were encouraged to compare and contrast the host communities with their own. Their tasks included mapping the various communities and exhibiting findings for community elders and facilitators, highlighting significant sights, resources and places understood to be relevant by the young people themselves.

These maps were also displayed in farewell celebrations and ceremonies and recorded for the EU-LAC Museums Bi-Regional Youth Exchange video documentary produced by John Large. They are available on the project's social media platforms: Facebook, Twitter and YouTube (Community Museums, 2019). The documentary was screened at the Byre Theatre in St Andrews, Scotland, in May 2019, to celebrate International Museums Day. There was also a community screening in Boruca, Costa Rica, during the La Red de Museos Comunitarios de América community facilitators exchange workshop in June 2019, which explored the world of 'community museums' in Europe and Latin America (Byre Theatre, 2019).

Another example of youth engagement in the programme involved 'Community Festivals' that were organised in each community to celebrate the exchange and to showcase the young people's commitment and work. These gave the opportunity to showcase traditional dances, such as the *Fiesta de los Diablitos* (Festival of the Little Devils) in Boruca, *cèilidh* dancing in the Isle of Skye and *ranchos* in Porto, as well as to share traditional food such as *tamales* (chicken in dough steamed in a corn husk, Costa Rica), haggis (innards of a sheep with spices from Scotland) and *francesinha* (a traditional sandwich from Porto made with roast meat, cheese and tomato sauce), as well as folk stories from Boruca, the Isle of Skye and Porto. Young people were encouraged to produce a mural of each community that highlighted their journey and progress as a group. In particular, the Costa Rican and Scottish communities hosted young people so they could experience everyday life in their respective communities by visiting community elders and meeting host families, where they tried traditional food and music in their homes.

Working under the theme of 'Community Crafts and Collective Memory' further helped young people to understand the importance of memory for each community, promoting debate about young people's identities. For example, various workshops were held, facilitated by community elders and artisans, to involve the young people in, and inculcate an appreciation for, the unique artisanal skills and industries in each community. These included

- masks and textiles at Boruca and Rey Curré;
- pottery at San Vicente de Nicoya;

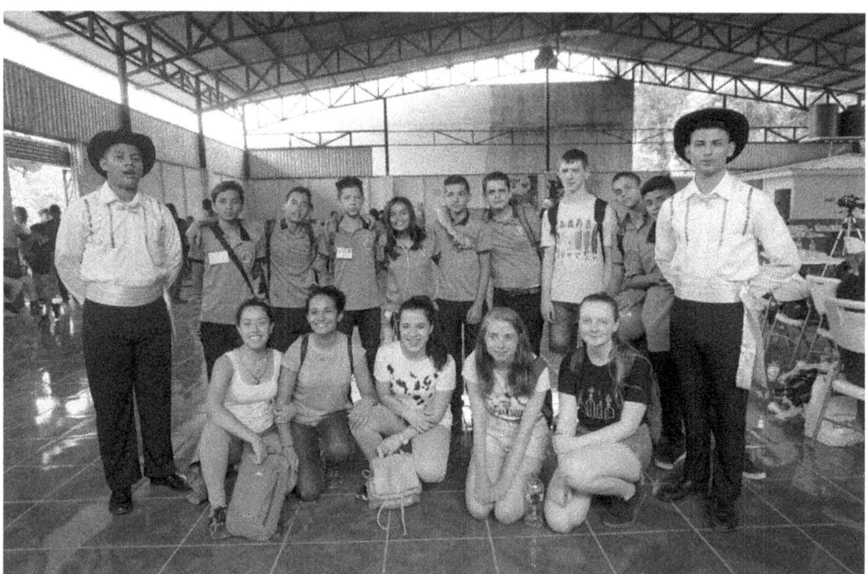

FIGURE 2.1 Community Festival, Boruca, Costa Rica, August 2017. © Jamie Allan Brown

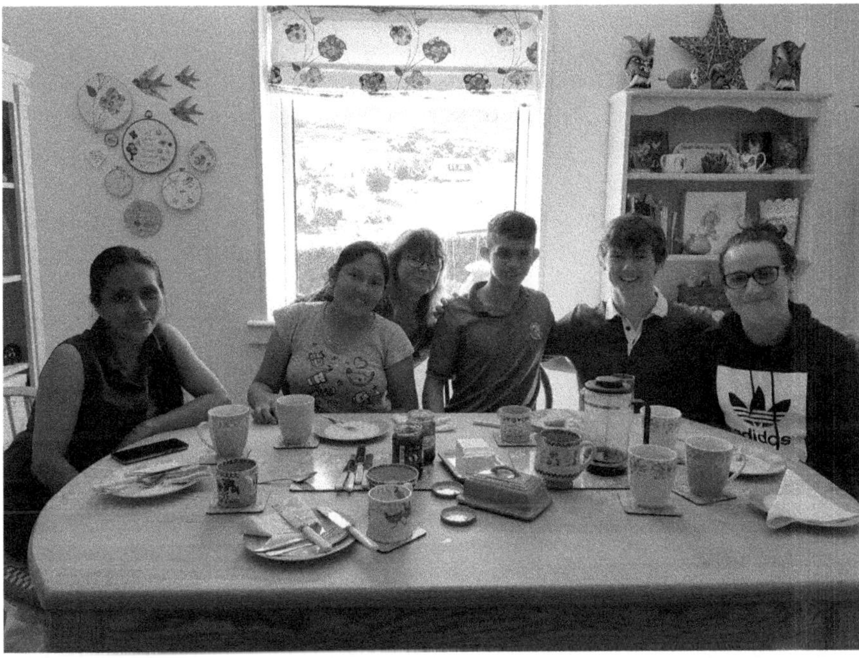

FIGURE 2.2 Home visit, Isle of Skye, Scotland, July 2018. © Karen Brown

- crofting, textiles and fishing on the Isle of Skye;
- pottery at Barcelos and Penafiel and
- hats, shoes and pencils at São João da Madeira.

Through these activities, young people developed a greater appreciation of their natural and cultural landscapes and of traditional crafts and practices. As a result of the workshops, relationships between nature and culture became more fluid and open for the young people, and they left the programme with a more profound sense of place.

A commitment to hold on to tradition and community identity was also deepened through the Youth Exchange process. One aspect shared by the young people from Scotland and Costa Rica was a concern about the issue of over-tourism and how it relates to questions of environmental, economic, social and cultural sustainability in the context of the Isle of Skye. In Skye, the district of Staffin, where the ecomuseum is located, has been appointed a National Scenic Area and is protected by the government. This poses a problem for the mass tourism that has grown on Skye, which has been made popular by not only the ecomuseum's marketing but also the filming in the locale of popular movies such as *Prometheus* (2012), *Macbeth* (2012) and *The Big Friendly Giant* (2016). In Costa Rica, tourism issues are slightly different: they concern the ethical economic development of their craft industries of mask-making, textiles and pottery (see Brown, 2017). Both locations—Scotland and Costa Rica—were seen to face threats to their traditional ways of life.

Linked to this discussion were workshops held on disaster resilience suitable for community and ecomuseums. In Skye, Samuel Franco led a theoretical and practical workshop with the young people on disaster preparedness and community resilience. The group engaged in a range of activities, including writing a report on risk-assessing their building and discussing what areas of the ecomuseum were currently most at risk from erosion (caused both naturally and by humans) due to the heavy footfall of tourism on the island.

A similar workshop was held in Rey Curré and neighbouring UNESCO site Finca 6, also led by Samuel Franco in association with the National Museum of Costa Rica and ICOM Costa Rica (then chaired by Lauran Bonilla-Merchav), and supported by the University of St Andrews Scottish Funding Council Global Challenges Research Fund project, Community Crafts and Culture (2016–2021). In this case, local first responders, such as firemen, police, first-aiders and community elders, were recruited to take part in a workshop, initiated in direct response to the severe floods that affected the community as a result of Hurricane Nate in 2017.

Rey Curré is located in a pre-Colombian settlement close to the Térraba River in the south of Costa Rica, close to the Panama border. Called Diquis ('great river') in the Borucan language, the Térraba is the largest river in the country. In the first week of October 2017, Rey Curré suffered torrential flooding. Many lower-lying family homes close to the river were ruined, and the school and community museum with which we had worked so closely and which had been constructed from traditional materials and using traditional techniques were flooded to roof height. The community suffered degrees of disruption and trauma that affected community life and indigenous systems of governance. However, their response showed enormous community resilience, as the entire community helped to rehabilitate the facilities and assist neighbours whose homes had been most affected by the disaster. Thus, they turned the challenges into opportunities through adaptability, flexibility and innovative approaches. Heritage preservation and safeguarding in this context was not always the priority of the community, and so the project needed to adapt to assist the community on its road to recovery and preparedness for the future. Adhering to the Blue Shield standards and promoting international standards in risk assessment and management, the programme addressed how to coordinate the preparation work needed to meet and respond to emergency situations and how to recover and store cultural objects until they could be rehomed. Focusing on practical tasks, the training gave participants an opportunity to reflect on past experiences and think of ways to secure their own community museums, especially to preserve local sites and artisan workshops.

In the monthly Youth Exchange workshops, young people were also encouraged to analyse and reflect on *Community Resilience* by assessing the strengths and weaknesses of their heritage communities, along with any future opportunities and threats they face. Prompted by community facilitators, young people led discussions about what a community is, what a community museum is, its role within the

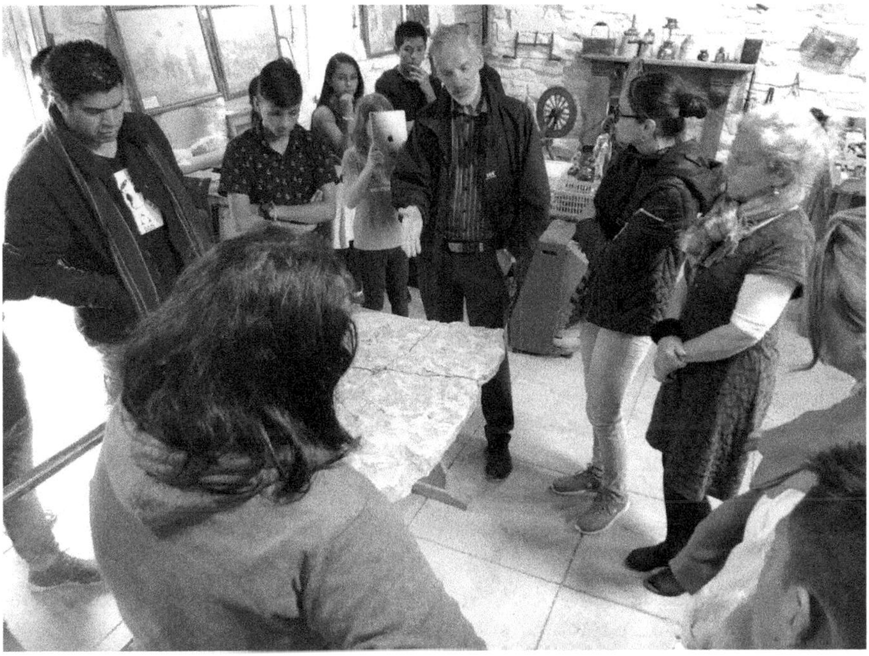

FIGURE 2.3 Community presentation, Isle of Skye, Scotland, July 2018. © Jamie Allan Brown

community and how communities can be empowered to find solutions to problems. Building on their previous workshop activities of interviewing community elders, the young people debated community resilience and identity, with a view to determining how this could be achieved in an ever-more globalised world.

During reflection workshops, the young people were encouraged to document their personal journey in an online blog, thereby building and enhancing their critical reflective skills. Blog tasks were set with various themes relevant to the topic at the time. The group acknowledged that blogging for an international audience, rather than for their local peers, community or school, boosted their creativity as they thought critically about what they wrote and how they presented themselves online through their communication and language skills.

The young people's online blogs can be found here:

- Costa Rica: *Blog de la Juventud de Costa Rica* (https://eulacmuseumsyouthcostarica.wordpress.com/)
- Portugal: *Blogue de 6 jovens de Portugal participantes* (https://eulacmuseumspt.tumblr.com/)
- Scotland: *Òigridh Air Iomlaid na Alba* (https://eulacmuseumsyouthscotland.wordpress.com/)

Accreditation and international recognition

The young people taking part from Scotland were accredited through the Scottish Government's Saltire Award volunteering certification programme in accordance with the framework of the Scottish Curriculum for Excellence, which is intended to encourage young people to gain knowledge, skills and attributes needed for learning, life and work in the 21st century (Scottish Government, 2008). This award highlights their commitment to the project and their rural community and is signed by the First Minister of Scotland. It can be used in university or further education applications and fully adheres to the national high school curriculum. All six young people have listed their Saltire Award in university or further educational CVs.

Large EU-LAC Museums project meetings were taken as opportunities to develop and share the experiences of the Youth Exchange, and further develop the young people's skills of public speaking with confidence. The Saltire Awards were presented to the group at the project's Second General Assembly and International Committee for Museology of the International Council of Museums conference on 'Defining the Museum of the 21st Century', held in St Andrews, Scotland, in November 2017.

The EU-LAC Museums *Itinerant Identities: Museum Communities/Community Museums* conference in 2018, coordinated by the project's Caribbean partner, the

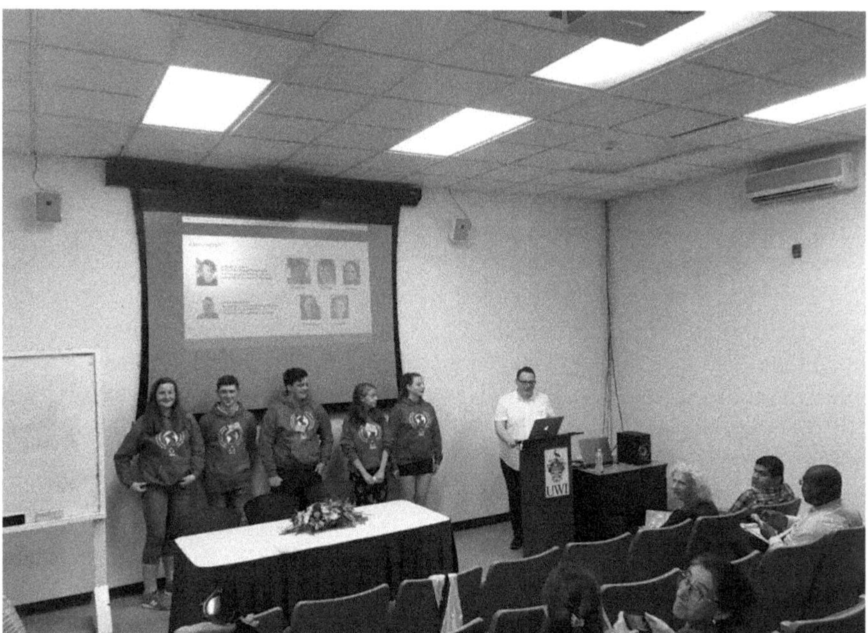

FIGURE 2.4 Presenting at the *Itinerant Identities* conference, the University of the West Indies, Barbados, November 2018. © Karen Brown

University of the West Indies and by the Museums Association of the Caribbean, explored critical issues researched within the project and beyond. These included gender and migration, sustainable development, the role of youth and the role of new technologies. Youth Programme Worker Jamie Allan Brown presented the Bi-Regional Youth Exchange alongside five of the Scottish young people. The presentation stimulated interactive discussions from the international audience and highlighted the plight of young people's daily lives within rural and island communities, touching on the cultural and colonial legacies between Scotland and Barbados as well as Spain and Costa Rica. The conference also offered an opportunity to run additional disaster resilience workshops, with Samuel Franco leading a session in the Barbados Museum and Historical Society for the museum's 'Young Curators', the Youth Exchange participants and museum professionals.[4]

It is hoped that the Youth Exchange will serve as a model to be applied in other contexts and communities, such as at the parish of St Andrew in Barbados. To this end, during the General Assembly, Kaye Hall from the Barbados Museum and Historical Society arranged for Jamie Allan Brown and Karen Brown to visit a new area of the island targeted for local development, with Jamie Brown also visiting the local high school to initiate early discussions.

Finally, during the 25th ICOM General Conference, held in 2019 at Kyoto, Japan, and entitled 'Museums as Cultural Hubs: The Future of Tradition', Jamie Allan Brown presented the Bi-Regional Youth Exchange to the ICOM Committee For Education and Cultural Action. He and Karen Brown also presented at the ICOM International Committee for Regional Museums before museum professionals, experts, educators and community leaders (see ICOM, 2019a).

During the general conference, ICOM members voted to adopt the EU-LAC Museums' proposed resolution on 'Museums, Communities and Sustainability' with its specific focus on building the capacity of ecomuseums and community museums, in order to remain sensitive to local and regional differences, and to demonstrate awareness of the geopolitical dimension of the concept of the museum, especially relating to the resource needs of community-based museums in low- to middle-income countries such as those involved in the project (see ICOM, 2019b). Future youth exchanges would be one concrete way in which this resolution could be taken forwards.

Final reflection

The Bi-Regional Youth Exchange was developed in collaboration with communities around thought-provoking ideas on the challenges facing young people living in rural communities across Europe, Latin America and the Caribbean. It was based on the broad EU-CELAC Action Plan themes expected by the funders, including social inclusion, sustainability, gender and addictions (European Council, 2015), but was adapted to each context, such that in Costa Rica, the 'Our Vision of Change' programme was implemented in full as the Red de Museos Comunitarios de America works to strengthen the community for the future. Activities sought to foster empathy between the young people and to nurture debate and critical thinking, creating a mindset in which our young people believe that they have the power to make a difference. In this way, it is in tune with our global project aim for long-term, sustained societal change based on lasting institutional partnerships, professional relationships and friendships beyond this project.

The programme was developed by the project but, most importantly, was led by the communities involved, who took part in all the decision-making processes. By creating a space for the community as a whole as well as the young people to share their thoughts, debate solutions and physically visit each community, EU-LAC Museums has fostered mutual understanding, and cultivated shared experiences and knowledge among the regions through intergenerational discussions, traditional demonstrations and the empowerment of the young people for each community's sustainability in an ever-more globalised world.

Here is what the young people themselves had to say.

Jonathan Smith (2018), Scotland:

> Though many miles lie between them and us we all are brought together through our shared passion for music, dance, art and community spirit. The exchange changed me as a person in so many ways. It made me proud of my island background[,] improved my confidence and gave me skills which will stay with me forever. I want to stay in Skye, really make a difference to the island, challenge tourism and retain our way of life for both locals and visitors, like the way the Boruca community does.

Yunieth Quirós (2018), Costa Rica:

> It's not every day you get to stay with people on the other side of the world, who show you that they're proud of their community, that they work as a team, and it moves you, because you start to understand that you can do the same in your own community. You can become more interested in your own culture, your own heritage and in your own language.

Rita Leite (2018), Portugal:

> I am very grateful for the friendships that resulted through the exchange. Of the things I learned, I feel that the most important is to value and respect the different ways of life in all our communities. Our history, our heritage is different, but we are all the same, facing the same problems.

And here are some responses from the professional museum world.

Teresa Morales, Co-Director of La Red de Museos Comunitarios de América:

> The testimonies of the young people who participated in this programme are evidence that it is possible to share concepts, methods and experiences in Europe

> and Latin America, in ways that enrich the practices of community museums and ecomuseums in both regions.
>
> *(cited in EU-LAC Museums, 2019a)*

Adriana Morales (2019), Community Museum Association leader:

> For the Boruca people, it was very valuable for the school, the community, very important because we have seen the change in them [young people]. We have seen that they are more focused in culture, histories, and our memory. It has been very important [...] for us, it has been an achievement as a museum, as a community and also personally [...] [We are] very grateful that we had the opportunity to strengthen our relationship and know that we can work together and knock on doors so it doesn't end here and this legacy grows and more young people get involved and carry on this path.

In conclusion, young people in Europe and Latin America—as in the rest of the globe—face many challenges today. The teenage years are an especially challenging period of transition in which young people seek to understand their family, their heritage and their place within their ever-changing community and the wider world. The EU-LAC Museums Bi-Regional Youth Exchange offered an alternative way to empower young people through shared heritage.

Museums need to be seen as forward-looking and innovative, rather than as institutions that look only to the past. In re-thinking the role of regional museums, we would suggest that encouraging the sustainable use of cultural and natural heritage must include—if not begin with—youth and intergenerational transmission of knowledge. Such an investment will promote positive cultural attitudes towards the environment and re-interpret social and/or ecological issues with fresh eyes and sustain communities with new ideas.

Acknowledgements

This project has received funding from the European Union's Horizon 2020 research and innovation programme under grant agreement No. 693669. For further information, please visit: https://www.eulacmuseums.net/.

Notes

1 For the purposes of this essay, we are working with the ICOM (2007) Museum Definition.
2 The goal of Horizon 2020 is to ensure that Europe produces world-class research, removing barriers to innovation, thus making it easier for both the public and private sector to collaborate delivering innovation (see European Commission, n.d.).
3 The dedicated youth leader from Costa Rica is Ronald Martínez Villareal from MNCR, and, from Portugal, Paula Menino Homem from Porto.
4 The Scottish youth group further presented at the "International Conference on Community Heritage" funded by the Royal Society of Edinburgh in 2019, organised by the University of St Andrews, which provided a platform for grassroots initiatives

and organisations to better understand the community heritage landscape of Scotland (Brown and Caezar, 2020).

References

Brown, K. (2017). 'Paradigm or predator? Eco- and community museums in Scotland and Costa Rica', *ICOFOM Study Series*, 45, pp. 23–36.

Brown, K. and Caezar, A. (eds.) (2020). *Proceedings of the International Conference on Community Heritage*. St Andrews, 6–8 November 2019. St Andrews: Proceedings of the International Conference on Community Heritage. Available at: http://communityheritage.wp.st-andrews.ac.uk/files/2020/06/Community-Heritage-Proceedings-Nov-2019.pdf (Accessed: 15 March 2021).

Brown, K., Davis P. and Raposo, L. (eds.) (2019). *On community and sustainable museums*. Lisbon: EU-LAC Museums. Available at: https://eulacmuseums.net/index.php/detail-2/detail-4/207-compedium (Accessed: 5 April 2021).

Byre Theatre (2019). *EU-LAC: Documentary screening of the Horizon 2020-funded EU-LAC Museums Project Youth Exchange*. Available at: https://byretheatre.com/events/eu-lac-documentary-screening-of-the-horizon2020-funded-eu-lac-museums-project-youth-exchange/ (Accessed: 20 April 2021).

CEPALSTAT (2020). *Regional panorama of Latin America and the Caribbean: Selected indicators*. Available at: https://estadisticas.cepal.org/cepalstat/portada.html?idioma=english?string_busqueda=poverty (Accessed: 5 April 2021).

Community Museums (2019). *Bi-Regional youth exchange experiences*. [Online video]. Available at: https://www.youtube.com/watch?v=BBP25BQoTtI (Accessed: 20 April 2021).

Davis, P. (2007). 'Place Exploration: museums, identity, community', in Watson, S. (ed.) *Museums and their communities*. London: Routledge, pp. 53–75.

Davis, P. (2011). *Ecomuseums: a sense of place*. 2nd edn. London: Continuum.

EU-LAC Museums (2019a). *Experiences from the Bi-Regional Youth Exchange*. Available at: https://eulacmuseums.net/eulac_museums_docs/Horizon2020_693669_eulacmuseums_Experiences_Oct2019_LOWRES2.pdf (Accessed: 20 April 2021).

EU-LAC Museums (2019b). *Select ecomuseums bibliography*. Available at: https://eulacmuseums.net/index.php/bibliography/details/1/89 (Accessed: 20 April 2021).

EU-LAC Museums (2021a). *International advisors*. Available at: https://eulacmuseums.net/index.php/international-advisors (Accessed: 5 April 2021).

EU-LAC Museums (2021b). *Project partners*. Available at: https://eulacmuseums.net/index.php/project-en/partners-2 (Accessed: 20 April 2021).

European Commission (n.d.). *Horizon 2020*. Available at: https://ec.europa.eu/programmes/horizon2020/en (Accessed: 20 April 2021).

European Council (2015). *EU-CELAC action plan*. Available at: https://www.consilium.europa.eu/media/23757/eu-celac-action-plan.pdf (Accessed: 5 April 2021).

European Parliamentary Research Service (2018). *Ten issues to watch in 2018*. Luxembourg: Publications Office of the European Union. Available at: https://op.europa.eu/en/publication-detail/-/publication/8d7ecf50-f74d-11e7-b8f5-01aa75ed71a1 (Accessed: 5 April 2021).

European Youth Forum (2015). *Youth Forum addresses European, Latin American and Caribbean leaders on decent employment*. Available at: https://www.youthforum.org/youth-forum-addresses-european-latin-american-and-caribbean-leaders-decent-employment (Accessed: 5 April 2021).

Eurostat (2021). *Young people – Social inclusion*. Available at: https://ec.europa.eu/eurostat/statistics-explained/index.php?title=Young_people_-_social_inclusion#Young_people_at_risk_of_poverty_or_social_exclusion (Accessed: 5 April 2021).

Eurostat (n.d.). *Youth database*. Available at: https://ec.europa.eu/eurostat/web/youth/data/database (Accessed: 5 April 2021).

Ghilleasbuig, S. (2019). Interviewed by Karen Brown and Jamie Allan Brown. 16 February, Skye, Scotland.

ICOM (2007). *Museum definition*. Available at: https://icom.museum/en/resources/standards-guidelines/museum-definition/ (Accessed: 5 April 2021).

ICOM (2019a). *ICOM Kyoto 2019: 25th ICOM General Conference*. Available at: https://icom-kyoto-2019.org/index.html (Accessed: 20 April 2021).

ICOM (2019b). *Resolutions adopted by ICOM's 34th General Assembly, Kyoto, Japan*. [PDF]. Available at: https://icom.museum/wp-content/uploads/2019/09/Resolutions_2019_EN.pdf (Accessed: 20 April 2021).

Janes, R. R. and Sandell, R. (eds.) (2019). *Museum activism*. Routledge.

Leite, R. (2018). Workshop contribution, *First EU-LAC physical exchange*. Boruca, Costa Rica, 1 August.

Mac Devitt, A. (ed.) (2017). 'Museums and cultural landscapes [Special issue]. *Museum International*, 69(1–2), pp. 3–5.

Morales, A. (2019). Workshop contribution, *EU-LAC Museums community leaders' workshop*. Boruca Community Museum, 6 July.

Museos Comunitarios (n.d.). *Red de Museos Comunitarios de América*. Available at: https://www.museoscomunitarios.org/redamerica (Accessed: 20 April 2021).

Museums Association (2020). *Declaring a climate emergency*. Available at: https://www.museumsassociation.org/campaigns/climate-crisis/declaring-a-climate-emergency/ (Accessed: 5 April 2021).

Quirós, Y. (2018). Workshop contribution, 'Memory and community museums: the right to self-determination and the role of youth', *EU-LAC Museums Conference*. Boruca, 9–10 March.

Riva, R. (ed.) (2017). *Ecomuseums and cultural landscapes*. Santarcangelo di Romagna: Maggioli Editore.

Smith, J. (2018). Workshop contribution, 'Itinerant identities: Museum communities/community museums', *EU-LAC Museums conference*. Bridgetown, 8 November.

UN DESA (Department of Economic and Social Affairs) (2016). 'Identifying social inclusion and exclusion', in *Report on the world social situation 2016*. New York: United Nations, pp. 17–32. Available at: https://www.un.org/esa/socdev/rwss/2016/chapter1.pdf (Accessed: 5 April 2021).

Youth Policy Labs (2007). *Latin America and the Caribbean: youth facts*. Available at: http://www.youthpolicy.org/mappings/regionalyouthscenes/latinamerica/facts/#FN18 (Accessed: 5 April 2021).

3

PASSION AS A MOBILISING TOOL FOR COMMUNITY-BASED MUSEUMS

Case studies from southern Chile[1]

Karin Weil, Bárbara Elmúdesi Krögh, Laura Fúquene Giraldo and Javiera Errázuriz Contreras

The role of passion and affection in the making of community-based museums

> Community museums are initiatives in response to many communities' urgent need to defend their identity and their very existence in this situation [...]. They are a response to the imposition of oblivion, an expression of the firm decision to remember, to preserve memory, to build self-determination and dignity.
>
> *(Camarena and Morales, 2016, p. 10)*

Community-based museums emerge from communities, from the people who inhabit a territory and share a collective memory produced through dialogue and interaction. Meaning, which is constructed from personal communication that is open to plurality, is expressed through different languages and activities in museum exhibitions, in the selection of the objects that represent community and trigger those memories, and through the ongoing participation of the communities themselves.

Within the framework of the Europe-Latin America and the Caribbean (EU-LAC) Museums project, an interdisciplinary team from Universidad Austral de Chile studied the social role of community-based museums and their impact on the sustainability of their communities, heritages and territories. The inquiry revealed the central aspects of this social role in the traditional museological sense, both from a theoretical standpoint and from the actual experiences and memories of visitors—that is, those who bring community-based museums to life.

The main value of community-based museums lies in the ideas, actions, decisions and emotions of the community members, along with their personal and collective relationships. In such museums, the people who inhabit and share a territory construct meaning from their collective memory, preserving and presenting

DOI: 10.4324/9781003288138-5

their own interpretations of their histories. They provide spaces for the collective expression of subjective experiences, which may include uncomfortable narratives often glossed over by official institutions. In so doing, they demonstrate the community's determination and commitment to legitimise their own stories, rendering them visible once more.

We investigated the role of five museums in Los Ríos, Chile, over the course of three years. Through case studies, comparative analysis and a review of bibliographic sources, we have found that social construction in community-based museums takes place through the affective bonds and enthusiasm of those community members who bring these spaces to life and sustain them.

We have also been able to show, and thus problematise, the direct relationship between, on the one hand, the sustainability of community-based museums and the community, territory and heritage they protect, and, on the other, the implicit motivation of those who bring them to life. In this relationship, affection and enthusiasm for one's own community play fundamental roles in the 21st-century human need to restore community life, in the political-cultural struggle for social change and in meeting the challenges and objectives of sustainable development. In this regard, issues such as gender equity, inclusion and equal representation, free access to information and uncensored communication of local history are relevant.

Our research showed that the strength of community-based museums comes from the passion and commitment of each member of the community; they are bearers of knowledge, the ones who document the history and feelings attached to the creative process as they decide what is to be told, how and for whom. What gives community-based museums their identity and value is *not* related to museography—that is, local themes, such as community characteristics and territory—nor to regional or local funding. Rather, their identity and value are rooted in the fact that they tell life stories and share history and community memories; beyond the topics they address, their existence is a testament to community life, to actual living, breathing cultures. They exemplify the appropriation of an institutional condition onto which a deep historical, political and cultural meaning is imprinted. The perseverance of a community willing to create a space that allows them to share their memories, stories and meanings, as well as to transform it into a counter-hegemonic movement to raise consciousness, demonstrates that community's enthusiasm to build a sense of belonging.

From *museion* to community-based museum: passion as a mobilising tool

> Neltume is a museum characterised by writing what people have said as they say it, how they think and feel it. We do not put the words forward from another more enlightened field, but only as people speak and feel it; [thus] it is transmitted.
>
> —*Angélica Navarrete, Centro Cultural Museo y Memoria de Neltume [Museum and Memory Cultural Center], Neltume, Chile, May 2017*

Museums have long had the reputation of being temples of knowledge, repositories of history and guardians of the past, spreading educational and hegemonic monologues. Today's museums are a product of the 'new museology' and trends that emerged in the 1960s. As such, they aim to collect the past and the present, engage in dialogue (rather than pronounce in monologue) and encourage debate and interactive learning. Communities are at the core of their mission and their institutions.

Taking on board the guidelines adopted at the 16th UNESCO General Conference, a transformation in museum practices took place in Chile with the so-called Round Table of Santiago de Chile, 1972 (Mostny, 1972a, 1972b). The General Conference took place at the GAM (Gabriela Mistral Cultural Centre), a building constructed between 1971 and 1972 to host the Third United Nations Conference on Trade and Development (UNCTAD III) held between 13 April and 21 May 1972. The General Conference was marked by the presence of delegates from the developing countries who demanded fair global trade, as well as world solidarity with the Chilean process led by President Salvador Allende. Allende's government invited a group of museologists from different Latin American countries to discuss 'the importance and development of museums in the contemporary world' (Mostny, 1972a, p. 5). It was not a coincidence that a democratic socialist government such as Allende's took on the principles of the new museology to propose an inclusive Latin American response. The gathering, which took place over ten days, was characterised by the participation of specialists from different disciplines—urban planning, agriculture, education, and science and technology, with an approach focusing especially on the region's needs. The meeting reaffirmed the perspectives already considered at the ICOM (International Council of Museums) General Conference in Grenoble the previous year, deepening ICOM's intention to strengthen the commitment of Latin American museums to the social development of their respective communities, and expressing the importance of the museum, as well as its potential to contribute culturally to the transformative action of these communities.

The Round Table's main conclusions were articulated around the notion of the 'integrated museum', through which participants proposed a new image for the institution, which should be closely linked to the present and future of the community (Varine, 2012). It ended by outlining ideas to subvert museum practices at their very foundations, decentring the traditional role of museums and relocating their value in their potential to reinvent themselves. Museums were to contribute as developmental tools to provide a 'social function', thus anticipating today's museologists' professional practice regarding their political and cultural responsibility (Varine, 2012, p. 98).

In this way, the museum's social role and its relevance for the present and future of its communities not only upends the museum (considering its etymological origin, museion) and its nationalist burden but also powerfully connects the museum to the present aspirations, challenges and pain of the communities it represents (González, 2016). This social function involves affection and enthusiasm as mobilising tools for museums, especially those with a community base, which restores the agency and vibrant materiality of museum objects (Escobar, 2016). This approach challenges the supposed inert nature of objects and turns them into social agents that promote experiences in the social network, connecting with those people who create, share, criticise or become part of the object´s meaning. (Simon, 2010).

Affection effects

> There is nothing more fascinating than examining how memory returns; for some the trigger is a smell, for others, the line of a face, the colour of an object or even a word that will make sprout an infinity of small moments that will confirm the whole of memory.
>
> *(Brousseau, 1991, p. 10)*

The social role of community-based museums is imbued with not only history and memory but also emotion and deep affection, since these museums are deeply rooted in their own contexts, in what is shared, what unites them and what differentiates them.

When we speak of community-based museums, we are talking about a group of people who recognise themselves in others, and who have shared needs, desires, territory, identity, affinities and so on. However, sharing certain attributes alone is not enough to create a community; the relative strength of the connection between members of the community is also important. We agree with John H. Falk (2009, p. 147) who states: 'the stronger the emotional "value", the more likely sensory information is to pass this initial inspection and be admitted into memory; and interestingly, pleasant experiences are strongly favoured over unpleasant ones'.

In the context of our research, and especially with regard to community-based museums, emotion and affection are historically and culturally contingent. Linked to power and political relations, they are starting points when discussing issues of heritage and local territory. Here, we follow Laurajane Smith, Margaret Wetherell and Gary Campbell's (2018, p. 19) assertion that:

> heritage, as a practice of making meaning, draws heavily on affect/emotion to legitimise the meanings and narratives produced and propagated. Heritage's emotional force is part and parcel of the power of heritage to stand in for and legitimise claims to inclusion or exclusion based on identity, nation and citizenship.

In this sense, the power struggle against hegemonic narrative discourses aimed at decolonisation and equality in relation to indigenous peoples, gender equality, human rights, guerrillas and natural resources, among others, materialises mainly because of two great forces: (1) emotions that transcend individuality, commitment and the need to give meaning to past and present situations and (2) the need to establish membership of, and identification with, a community.

In this way, museums are vital to the societies in which they belong. By museums, we mean both the spaces (the museum as an institution at community, rural, small or medium scale) and the people who give life to them. The latter includes the visitors or users, and above all the museum workers themselves.[2] Often, these workers are those who not only manage the museum but have also played important roles in its foundation. They have made the museum's work their own, and their involvement goes beyond its management. The museum's story is their own story; they are part of the narrative. A truly community-based museum is marked by the affections and enthusiasm of the people who created it. We should understand that:

> affection and emotion have a consequence for the way people understand and experience the world in which they live [...] Emotions are not actions, but they provide an inner energy that propel us towards an act, they provide the energy for cognition and evaluation.
>
> *(Smith and Campbell, 2015, pp. 15–16)*

Within the framework of the international research project 'EU-LAC Museums: Museums and community: concepts, experiences and sustainability in Europe, Latin America and the Caribbean', we have been able to share experiences around museology specifically in community-based museums, and how they developed from their own territorial and historical contexts to become spaces that encourage dialogue, add value and make sense of their 'places'. In each of our case studies, the collections, narrative, research, actions and exhibits highlight the centrality and reach of each community-based museum's social role as fundamental to its sustainability, linking it directly to the motivation, experiences, memory and affections of the people in charge and those who participated in the process.

Each of these museums or community spaces illustrates the complexity and diversity of individual and communal emotional relationships established with the past. A significant insight offered by Margo Shea (cited in Smith, Wetherell and Campbell, 2018, p. 21) is that 'the contemporary social meaning constructed in affective commemorative practices may at once be complex, contradictory and ambiguous', causing internal conflicts. Dynamic museum projects tell the story of the individual and the social context of community-based museums, of how personal and communal affections must be collectively negotiated, and how the kind of consensus achieved through such practices reconstructs the meaning of the past.

The passions behind the construction of the Neltume, Malalhue and La Aguada museums

> They [the Mapuche people] see us as a valuable addition, given the knowledge that many of them in their homes are not able to continue replicating history, so

> this space was built for that purpose [...] this is the basis of this museum. We are not digging, buying or asking for things: they come on their own.
>
> —*Nerys Mora and Isabel Riveros, Museo Despierta Hermano, Malalhue, May 2017*

Within the framework of the EU-LAC Museums project, we investigated five cases of community-based museums in the Los Ríos region. Based on interdisciplinary work and exhaustive methodologies, we selected these cases for their diversity in terms of locality, resources, theme, size and social role, among others. Shortly after moving forwards with the study, we realised that the evolution of their social role was marked by their founding principles.

For Museo Despierta Hermano de Malalhue (Museum awakens brother of Malalhue), the founding principle was discrimination; for Museo Escolar Hugo Günckel de la Aguada (Hugo Günckel School Museum of La Aguada), it was environmental conflict; and in the case of Centro Cultural Museo y Memoria de Neltume (Cultural Center Museum and Memory of Neltume), it was human rights violations. By contrast, the genesis of Museo Tringlo de Lago Ranco (Lago Ranco's Tringlo Museum) and Museo Histórico y Antropológico Maurice van de Maele (Maurice van de Maele Historical and Anthropological Museum) reflected extrinsic demands imposed over those of the community in a top-down, rather than bottom-up, approach (Weil et al., 2018). The collections on which these five museums were formed were officially legitimised as heritage, that is, recognised as traditionally

TABLE 3.1 Case studies summary

Museum	*Source*	*Environment*	*Institution*	*Theme*
Centro Cultural Museo y Memoria Neltume	Community	Andes mountains	Autonomous	Socio-political Agrarian reform Human rights
Museo Despierta Hermano Malalhue	School and community	Andes foothills	Autonomous	Intercultural
Museo Escolar Hugo Günckel Corral	School	Coast	Municipal Public School	Environmental education
Museo Histórico y Antropológico M. Van de Maele	University	Urban Valdivian jungle	Universidad de Austral, Chile (UACh)	Hispanic and German colonisation Native people
Museo de Tringlo Lago Ranco	Municipal	Lacustrine valley	Municipal	Prehispanic and colonial ceramics

Source: Karin Weil.

'valuable', and worthy of preservation and care. In these latter two museums, the social role has historically been less obvious.

Among the five museums that make up the case studies, we can highlight the coherence with which three of them create and recreate their social role. Despierta Hermano de Malalhue, Centro Cultural Museo y Memoria de Neltume and Museo Escolar Hugo Günckel de la Aguada play a prominent role in this analysis because they are museums that, of their own volition, have taken over the socio-cultural, political, environmental challenges and conflicts in their respective communities. From the beginning, this function has given them deep roots in their territories, and they have found solutions to the problems that affect them by using the language of museology.

Today's community-based museums, often located in marginal situations, are spaces that seek to protect and offer relief to their communities through their own memories and shared contexts. As institutions that carry threads from the past through the present and to the future, they provide safe spaces for reflection in resistance to the dominant powers and hegemonic institutions. The three museums discussed here do this in different ways.

The Museo Despierta Hermano de Malalhue was established to emphasise diversity as a value in a territory with a significant Mapuche population. This was a reaction to the significant discrimination suffered by the children of those indigenous people. As Maori author Linda Tuhiwai Smith (2012, p. 108) suggests, '[...] the activities of indigenous peoples can be understood at one level simply as an indigenous social movement [...] [that] developed simultaneously out of the survival strategies and cultural systems', values and beliefs that have nurtured these communities in the past. Processes, like the ones referred, began under the radar of the dominant society. Examples of this exist throughout 500 years of anti-colonial struggle. However, in the museum context, the new museology and its development from the 1960s has generated new spaces in which this movement has found new tools to express itself. Indigenous communities have been turning these often isolated and marginalised spaces into places steeped in spiritual meaning and indigenous identity, generating content from the communities, and traditional knowledge from oral testimonies and stories. In the case of the Museo Despierta Hermano de Malalhue, the community's active participation, based on their commitment and passion, validates new methodologies, celebrating survival through resistance and a process of regeneration that creates connections among people and represents diversity. The aim is to protect the people, communities, languages, customs, beliefs and natural resources of the region, and to validate, value and recognise traditional knowledge in the context of the 21st century. This has become a necessity in order to recover the values of the community, along with its social relations, sense of well-being and balance within the ecosystem.

The Centro Cultural Museo y Memoria de Neltume, located in a Cordillera town that was established in the 1940s as a result of forestry, aims to reconstruct the workers' social history and human rights during the Pinochet dictatorship.

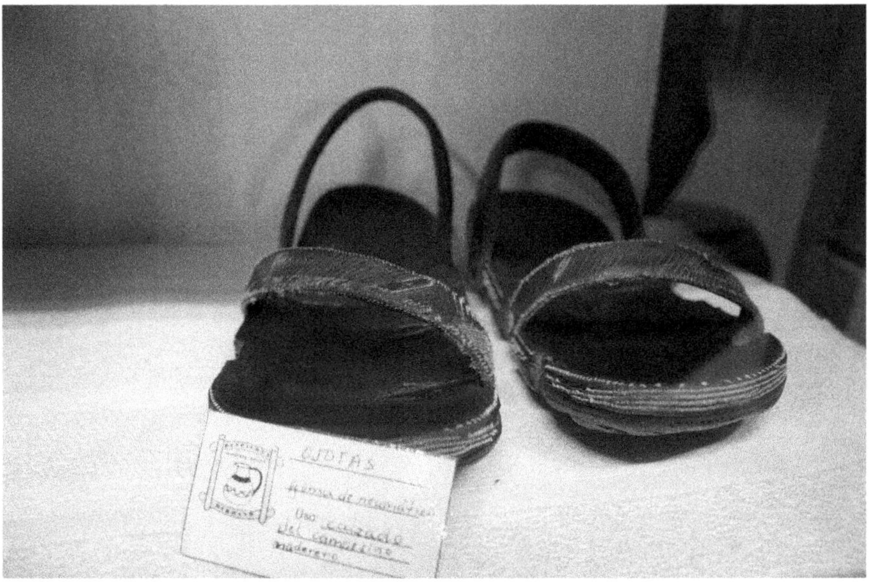

FIGURE 3.1 Ojotas. Footwear made from tyre rubber used by wood farmers. Part of the Museo Despierta Hermano collection. © Claudia Ordóñez for EU LAC Project

Freud has affirmed the formidable existential power of the past, which we cannot evade, and which continues to influence our mental processes and reality (Anton, 2014). The world is full of memories of repeated wars, genocides and violations of individual and community rights, which we often repress or deny so that the group can continue to move forwards into the future. However, hegemonic narratives often manage to avoid the memory of trauma. Although we try to replace those memories considered negative with more benign and tolerable memories, the inescapable presence of trauma impairs our vision of the present and prevents the future from coming to life. It is not possible, from a psychological point of view, to displace trauma altogether; people will at some point have to face and understand the traumatic events of the past. They can do this by recovering their own experiences and memories, by rendering trauma visible, and by working on their suffering to transform and overcome it, thus preventing it from happening again and developing community resilience (Veneros Ruiz-Tagle and Toledo, 2009).

Finally, the Museo Escolar Hugo Günckel seeks to generate critical reflection regarding the socio-environmental conflicts affecting a territory heavily degraded by different extractive industries, such as forestry and siderurgy. The new museology and the movements that emerged in the 1960s are committed to this kind of integral museum, that is, one that considers the totality of society's problems and establishes itself as a dynamic instrument of social change. Setting aside traditional museum roles, such as collecting and conserving, it proposes a flexible and dynamic

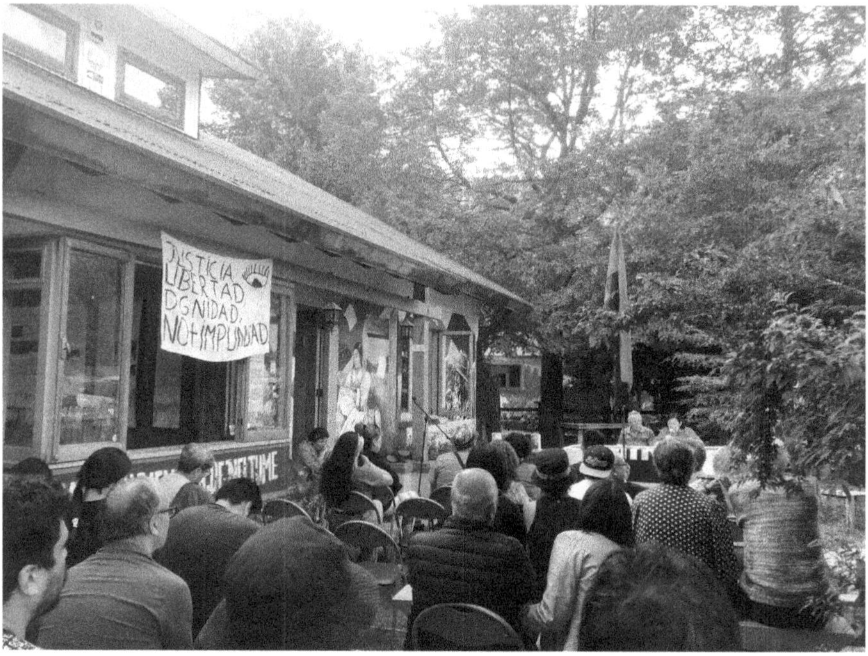

FIGURE 3.2 Presentation of the book *Flor de la Higuera* by Omar Rubio in the context of the Memory encounter at the Centro Cultural Museo y Memoria, Neltume, February 2020. © Karin Weil.

concept of heritage that is managed by the communities themselves. The committee of experts who took part in the Round Table of Santiago de Chile concluded:

> It is recommended that the museum intensifies its role as an excellent element in the education of the community in general using all means of communication. It is recommended that museums should intensify their function as the best possible agent of permanent education for the community in general by making use of all the communication media.
>
> *(Fernández Guido, 2012, pp. 229–30)*

School museums, in particular, seek to make students aware of the value and necessity of conserving and recovering the ecosystems and biodiversity of their own territories. The museum is thus transformed into a laboratory in which children gain knowledge of their environment, develop meaningful learning tools and are mobilised to protect and deepen their knowledge of local heritage through active and meaningful conservation. The space for democratic education provided by museums is thus linked to the pupils' own world, while at the same time taking advantage of the plurality of voices, latent in all classes, which enhances the educational potential (Dysth, et al., 2016).

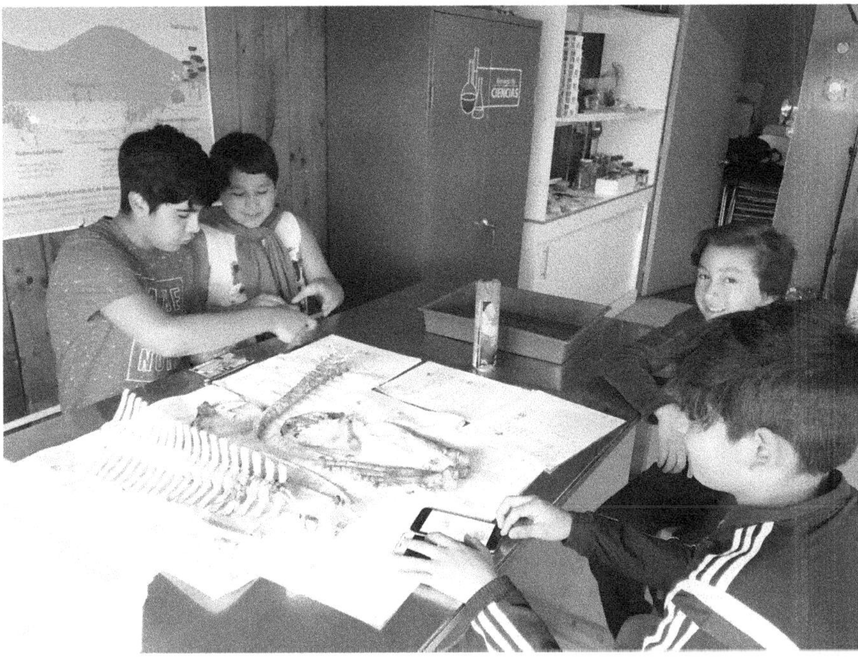

FIGURE 3.3 Museo Escolar Hugo Günckel, La Aguada, 2019. © Karin Weil.

Whether they are shaped or have been shaped by these museums, the people who support these museums have done so because of their enthusiasm for, and strong emotional connection with, the core principles behind them. The passion and commitment that drives their work—their yearning and pain—is channelled into preserving their communities, heritage and territory, and not only keeps them aligned with the cause but also becomes part of their story.

From a classical economic standpoint, these museums exist thanks to the unpaid efforts of the people in charge who feel a strong sense of commitment to the community. In Museo Escolar Hugo Günckel, the science teacher took over the running of the museum as part of his role, including more responsibilities to his activities and programmes. In the Neltume and Malalhue museums, this precariousness is more extreme, since their managers receive no salary at all.

Unfair passion: inequalities in community-based museums

> Here, I do not work to please the tourists, I work to meet the needs of my community. Here the tourist adapts to our activities.
>
> —*Lorena Carrillo, Biblioteca Municipal 332 BC1, Futrono, 2017*

Even though the enthusiasm and strong feelings of those who brought these museums to life have caused these vibrant spaces to flourish and become the cornerstones of their communities, this mobilising passion also promotes inequities and inequalities. For example, they have perpetuated the marginalisation of these spaces, along with already existing inequities and unequal gender roles. People who saw the need for spaces that might resolve social, political, cultural and environmental conflicts in their communities created the community-based museums that we investigated. These people were driven by enthusiasm and emotion rather than guided by the theoretical and practical knowledge associated with museum disciplines (which, for the most part, they do not possess). Thus, through tenacity and intuition, they have created spaces that are inadvertently marginalised from official considerations of what a museum is.

Researchers at a community-based museum often do not document and follow international guidelines and good practices; instead, they build the museum from the personal and collective experience of their community. They do not investigate from a place of academic or theoretical expertise, but from local histories and traditions. Their administration practices developed from knowledge of the community, its territory and heritage rather than a cultural management perspective. A community-based museum recognises the value of its heritage and work in the community, its experiences, its stories and memories. This work often sets such institutions apart from formal museum practice, but not from the museum's social role.

The marginalisation of community-based museums from national and international public policies has serious effects on the perpetuation of inequities and inequalities. Community-based museums exist outside the ICOM (2007) definition of a museum; therefore, they are beyond official recognition. This marginalises them from discussions on public policy, access to public funds and official recognition of their collections—a situation that puts their heritage at risk and makes official stories and memories even more dominant.

Researchers, curators and artists have been studying and bringing to light gender inequalities in the cultural heritage context for decades. Since the 1980s, the feminist group Guerrilla Girls has carried out many artistic interventions in the world's most popular cultural spaces and drawn attention to sexism in the world of the arts and museums.[3] Likewise, gender studies of heritage and museum culture have proliferated in recent years, revealing stereotypical representations of women and a lack of representation in museum management positions (Turner, 2002). Even though the largest number of museum workers are women, the highest leadership positions continue to be held mostly by men (Westerman, Schonfeld and Sweeney, 2019).

Our study into community-based museums shows an opposite situation: most of the managerial positions in the Los Ríos region are held by women. Although this is good news, it obscures the perpetuation of associated injustices and stereotypes in Western society in which the functions of caring and unconditional dedication are typically carried out by women. The passion that helped to create these museums is the same passion that has perpetuated gender stereotypes and injustices in the museums.

As Amaia Pérez Orozco and Sira del Río (2002) argue, in a system following the logic of accumulation, social sustainability is not a priority. In such contexts, women tend to act as agents of readjustment for the economic system, taking on unpaid labour to ensure that needs are satisfied and living conditions improved. Such gender inequalities have been maintained in the safeguarding of the identity, memory and history of communities, like Neltume and Malalhue. In both cases, Angélica Navarrete and Nerys Mora, respectively, have taken care of their community, rescuing and strengthening their heritage, nurturing the relationships between different actors, protecting traditions and keeping the community memory alive. This 'caretaker' role is understood by them and their communities as a natural and unconditional commitment. For years, the work they have carried out has not been remunerated. At home and in the community, the work of those who care for, protect, mediate and bring the community together seems taken for granted and unvalued; yet, without their work, the museums would be less wide-ranging and their communities less sustainable.

From a social sustainability perspective, in the Neltume and Malalhue territories, the effect of the community museums is undeniable. In Neltume, which was deeply affected by human rights violations during the dictatorship, the museum has pursued the healing of individual and collective trauma through various initiatives that have broken the silence around the region's historical trauma. During its early years, Malalhue addressed the discrimination against Mapuche children, and once that suffering had been healed, it dedicated itself to promoting intercultural values.[4]

These processes were only possible thanks to unpaid work. A sentimental approach might celebrate the fact that the sustainability of a community museum and its territory depends on passion and enthusiasm. From a more clear-eyed perspective, however, it can be seen as another expression of the sexual division of labour, where care roles have historically fallen to women while the production of goods for market is the domain of men. If care work is broadened beyond the domestic sphere and if the care of sick people is to be understood as 'activities for the maintenance of life and health' (Esteban and Otxoa, 2010), the work that Angélica and Nerys carry out fits perfectly into this category of strong emotional connection, invisibility and gratuitousness.

Considering that in geopolitically marginalised areas, such as Neltume and Malalhue, museums must often transcend their traditional museological role to bridge the social gap that neoliberal governments have failed to address, the role of caring for a neglected community is even more demanding and inescapable. In Malalhue, Nerys has alleviated the suffering of discriminated children, contributed to the cultural recovery of the Mapuche people and pioneered intercultural education practice. Angélica has created a space in Neltume where those who have never been able to speak about the torture and death of their relatives are finally able to do so; she has brought together a wounded community, vindicated the logging-worker identity of the former inhabitants and made the value of women's work in the mountains visible. In community museums where the museum's original role has been transcended, therefore, women can be seen to have transcended their stereotypical roles.

Besides not having intermediate objectives, such as obtaining benefits from those mediated by the market, care work involves an emotional, affective and relational component, as Pérez Orozco (2014) explains: as long as a care work service is provided, they include emotions. This emotional involvement makes it impossible to find market substitutes; it requires managing multiple simultaneous tasks and a versatility of knowledge (Pérez Orozco, 2014). In this sense, the gendered inequality we have been discussing can also produce a position of privilege when it comes to the construction of knowledge that has proven to be more in line with the continuity of life. For this reason, any definition of sustainability has to take its lead from women and take into account the culture of care (Pascual and Herrero, 2010).

The passionate communities mobilised by their museums

> One time there was an oil spill in the fishery, which affected the entire bay coastline. There, for example, we went out to the field, we took samples and linked them to the natural science class […]. This inquiry that was started by us, which passed through the children, then reached the homes.
>
> *—Diego Oyarzo, Museo Escolar Hugo Günckel, La Aguada, 2018*

Museums with a social role at their core are committed to their audiences; they are inseparable from their territories and, therefore, from their communities' concerns. As we have verified in this research, the synergy between the museum and its community arouses emotions that lead its members to commit deeply and passionately to these spaces. These passionate communities with their museums are, without doubt, promoters of sustainability, adopting practices and defending their territories, natural resources and community to promote harmonious coexistence and generate spaces of trust and reflection. There are many examples of this in the world, each with its own context. However, within the framework of this research, we have observed that although there are many differences between community-based museums, there is, at their core, an active commitment from the community that stems from emotion.

For example, within the frame of the EU-LAC Museums project, we have collaborated with Valencia, Spain, where the team had prepared a proposal to develop a project based on the Huerta de Valencia. The idea was to turn the territory, which in symbolic, aesthetic, economic and cultural terms largely defines the Valencian identity, into a museum. The Water Tribunal of the Plain of Valencia was included in the proposal, along with the Council of Wise Men of the Plain of Murcia, which has been defined as a relevant tourist attraction and was declared as Intangible Cultural Heritage in 2007, thus acknowledging its status as one of the most deeply rooted cultural traditions. The museum focuses on the distinctiveness of this regional tradition of water management. It deals with locality from an interdisciplinary approach, is the only space that can show the interaction between human culture and natural environment over time and space, and illustrates the interdependence of culture around a local model (Santacana i Mestre and Llonch Molina, 2008, p. 22). Today, real estate speculation in the Huerta de Valencia region is a threat to the traditions rooted in agricultural activities, the distribution of water and farming families. Industries have been closing down, and both human and animal migrations are evident. The joint action between the community and the social organisations has been key in making this threat visible, alerting the community and generating a necessary movement to protect the community's heritage. The local museum has been the foundation of the so-called culture of campanilismo, that is, understanding local history as 'microhistory'. These museums have been transformed into spaces of local struggle, starting with the recovery of the inhabitants' stories and memories in order that communities might understand the locality's transformation and respond to these changes. They provide meeting spaces, catharsis and local reflection.

As the case studies of southern Chile and Valencia demonstrate, museums do not exist in isolation from their territories. Rather, through active involvement in their local contexts, they become instruments of social cohesion, networking and vision for the future. They do this by taking into account the political and organisational ways in which new movements and identities are introduced, such as migration, cultural and social trends, and so on.

Turning to the case study of Peru (detailed in Chapter 6 of this edited book), the museums selected for this research are mainly those related to archaeological sites of great historical and heritage value, for which tourism and research are central. They maintain a commitment to their local communities that has guided their actions over the years, and although the situation in each institution is different, their commitment speaks of a clear intention to create and strengthen regional identities, to promote local development through education, and to enhance archaeological and cultural heritage. To understand museum sustainability, we must remember its relationship to how museums carry out conservation, research and the dissemination of cultural heritage. Their work is valued by local communities in a way that helps them contribute to local development through direct action, which has an immediate and tangible effect on the territory.

For our five selected museums from Chile, achieving sustainability is part of a daily challenge to reconcile their responses to the challenges their communities face, on the one hand, with their mission to preserve, communicate, investigate and teach the various aspects of culture and local heritage, on the other. They can be understood as agents of sustainable development and promoters of cultural change in their respective contexts.

Emotions are expressed in the exhibits where objects effectively form a bond with the community and have been established through different mechanisms, including donation, the transference of stories and meaning, and the preparation or documentation of objects. Our case studies illuminate this relationship. The

Museo Escolar Hugo Günckel in Aguada, Corral, works with science classes: the students first encounter taxidermy exhibits of species, and then reflect on the animal, its habitat and value to the environment. The children themselves are guides in the museum and are provided with technical knowledge related to conservation or associated socio-environmental risks. In the end, they deliver a story with high emotional content.

At the Museo Despierta Hermano in Malalhue, donations from the community comprise the exhibits. Once someone makes a donation, the guide, Isabel Riveros, registers not only its historical and technical characteristics but also the 'feeling with which the object arrives'.[5] This is then transmitted to the public once the object goes on display.

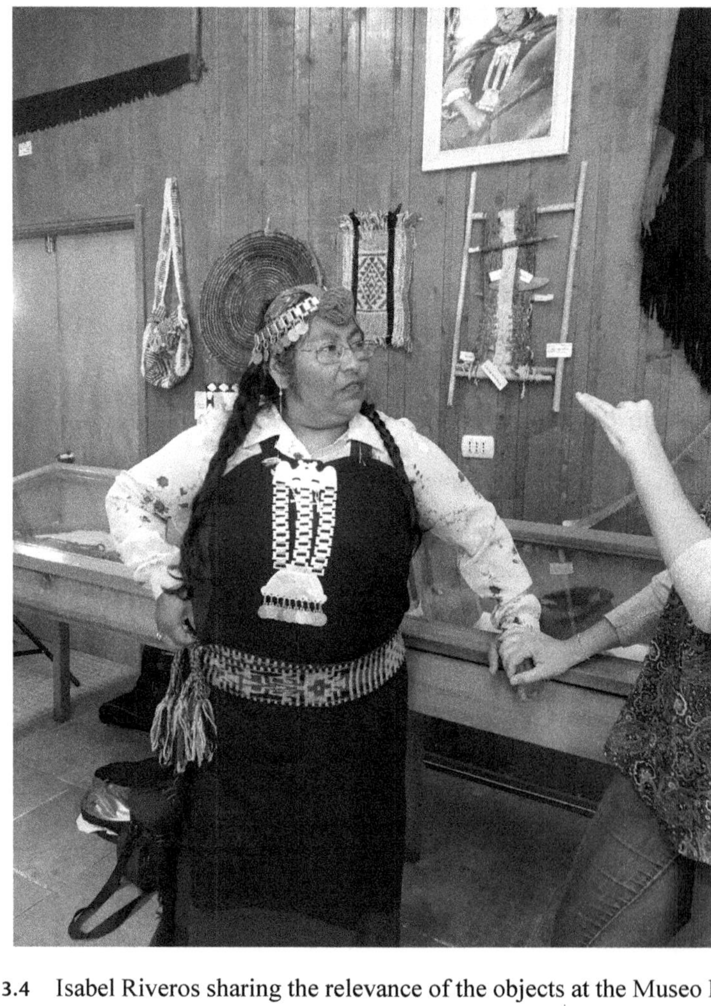

FIGURE 3.4 Isabel Riveros sharing the relevance of the objects at the Museo Despierta Hermano, Malalhue, January 2019. © Karin Weil.

The community considers the museums that derive from the current wave of new museology highly relevant, especially for living people and future generations. People donate objects to these trusted spaces, allowing them to build collections for the future. In response, these spaces support their communities, displaying and valuing their significant objects. In February 2020, like every year, the Centro Cultural Museo y Memoria de Neltume held a meeting of local guerrillas. This year the meeting was particularly significant:

> [the] first activity was an emotional reception ceremony, by the Centro Cultural Museo y Memoria de Neltume, of some material fragments belonging to the 'Pablo' guerrilla, Raúl Rodrigo Obregón Torres, donated by the family to remain in the museum's memory and resistance room.
>
> *(Correa, 2020)*

These are simple examples of how museums' missions have changed profoundly. Moreover, the traditional way of exhibiting objects and inviting the viewer to contemplate and inspect them has changed. What we observe in a museum today is not unequivocally an object; rather, objects have been reconstructed as experiences in themselves. Exhibits with an emotional value invite others to experience the communities' passions, promoting sustainability and the museum's social function—affection, enthusiasm and emotion encapsulated in a display case. However, this practice also raises questions about the value of objects, their legitimacy according to different experiences. Today, museums proudly embrace an expanded educational mandate to stimulate and provoke curiosity (Hein, 2014, p. 6).

Those Latin American museums that are a product of the new museology therefore fulfil a social role that also involves a degree of agency. In Norman Long's (2007, p. 50) terms, such agency is evident 'when particular actions make a difference in a pre-existing state of affairs or course of events', with the understanding that 'all forms of external intervention necessarily enter the ways of life of the individuals and social groups affected, and in this way, they are mediated and transformed by these same actors and their structures'. Thus, one would view a museum as an agent of change and its related community as actors—'active participants who receive and interpret information and design strategies in their relationships with various local actors, and with external institutions and their personnel' (Long, 2007, p. 43). The museum can then be understood as a place from which to observe and analyse the structural dynamics at play in certain territories, and the social outcomes delivered by museology. The case studies we have discussed lead us to conclude that sustainability refers to the quality of living systems. Moreover, these museums, examined using a new museological framework, represent the heterogeneity, principles and practices promoted by the New Museum movement. In effect, they reveal a social function empirically developed: one which is not perfect, but instead acknowledges the fragility of human experience.

This fragility, dependent on the affection and enthusiasm of its founders, becomes a threat to the sustainability of community-based museums whose communities are not as passionate as they are. The museums discussed in this chapter were built on the stories and personal memories of those who brought them to life. Their connection with their managers is not merely a professional relationship, therefore, but a deeply personal one in which the managers' own histories and personal experiences play a leading role in sustaining these spaces. What happens when the future generations do not connect with the enthusiasm and the social needs that gave life to the museum in the first place? What happens, furthermore, when the museum reflects the emotions and passions of a generation, but not a community? In short, the museum becomes weaker when its foundational managers are no longer involved; this brings into question the museum's endurance over time. We suggest that when that happens, its social function will be lost, and the museum will cease to contribute to the sustainability of its community, territory and heritage.

Transmitting and spreading enthusiasm for the sustainability of community-based museums

> We are our memory,
> we are that chimerical museum of inconstant forms,
> that pile of broken mirrors.
>
> *(Borges, 1974, p. 125)*

Humans have a deep need to belong to and to identify with a community; they have a need to share feelings. Museums, mainly small- to medium-scale institutions or those in marginal sectors—including community museums, social museums, local museums or community-based museums—develop local heritage initiatives with a strong communal commitment to create their own narratives to reconcile the past and the present. Each of these museums propose unique forms of participation seeking to co-construct memories associated with the contexts in which they are located, giving meaning to their territories and strengthening the polyphonic networks of history. Working with history, heritage and memory is, above all, an act of interpretation. Determining and communicating the 'truth' of what has happened is also an exercise of power over the past and future.

The invitation extended by the Round Table of Santiago de Chile in 1972, along with the recommendations developed by the New Museum movement, called for a commitment from the territories and their communities to respond to social and political change. This call invested these museums with a social function, a cross-sectional purpose to their activities, helping them to contribute to the building of a better life for everyone.

The relevance of invoking the Round Table and its declaration about 'the importance and development of museums' lies in conceiving and positioning the museum as a living, dynamic space at the service of sustainable development. This is necessary and urgent since, beyond the enthusiastic and committed efforts of those who establish and run them, museums are subject to precarious technical, institutional and financial conditions that undermine their ability to fulfil the noble purposes laid out in Santiago in 1972. As we have argued, the same affections and passions that give life to these spaces often render them uncomfortable for certain sectors of the ruling classes because they are outside the official sphere. This discomfort can prompt those with power to make these museums invisible, thus negating their social role and the effort, enthusiasm and dedication that the managers use to keep their communities' voices and memories alive.

Yet, even as affections and passions can pose a risk to the museum and its social role, they also save them from oblivion, from disappearance. As these spaces spread and transmit the passion that engendered them, the link between the community-based museum and its community is renewed and strengthened. The museum's social role, as dynamic as societies and cultures themselves, is reaffirmed in the affections of each community, in the memories that create and recreate history and in the enthusiasm and sense of belonging that unite them. It is through these dynamics that the centrality and reach of the museum's social role is maintained, and a museum becomes sustainable, as well as an agent of change in a territory.

The challenge, ultimately, is permanence. The mission of a community-based museum must be constantly reviewed to incorporate means of sustaining and renewing the bond with its community, and to consider the present as well as the past. It must bring everyday experiences and challenges back to life, chart the history of the present, involve new generations of the community and remain open to 21st-century solutions and formats. It should provide a space for dialogue and reflection that might enable communities to face the 'instant of danger', that is, to meet the challenges and commitment of sustainable and inclusive development. These objectives should be central to the museum and its practices. Indeed, a museum project's success in this regard will determine the transformation of its environment and its social sustainability.

Thinking about sustainability parameters such as economic, social and environmental, alone may render invisible those mechanisms that have allowed community museums to be sustained over time. Seeking to fulfil such criteria in isolation would present many challenges since, by definition, the sustainability of these museums is the product of shared interests and collaboration between communities and managers of natural, material and intangible heritage. Therefore, there should be as many ways to define and measure sustainability as there are situations in real life that perfect the art of living, rather than the march of progress.

Notes

1 Adapted from the authors' presentation at *Itinerant Identities*, the International Museum Conference, 7–9 November 2018, University of the West Indies, Barbados.

2 Audience has been a very well-researched topic within museum studies in recent years. See McSweeney and Kavanagh (2016), Lang and Reeve (2017) and Longair (2015).

3 Their projects include their iconic and well-known interventions, 'Do women have to be naked to get into the Met. Museum?' (1989) and 'The advantages of being a woman artist' (1988), which can be seen here: https://www.guerrillagirls.com/projects [Accessed 23 August 2020].
4 For more information, see the video *Museo Despierta Hermano* here: https://www.youtube.com/watch?v=CDhzqQbmdN8 [Accessed 23 August 2020].
5 Isabel Riveros, conversation, 2018.

References

Anton, M.C. (2014). 'Pasado e historia en psicoanálisis', *Perspectivas en Psicología: Revista de Psicología y Ciencias Afines*, 11(2), pp. 71–4. [Online]. Available at: https://www.redalyc.org/pdf/4835/483547666008.pdf (Accessed: 11 November, 2021).

Borges, J. (1974). *In praise of darkness*. New York City: Dutton Adult.

Brousseau, L. (1991). 'La exposición', *Museos*, 13(2), pp. 10–11. [Online]. Available at: http://doc.ocim.fr/opac_css/index.php?lvl=author_see&id=1763 (Accessed: 11 November, 2021).

Camarena Ocampo, C. and Morales Lersch, T. (2016). *Memoria: Red de Museos Comunitarios de América. Experiencias de museos comunitarios y redes nacionales*. Oaxaca: Red de Museos Comunitarios de América.

Correa, G. (2020). *Neltume 2020: uniendo la memoria combativa de la montaña con las luchas rebeldes del presente*. [Online]. Available at: https://www.elclarin.cl/2020/03/29/neltume-2020-uniendo-la-memoria-combativa-de-la-montana-con-las-luchas-rebeldes-del-presente/ (Accessed: 11 November, 2021).

Dysthe, O., Bernhardt, N. and Esbjorn, L. (2013). *Enseñanza basada en el diálogo: El museo de arte como espacio de aprendizaje*. Copenhagen: Skoletjenesten.

Escobar, A. (2016). *Autonomía y diseño: la realización de lo comunal*. Popayán: Universidad del Cauca Sello Editorial.

Esteban, M.L. and Otxoa, I. (2010). 'El debate feminista en torno al concepto de cuidados', *Boletín ECOS*, 10.

Falk, J. (2009). *Identity and the Museum Visitor Experience*. California: Left Coast Press.

Fernández Guido, H. (2012). 'Final report: round table on the development and the role of museums in the contemporary world', in Nascimento Júnior, J. do, Trampe, A. and Santos, P.A. dos (eds.) *Mesa Redonda de Santiago de Chile 1972, Vol. 1*. Brasilia: Ibram, pp. 210–31.

González, D. (2016). 'La función social del museo. Sus límites y posibilidades', *Reflexiones marginales*, 34. [Online]. Available at: https://reflexionesmarginales.com/blog/2016/07/31/la-funcion-social-del-museo-sus-limites-y-posibilidades/ (Accessed: 11 November, 2021).

Hein, H. (2014). *The Museum in Transition: A Philosophical Perspective*. Washington: Smithsonian Institution.

ICOM Statutes (2007). Adopted by the 22nd General Assembly (Vienna, Austria, August 24, 2007). [Online]. Available at: https://www.icom-ce.org/documentos/ (Accessed: 11 November, 2021).

Lang, C. and Reeve, J. (2017). *The responsive museum: working with audiences in the twenty-first century*. London: Routledge.

Long, N. (2007). *Sociología del Desarrollo: una perspectiva centrada en el actor*. San Luis Potosí: Editorial El Colegio de San Luis.

Longair, S. (2015). 'Cultures of curating: the limits of authority', *Museum History Journal*, 8(1), pp. 1–7.

McSweeney, K. and Kavanagh, J. (eds.) (2016). *Museum participation: new directions for audience collaboration.* Edinburgh: MuseumsEtc.

Mostny, G. (1972a). 'El desarrollo y la importancia de los museos en el mundo contemporáneo', *Noticiario mensual del Museo Nacional de Historia Natural de Santiago*, 16(190–1), pp. 3–4.

Mostny, G. (1972b). 'Mesa Redonda sobre la importancia y el desarrollo de los museos en el mundo contemporáneo. Resoluciones', *Noticiario mensual del Museo Nacional de Historia Natural de Santiago*, 16(190–1), pp. 5–7.

Pascual Rodríguez, M. and Herrero López, Y. (2010). 'Ecofeminismo, una propuesta para repensar el presente y construir el futuro', *Boletín ECOS*, 10.

Pérez Orozco, A. (2014). *Subversión feminista de la economía. Aportes para un debate sobre el conflicto capital-vida*. Madrid: Traficantes de Sueños.

Pérez Orozco, A. and del Río, S. (2002). *La economía desde el feminismo: trabajos y cuidados.* [Online] Available at: https://www.ecologistasenaccion.org/13104/la-economia-desde-el-feminismo-trabajos-y-cuidados/ (Accessed: 11 November 2021).

Santacana i Mestre, J. and Llonch Molina, N. (2008). *Museo local: la cenicienta de la cultura.* Gijón: Ediciones Trea.

Simon, N. (2010). *The participatory museum.* Santa Cruz: Museum 2.0.

Simon, N. (2016). *The art of relevance.* Santa Cruz: Museum 2.0.

Smith, L. and Campbell, G. (2015). 'The elephant in the room: heritage, affect, and emotion', in Logan, W., Nic Craith, M. and Kockel, U. (eds.) *A companion to heritage studies.* Chichester: Wiley Blackwell, pp. 443–460. [Online] Available at: https://www.researchgate.net/publication/299410843_The_elephant_in_the_room_Heritage_affect_and_emotion (Accessed: 23 June 2021).

Smith, L., Wetherell, M. and Campbell, G. (2018). *Emotion, affective, practices, and the past in the present.* New York: Routledge.

Tuhiwai Smith, L. (2012). *Decolonizing methodologies: Research and Indigenous peoples* New York: Zed Books.

Turner, V. (2002). 'The factors affecting women's success in museum careers: a discussion of the reasons more women do not reach the top, and of strategies to promote their future success', *Journal of Conservation and Museum Studies*, 8, pp. 6–10. [Online]. Available at: The Factors Affecting Women's Success in Museum Careers: A Discussion of the Reasons More Women Do Not Reach the Top, and of Strategies to Promote their Future Success (Accessed: 11 November, 2021).

Varine, H. de (2012). 'Alrededor de la mesa redonda de Santiago', in Nascimento Júnior, J. do, Trampe, A. and Santos, P.A. dos (eds.) *Mesa Redonda de Santiago de Chile 1972, Vol. 1.* Brasilia: Ibram, pp. 97–8.

Veneros Ruiz-Tagle, D. and Toledo Jofré, M.I. (2009). 'Del uso pedagógico de lugares de memoria: visita de estudiantes de educación media al Parque por la Paz Villa Grimaldi (Santiago, Chile)', *Estudios pedagógicos (Valdivia)*, 35(1), pp. 199–220.

Weil, K., Elmúdesi, B., Fúquene, L., Blanco, G., Godoy, M., Urbina, S. and Barbosa, O. (2018). *Demonstration sketch of videos and inform of close analysis.* Universidad Austral de Chile. Unpublished.

Weil, K., Elmúdesi, B., Fúquene, L. and Godoy, M. (2018). *Final report on sustainability in community-based museums*. Universidad Austral de Chile. Unpublished.

Westerman, M., Schonfeld, R. and Sweeney, L. (2019). *Art museum staff demographic survey 2018.* [Online]. New York: The Andrew W. Mellon Foundation. Available at Art Museum Staff Demographic Survey 2018 (Accessed: 11 November, 2021).

4

MUSEUMS AND COMMUNITY ENGAGEMENT IN BELIZE

Case studies for protection and active participation of knowledge

Sherilyne Jones

Belize is a former British colony, geographically located in Central America, but has its roots in the Anglophone Caribbean. With Mexico as a northern neighbour, Guatemala to the west and the Caribbean Sea to the eastern border, the country has a unique historical context. Historian Nigel Bolland (2003) asserts that Belize's history is intrinsically linked to the Caribbean, especially Jamaica, and that

> the settlement at Belize was so small and dependent on Jamaica and the rest of the British Colonial system, that its version of slavery was inevitably affected by the cultural traditions and legal structures which has been well established on the islands.

This perspective, combined with a recent surge in immigrants to the country, has significantly altered the multicultural society, challenging its discourse on national identity. As the country embarks on its fourth decade of independence from Great Britain, in the context of nation building, it is now home to an uncommonly large number of very distinct ethnic groupings. It must therefore grapple with the difficult challenge of integrating those differences into a single, homogenous society. Watson asserts that

> As nations become more multicultural and governments struggle to balance the need for national unity with the idea of the toleration of the differences, museums become arenas where concepts of what it is to belong with a nation are explored and made explicit.
>
> *(2007)*

This also applies to Belize, where there is a perceived erosion of cultural heritage and national identity. The values upheld by its residents are based on how

DOI: 10.4324/9781003288138-6

museums' work contributes to historical consciousness, and attempts to construct a nation are currently stymied.

A critical understanding of how best to curate its history through museums, achieving a balance between cultures, nation building, and mitigation of the global ecumene, becomes critical for the preservation and development of Belize's cultures. According to Watson, 'National Identity has been one of the preferred and privileged forms of identity in the modern world and continues to exert a powerful influence in the way museums represents communities' (2007). Communities and the way the nation identifies, acknowledges and collaborates with these groups becomes a critical factor which, in effect, determines the success of the museum programmes and activities. Anderson alludes to this by suggesting that 'Museums are tools by which nations "imagine themselves"' (1983). Richard Fortey, the British palaeontologist and writer, stated in his book, *The Secret Life of the Natural History Museum*, 'I believe profoundly in the importance of museums; I would go as far as to say that you can judge a society by the quality of its museums'. From my experience, I concluded that museums validate the quality of our societies and are repositories that assist humanity in its regressive reflection and self-validation, while providing the ability to not only appreciate our past but also to hope for a brighter and unified future.

The Oxford Advanced Learners' Dictionary defines culture as '... the customs and beliefs, art, way of life and social organization of a particular country'. What we refer to as culture, in the simplest sense, is essentially an account of people and their collective values. With an increase in the immigrant population, added to its unique collective history, Belize was forced into an identity crisis and culture dilemma. Shoman (1995) writes,

> our national culture is not defined simply by the fact that we are a people made up of various ethnic groups who came to this land at different times in response to various historical processes, but it is a direct result of defining effects of colonialism and its citizenry's reaction to colonialism.

He goes on to assert that 'the colonial strategy of a divide, and subdivide rule ensured, that on the one hand, our various cultures remained largely isolated and on the other hand the "colonizer's" culture remained dominant'. The need for change, transformation and movement to transcend from a colonial society to that of an independent nation would only be accomplished through the implementation of an engaging educational and cultural programme. In this context, focus on the ongoing narrative of how people lived, interacted with each other and expressed themselves in a post-independent Belize became paramount. Culture and the communities in which we lived defined the new citizens.

Prior to independence, avenues for cultural and artistic expression were mostly missing in Belize. While a small number of Belizeans would rise to prominence during the artistic revolution of the 1970s, this surge would be short lived. After independence, the demands of nation building, addressing the social needs of the country, poverty alleviation, providing housing and health care for people, overshadowed further investments in the arts. With the introduction of television in 1983, 'the degeneration, devaluation and de-evolution of authentically Belizean culture' occurred (Shoman, 1995). Prior to independence, the country's historical records reflected the views of the coloniser, and post-independence, little or no Indigenous art was available or on display for the citizenry to appreciate. Television brought a new genre of entertainment for the citizens of the young nation of Belize. The Western way of dressing, speaking, consumerism and proliferation of a capitalistic society streamed into living rooms across the country. Traditional entertainment, such as cultural and artistic representations at the annual Festival of Arts, plays portraying everyday life and cricket games, was replaced with leisure activities and TV programmes coming from the United States. The inundation of US television programmes was swift and prolific, at a time when Belizeans would have benefitted greatly from coming together to develop a national identity. Youths began to emulate the MTV culture of North America, and little or no effort was placed on safeguarding Belize's authentic culture, nor were there community engagement programmes for people to participate in.

The adverse effect of television would go unabated during the 1980s and 1990s, until cultural pioneers started teaching drama in schools, collaborating with the Ministry of Education to include art education in the school curriculum. In essence, community engagement in Belize began in this rudimentary form given the fact that the country did not have a museum. According to Dodd and Sandell, 'Museums can deliver benefits to communities in specific neighbourhoods and locations, as well as individuals. The outcomes in this area include community capacity building, whereby communities learn competencies and develop both the ability and confidence to change' (2001).

Administering culture without the mechanisms in place would prove challenging but the individuals driving the cultural revitalisation would be undeterred. Through initiatives like the development of the Museum of Belize, communities became once again empowered to effect change in their society and the appreciation of culture increased into the late 1990s.

Museums play an important role as custodians of cultural heritage. Traditionally their prime function has been to gather, preserve and study objects. Community engagement then enables museums to get access to the community and become sustainable through feedback, ideas, views, new insights and relevance. It provides an opportunity for people living in such communities to find out about their own heritage and to help them realise that their active participation in museum activities helps to keep it alive for future generations. Museums are a relatively new concept in Belize, with the National Museum of Belize opening in 2002 only. In fact, given the low visitation to the museum and by extension the country's archaeological sites, the perception is that Belizeans are not in the habit of or accustomed to visiting cultural spaces.

Unlike its Caribbean counterparts, who were able to build on the foundations of early preservation and museums establishment, Belize is in a precarious position from which to develop the museum institution and guide the discourse necessary for future museum developments in the 21st century. The pervasive 'melting pot dialogue', questions of 'cultural identity' and 'loss of heritage assets' are a continuous challenge for the young nation. Recognising the disparity of ideologies and growing disconnect with the community, it became critical that avenues for cultural and heritage expression be holistically addressed. This was achieved with relative success in 2016 with the development and implementation of Belize's National Cultural Policy (NCP), whose objective is 'fostering and providing an enabling environment for the development of a cultural industry that can be harnessed to address issues of national identity'. Ultimately, for cultural revival to be successful, the process had to begin with an appreciation of one's own unique culture. This can be achieved by raising awareness among Belizeans of the importance of their cultural heritage. This will lead to the development of values and allows for systematic safeguarding of intangible cultural heritage, promotion and preservation of heritage assets.

In an effort to foster a cohesive environment for cultural development, the Belizean government conceptualised the idea of Houses of Culture (HOCs) to complement the national museum. HOCs are small eco-museums, which provide open and accessible cultural spaces in district towns, where the responsibility of preservation, transmission, promotion of art exhibitions and cultural characteristics is community driven. HOCs act as intermediaries for the Museum of Belize, which allows for community participation, access and engagement with resources that promote cultural development, which may otherwise be inaccessible to local communities. These spaces engage sectors of society who would not ordinarily visit a museum or even understand or appreciate the traditional concepts. Due to the grassroots nature of engagement, their level of support can vary, according to community needs. With inspiration from the Cuban model of 'Casas de Cultura', HOCs provided cultural spaces that were of benefit to the community and became neutral spaces in which the constituents and cultural stakeholders accessed free of charge arts and culture in all is forms. The mandate and mission of these HOCs is to inspire artists, artisans and individuals to explore their creativity in the fields of music, and creative and visual arts.

Similarly, the Gulisi Garifuna Museum is considered a cultural landmark for the Garifuna people and the people of Belize. Garifunas are the descendants of an Afro-Indigenous population from the Caribbean Island of St. Vincent, who were exiled to the Honduran coast in the 18th century. Following their exile from St. Vincent[1], Garinagus[2] migrated to the mainland of present-day Honduras, arriving at the coastal fort of Trujillo, Atlantic coast of Honduras, and continued to populate the coastline of Guatemala, Belize and Nicaragua. In Belize, the original settlements were established south of the Sibun River, which until 1859 was the southern border of the British Settlement at the mouth of the Belize River. By the 1950s, there were five Garifuna settlements, including the towns of Dangriga and Punta Gorda, and the villages of Barranco, Seine Bight and Hopkins, which have remained predominantly Garifuna. In 1981, a non-governmental organisation called The National Garifuna Council of Belize (NGC) was established. The NGC promotes economic growth and opportunity for the Garinagu people in Belize while also working to preserve, maintain and develop the Garifuna culture. To protect and advocate on behalf of Garifuna people in the country, the mission of the NGC is 'to promote the cultural identity, economic development and general well-being of the Garifuna People as well as interracial harmony, through means that ensure the sustainability of the organization, being mindful of the responsibility to protect the environment'. The NGC was instrumental in the development of the Gulisi Museum, whose core mission includes the collection of Garifuna objects, art, artefacts and literary materials that visually demonstrate the Garifuna history.

When the Gulisi Garifuna Museum opened in 2004, two years after the Museum of Belize, it was heralded as groundbreaking. Not only did the country now have a national museum but also the development of a Garifuna Museum was timely because the United Nations Educational, Scientific and Cultural Organization (UNESCO) had in 2001 declared Garifuna Language, Music and Dance 'Masterpieces of the Oral and Intangible Heritage of Humanity'. There was every expectation that this museum would bring new energy and vitality to the already rich culture and traditions of the Garifuna people in Belize. The museum would be a place to showcase Garifuna art, music and dance through exhibitions, workshops and summer courses in the Garifuna language; a place where Belizean youth could learn about Garifuna history.

The museum was named after 'Gulisi', the daughter of Paramount Chief Joseph Chatoyer, who arrived in Dangriga from St. Vincent via Roatan, Trujillo and Puerto Cortez. She was one of the first Garifuna women to arrive and settle in the Dangriga-Commerce Bight area. She came along with her 13 sons, who carried the surname Lambey, some of whom were the first to settle in the many Garifuna settlements in Belize. She would have arrived in the 1820s, and many family groups in all Garifuna communities descended from her. Her story was transcribed as oral history from her granddaughter's granddaughter, Felicita Francisco, by Dr. Joseph O. Palacio and was the inspiration for the naming of the Gulisi Museum in Dangriga. To date, the Gulisi Garifuna Museum stands as the latest addition to the expanding cultural landscape of Belize and a space accessible to all for expression and cultural understanding.

While the country did not have a national museum prior to 2002, it was not totally void of cultural offerings. In efforts to express their own identity in a multi-ethnic society, community members often explored and interpreted their history through their own lens and voices to safeguard what was most important to them. This was true of Luba Garifuna Museum, which was opened on 5 November 1999, by proprietor Sebastian Cayetano or 'Mr. Caye' as he is affectionately known. He is one of 13 siblings, is a retired historian, author, educator and linguist – fluent in five languages – a fact very few people are aware of. Retiring after 27 years as a teacher in 2004, his conceptual idea to develop a museum had taken root. The goal was to use his 30-year collection of Garifuna implements and artefacts he had amassed throughout his various travels abroad and in Belize.

The main goal of the museum was to provide the wider populace with knowledge about Garifuna culture and to ensure that Garinagus keep their language and origins alive given the growing multi-ethnic society. In visiting the tiny museum space, which is situated on the ground level of his home, Mr. Cayetano passionately relays stories from the elders about the Garifuna's rich cultural history, and the resilience, strength and perseverance of his people in the face of cultural genocide. This obscure museum informs visitors of Garifuna culture through the display of items used in a typical Garifuna household, effectively safeguarding their history by actively engaging the community through grassroots activities, without any financial support from the State. The success of Luba Garifuna Museums affirms what Karp (1991) asserts: it is the individuals that are responsible for a museum's mission, core values and work. Mr Cayetano's emic perspective and passion for his culture plays a significant role in what is exhibited for public consumption and presented to the community.

This success is in contrast with the impact of the Gulisi Garifuna Museum. While government funding was provided for showcasing the rich cultural heritage of the Garifuna people, this museum has failed to capture the interest of the general public who believed that the museum 'appeared static, cold and uninviting, with the same displays from its initial opening in 2004'.[3] While the museum does provide an educational outlet to students on school visits or by Belizean diaspora, the majority of the populace do not see its relevance to their lives, and many people in the other district towns and villages have never visited. The location on the outskirts of the town of Dangriga also created a barrier for accessibility for individuals who did not have transportation. Ultimately, for the Gulisi Museum a new relationship with visitors and the community needed to be developed, from a top-down model of management to a more community-driven partnership. While the Gulisi provided exhibitions from the expert perspective, the community stakeholders were missing from its curatorial practices. As the catalyst for regeneration initiatives, the agency responsible for culture partnered with Gulisi to establish a HOC within the museum. The Stann Creek's HOC enabled an 'open dialogue between people and engage their enthusiasm and commitment to a shared redevelopment process', while providing the necessary tools to successfully assist in 'developing the capacity of local communities to address their own needs' (Sandell Karp 1992: 12, 2007). The HOC created an enabling environment for community members to provide input into the programmes that were important to them, as well as collaborating in heritage and preservation initiatives.

This model of community engagement demonstrates that community members are actively involved in safeguarding their intangible and tangible heritage in alignment with the Belize's cultural policy, developed in 2016, which states that the government shall 'Provide information and facilitate access by national and community organisations, custodians and practitioners of intangible cultural heritage to the International Fund for Safeguarding Intangible Cultural Heritage that provides international assistance towards effective safeguarding of the intangible cultural heritage' (Belize National Cultural Policy, 2006:7). The successful engagement and participation that are at the core of HOCs infer that based on their individual characteristics, they are more effective than the mainstream Western concept of museums. While each HOC is unique in its conceptual development and community engagement, as each represents different areas of the country and the varying ethnicities of the community, they remain close to the eco-museum concept developed by Hugues de Varine in the 1960s. Luba Garifuna Museum and the HOCs have successfully empowered community leaders to be more active in the dissemination of cultural traditions to younger generations. Theodore Low suggests that 'museums must be willing to alter and to modify their internal structures and their ideas to fit a changing world and the advances of social thought' (2004).

In the Dangriga, known as the 'culture capital of Belize', the Gulisi Museum along with the HOC have been successful in preserving these cultural traditions and creating a welcoming environment for the community, by working in tandem with the Stann Creek HOC. This collaborative effort enabled the development of empowerment initiatives for the community spearheaded by elders by facilitating 'an enabling, creative, perhaps less threatening forum through which community members could gain the skills and confidence required to take control and play an active, self-determining role in their community's future' (Sandell, 2007). The museum helps sustain Garifuna heritage through drumming and dance programmes, especially during summer camps and specialised workshops targeting youths. The fact that these activities can take place outside the museum space aids heritage preservation, in collaboration with the community, and increases the successful retrieval and conservation of the local heritage, while maintaining traditional knowledge. These extra-mural activities are supported by the presence of Gulisi Garifuna Primary School which is located adjacent to the museum and Garifuna monument grounds. The Gulisi Community Primary School is trailblazing and unique, as it is the first school in Belize to combine traditional knowledge-based programmes with the Curriculum of the Ministry of Education. This is done through intercultural trilingual education programmes that teach English, Spanish and Garifuna to students, which allows for active transmission of Garifuna cultural heritage with that of the structured government system.

In these examples, by relinquishing control of how Gulisi's programming is administered, the State, through the Museum of Belize, grants community leaders 'decision-making opportunities' to promote their history and provides opportunities where they became the collaborator rather than the agency of authority and allow for structured guidance rather than control of community initiatives. The action of teaching the Garifuna language and its cultural traditions to Garifuna youths has become a source of community pride and will help to guarantee the safeguarding of this heritage for generations to come. Other successful examples coming from the Gulisi, and the HOC collaboration, saw the introduction of drum making and drumming lessons to the children by the elders in the community (Figure 4.1). A significant part of Garifuna life and culture is music, song and dance, and this is demonstrated in Garifuna celebrations and festivals. Most Garifuna songs and dances illustrate their history and culture. The drum is especially significant because it is directly connected to their African ancestry and consists of two main types: the Primero (tenor) and Segunda (bass). Roy Cayetano, Garifuna Linguistic Anthropologist and former Chief Executive Officer in the Ministry of Rural Development and Culture (2001–2007), explains that:

> In ritual, we have three drums, and those drums symbolize the totality of Garifuna life: past, present and future. And they play their part in representing the Garifuna world view, the Garifuna cosmology, if you will. In secular life we generally use the Primero and the Segundo to play the various types of music that we use for festive occasions. And when I say festive occasions of course I don't just mean fun times because we also mask our grief, our sorrows, our burdens in song and dance. [...] The present incorporating the past and the future emerging from the present.

These models demonstrate the direct correlation between the community and the museum. The community saw the interest in developing these cultural revival initiatives, and actively worked and campaigned to ensure that elders were available to teach not only how to make the drums but also the importance of drumming for specific activities in their culture. This intangible culture is showcased through events, outreach programmes, activities or museum displays in which the materiality of the objects is downplayed, and their meaning, value or use is given precedence (Varutti, 2013). This is also reflected in Luba Garifuna Museum where the main aim is cultural preservation through practising heritage. This is obvious in the relatively low priority given to the cataloguing or conservation of objects compared to other museum functions. While there are displays in all the HOCs countrywide, small eco-museums, like Luba, are focused on the stimulation of locally based cultural and creative initiatives that benefit the community, empower participants and encourage curiosity, openness and tolerance towards different cultures and traditions. Community engagement then bridges the gap between the government and the constituents they serve by allowing a more participatory approach to programmes and activities.

This symbiotic approach to managing authority and the community is demonstrated in the revival of the Habinahan Wanáragua Jankunu Festival in the Stann Creek District in southern Belize. This district has the largest concentration of active Garinagu communities. Green asserts that:

> today, the Garifuna exist as arguably the only people of African descent to escape the physical and psychological chattels of slavery, intermix with Native

> Americans in the Caribbean, develop and maintain an Indigenous language and corpus of postmortem propitiation rituals and dance-song genres, and endure numerous attempts at racial annihilation and cultural genocide.
>
> *(2018)*

The Habinahan Wanáragua, which translates to 'mask', is a dance rite ritual also known as Jankunu, usually performed during the Christmas season up until the Epiphany on 6 January. According to Green, it is a unique synthesis of three cultural traditions: (1) African harvest festivals, ancestor rituals and secret societies, (2) English mummer's plays and (3) Amerindian (Arawak and Carib) festivals.[4]

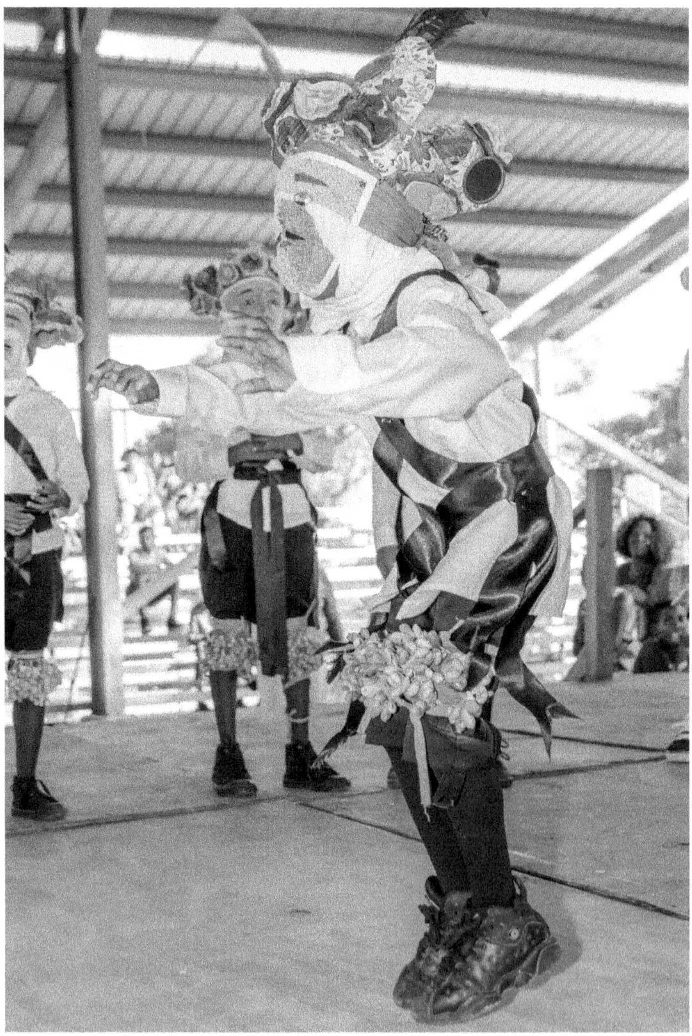

FIGURE 4.1 Habinahan Wanáragua Dancers. Image © JC Cuellar Photography

He further explains that this 'masquerade dance is of great social and festive importance and has evolved throughout the Caribbean for the last 200 years due to its pomp, pageantry, elegance and finesse' (Greene, 2018). The dance is traditionally performed by men, whose mask and costumes replicated British militia. Their costumes involve elaborate headdresses complete with feathers and mirrors, and they wear bands of shells around their knees with white shirts and black or white pants. Usually, black, green or pink ribbons cross their chests depending on the time of year the dance is done (Figure 4.2).

The Habinahan Wanáragua is one of the few dances where the drummers follow the dancer's movements, and not the dancer dancing to the beat of the drum, which

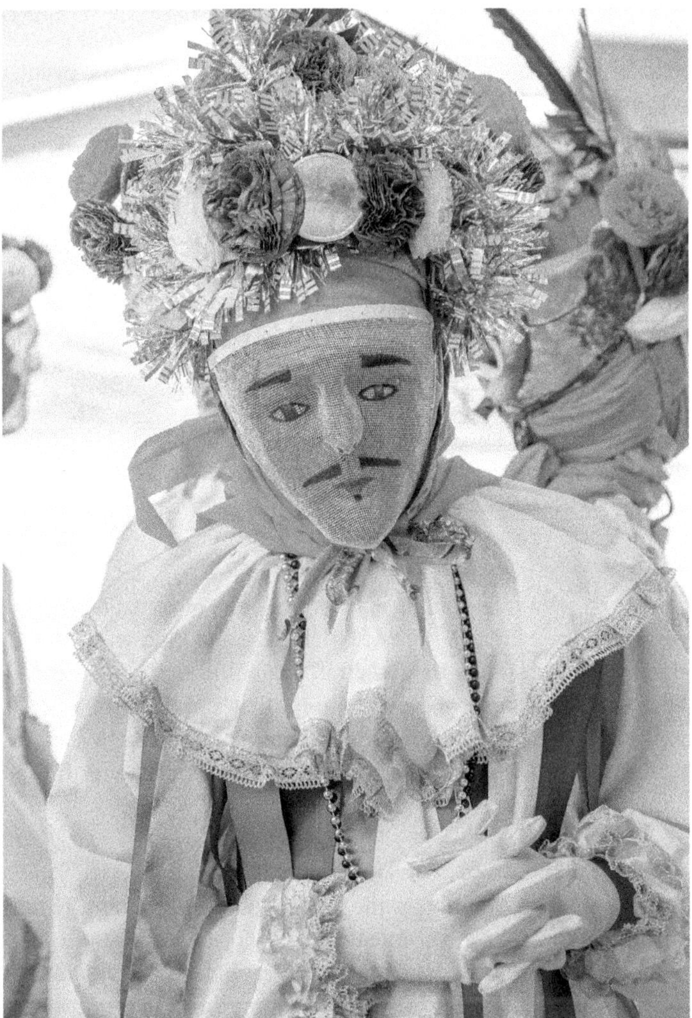

FIGURE 4.2 Male dancer at the annual Junior Habinahan Wanáragua Competition held in the culture capital, Dangriga Town. Image © JC Cuellar Photography

allows for an exciting show of skill by both the dancer and the drummer. According to an article by Rommen and Neely, it is the most difficult of Garifuna dance-song genres for both dancers and drummers. This significance was discussed by Roy Cayetano, who highlighted the impetus for the development of the Wanáragua festival in a conversation he had:

> When I approached the then President of the National Institute of Culture & History (NICH) in November 2010 regarding some support to establish Habinahan Wanáragua she acceded saying that NICH was encouraging the development of a major festival in each district.[5]

With funding from both the National Institute of Culture & History (NICH) and the National Garifuna Council, the inaugural edition was critical to the establishment of Habinahan Wanáragua, and the NICH has continued to fund the event. Cayetano goes on to illustrate how NICH was able to provide participatory opportunities in the community. He states,

> The case of Habinahan Wanáragua illustrates that NICH can and does have an impact on the development of Arts and Culture in Belize. It does this through a judicious and strategic use of the resources available to it. In this case, the art form or cultural phenomenon to be developed as well as the means to be employed for its development and promotion was determined by the community, not NICH.

In supporting the festival, NICH was successful in aligning its core values with its mission, thus making the organisation relevant and essential to the community it serves (Simon, 2010). Cayetano asserts that this type of civic engagement is the reason for which NICH was first conceived:

> This is in keeping with the conclusions of the series of consultations that eventually gave rise to the establishment of NICH. The thinking was that Government had no business determining the direction that cultural development should take and that it was the people, the community, that should do that. NICH would therefore be removed from direct government control and be established with a view to ensuring that the control, the choice, resided with the people, the community.

This example and others from the various HOCs throughout Belize illustrate the type of civic engagement that makes the Houses successful and true to the Cuban model. It demonstrates that government should not determine the direction that cultural development should take, but rather the people, and the community, should be responsible for that. The safeguarding programmes illustrate how the approach to museum management and community engagement differs in Belize compared to its Western counterparts. There is greater emphasis on intangible heritage, 'essentially the orality, transience, rhythm and vibrancy which overlays Caribbean cultures' (Cummins, 2012). Cummins advocates that 'the majority of signifiers of Caribbean identity, whether cultural or national, cannot be reflected through the material culture or physical heritage alone [...] but is perhaps most strongly expressed in the intangible' (2012). By decentralising access to culture, through the development of the HOCs and in supporting small museums like Luba Garifuna Museum and Gulisi Museum, government control was removed and the choice resided with the people and the community. The very fact that these museums exist and represent a marginally underrepresented portion of the population is significant. Sebastian Cayetano notes that 'while Belize has now fully embraced the Garifuna culture, there continues to be a strong need to address the economic plight of our people'. Through his museum, he urges Garinagu, 'never, ever to abandon their language and their roots and to continue – to teach the language and to teach the culture to all Belizeans and that way, we preserve everything for all of Belize and for the world' (Amandala Newspaper).

There is a varying degree of disparity as to the perception of what the museum's role should be in our society; definitions of the museum have evolved over the years, and the current view of the museum is not static and varies around the world. Christina Kreps remarks that the 'non-Western models of museums and curatorial practices have value as unique cultural expressions, and as examples of human cultural diversity. But they also have much to contribute to our understanding of museological behaviour cross-culturally' (2011). Kosinski expounds by stating that 'Museums and cultural institutions have long acted as a sort of glue, binding disparate individuals together through a shared interest in art, culture, and the humanities' (2020). Boasting high visitation no longer suffices, and museums must engage their constituents with innovative programming that is relevant to the communities they serve. In Belize, the HOC model and the autonomy of private small eco-museums demonstrates this with remarkable success, as community members actively participate in safeguarding the intangible and tangible aspects of their history for the specific ethnic group or municipality.

In these examples, community engagement began at the grassroots level, and there was a direct correlation with the community, the museums and the government's HOCs. Without the HOCs, a large portion of the population would not otherwise visit the museum or have any interest or see relevance to their lives. As museums move towards becoming more socially conscience agents, these examples provide a good illustration in which the visitor's role in the museum has transcended to that of 'community activists'. In this context, the term 'community' refers to a group of individuals with a shared history and interest, a network that offered a sense of belonging to its participants and is reminiscent of the 1960s, when social institutions were challenged by Indigenous Peoples to have their voices and views heard (Smith and Waterton, 2009: 13–14), and which gave rise to the new concept of community and their role within museums.

As with the term museum, 'community' has posed as much of a challenge in trying to determine a meaning. It is obvious that the meaning is difficult to define given the broad range of what communities represent. Aside from being an abstract grouping of people with shared characteristics, it also refers to a sense of belonging to a group, common experiences and specific characteristics of language, religion, ethnicity or other cultural markers (Crooke). Roy Cayetano considers community in the Belize context to mean the

> group of persons who have a common interest in the creation, preservation, promotion or enjoyment of a particular art form or cultural manifestation. There may be different groups/communities each with a different focus but combined they also make up the wider community of stakeholders.

Yasser Musa, a Belizean visual artist, teacher, poet, publisher and former President of NICH, is on the opposite end of the spectrum and views the community as

> a destination we want to reach, not a just a state of being. Let me make an analogy. As a teacher, I see a 13-year-old youth in front of me, but his state of mind, his spirit, and his attitude toward learning and life is not static. So, I am not just teaching him for the now, but for what he can become. So, in my mind, the real client in education is not the student, but the future student, the future self-directed, self-motivated learner. And this is how I see community and community engagement. We need to be open and expansive to the notion of community. It is a wide sea, where many boats must be launched into. And those floating vessels will best carry art into the imagination of society (YM).

Upon further examination, the term 'community' provides an enlightening exposé into what it means to be a part of community. No longer can community be considered simply as a homogeneous unit, that it is necessarily geographically based ('local'), that it has long established roots, with characteristics that are easily recognizable or even that communities are 'real' (Anderson 2006: 6). Instead, communities are more heterogeneous, ever-changing and dynamic. In a society, a person can belong to multiple communities simultaneously, knowingly or unknowingly and 'membership of a community may be fleeting, partial, or innate, lifelong, and unshakeable, often irrelevant of an individual's wishes' (Onciul 2013: 81). This is not to say that the term has no value but that the notion is not one to accept uncritically (Crooke 2011: 183):

> This range suggests that community is a word that alters in different contexts in an almost chameleon-like fashion. For some, this point would be grounds to dismiss the idea of community as outdated, vague, and of little use. However, no matter how forcefully an argument of redundancy may be presented, one cannot dismiss the frequency of the use of the word and therefore its importance.
>
> *(Ibid.: 173)*

This importance is underscored in the way museums apply the term community. Rather than narrowly categorising audiences on statistical data only (demographics of age, gender, local/tourist), museums must now apply community to a much broader investigation of audiences. Understanding the constituents that make up their audiences will assist museums in developing more targeted and inclusive programmes and activities. Luba Garifuna Museum is well placed to apply these principles as its mission is specifically geared to the transmission of knowledge and engagement to preserve the craft techniques, music and dance of the Garifuna people.

However, the knowledge of 'the community' does not necessarily make it easier to plan museum activities, as many visitors 'belong to many communities, often simultaneously', but if museums focus on specific communities, they will provide museum services that are better suited to specific target groups. This is far from a perfect model, and there is room for improvement, particularly in the manner museums interact with communities. Several museums tend to focus on specific ethnic groups while ignoring other, perhaps disenfranchised, local communities, thus creating an imbalance in the voices that are being heard (Karp et al., 1992: 12). Additionally, a major judgement error by museums occurs in the assumption that, in collaborating or co-curating with community members, those members or individuals represent the collective whole, when this may not necessarily be the case. Gulisi Museum addresses this through the partnership with Gulisi Community Primary School, where the curriculum instruction is done in both English and Garifuna, thereby supporting the language programmes and connecting the wider community with cultural knowledge and skills.

Given the Caribbean's ethnic diversity, it is challenging to effectively represent diverse communities invested in its museums. Rex Nettleford notes that:

> the encounter of Africa and Europe on foreign soil and these in turn with the indigenous Native Americans on their long-tenanted estates and all in turn with latter-day arrivants from Asia and the Middle East, has resulted in a culture of texture and diversity.
>
> *(2003)*

Cummins (1992: 37–38) asserts that the problem many museums in the Caribbean face is a lack of focus on the Indigenous cultural heritage, and failing to correlate that heritage with the descendants of the living communities. This relationship between museums and source communities must move beyond consultation and collaboration to that of partnerships in which both parties share power. In other words, museums must ensure that their programmatic activities are inclusive and pertinent and that their voices are reflected in the narratives. This will only be achieved if communities are consulted and provided a platform to contribute and provide input in the processes of exhibition development and safeguarding of their culture. Both Garifuna museums in Belize demonstrate that this relationship moved beyond consultations and illustrates the successes of individual and/or community interest in heritage preservation initiatives. The power relations within the museum space were redefined, and their approach included shared authority and a re-examination of ownership and curatorship. These spaces were not only places where heritage is kept, and disseminated, but also are spaces for the community to recreate their heritage and memorialise histories for future generations.

Notes

1 St. Vincent is known as Yurumein in the Garifuna language.
2 Oliver Greene notes that the term 'Garinagu' is a term which refers to 'the people as a whole, whereas the term Garifuna refers to the language, the culture, and a person in the singular form' (2002: 189). Today, Garinagu are commonly referred to as Garifuna in publications by outsiders.
3 Visitor comment cards – Museum of Belize (2011).
4 Ibid.
5 Dr. Roy Cayetano (2017). Email responses to Sherilyne Jones, 22 March.

References

Anderson, Benedict. 2006 (1983). *Imagined Communities: Reflections on the Origin and Spread of Nationalism.* London & New York: Verso.

Baltimore Sun Op-Ed. (Online). Museums Must Reach into Communities. https://www.baltimoresun.com/opinion/op-ed/bs-ed-museum-culture-20160203-story.html

Bolland, O. N. (2003). *Colonialism and resistance in Belize: Essays in historical sociology.* Jamaica: University of the West Indies Press.

Cayetano, I. (2019). 'The Art of Drum Making and its Significance to the Garifuna People'. Great Belize Production Channel 5 Belize, 18 November. Retrieved June 2020, from http://edition.channel5belize.com/archives/92605

Cayetano, I. (2017). Personal communication.

Cummins, Alissandra. (1992). Exhibiting Culture: Museums and National Identity in the Caribbean. *Caribbean Quarterly* 38.2: 33–53.

Cummins, Alissandra (2012). 'Memory, Museums, and the Making of Meaning: A Caribbean Perspective'. In: Michelle L. Stefano, Peter Davis and Gerard Corsane (eds.) *Safeguarding Intangible Cultural Heritage.* Woodbridge: Boydell Press, pp. 23–32, p. 27.

Dodd, Jocelyn, and Richard Sandell (2001). Including Museums: Perspectives on Museums, Galleries and Social Inclusion. University of Leicester. Report. https://hdl.handle.net/2381/34

Greene, O. N. (2014). Celebrating Settlement Day in Belize. In: T. Rommen and D. T. Neely, eds., *Sun, Sea, and Sound: Music and Tourism in the Circum-Caribbean.* New York, NY: Oxford University Press, pp.199–201.

Greene, O. (ed.). (2018). *The Garifuna Music Reader.* Solana Beach, California: Cognella Academic Publishing.

Karp, Ivan. (1992). Introduction: Museums and Communities: The Politics of Public Culture. In Ivan Karp, Christine Mullen Kreamer & Steven D. Lavine (eds) *Museums and Communities: The Politics of Public Culture.* Washington: Smithsonian Institute, pp. 1–17.

Kosinski, Dorothy (2016). 'Museums Must Reach into Communities'. *Baltimore Sun*, 3 February. Retrieved March 20, 2020, from http://www.baltimoresun.com/news/opinion/oped/bs-ed-museum-culture-20160203-story.html

Kreps, Christina (2011). 'Non-Western Models of Museums and Curation in Cross-cultural Perspective'. In Sharon Macdonald (ed.) *A Companion to Museum Studies*. Malden & Oxford: Wiley-Blackwell, p. 457.

Low, T. (2004). 'What is a Museum?' (1942). In: G. Anderson (ed.) *Reinventing the Museum: Historical and Contemporary Perspectives on the Paradigm Shift*. Lanham, MD: Rowman & Littlefield (AltaMira Press), p. 36.

Museum of Belize. (2011). Visitor Comment Cards. National History of Culture and History, Ministry of Tourism, Culture and Civil Aviation.

National Institute for Culture and History. (2016). Priority Policy Areas and Policy Interventions. In: *Belize National Cultural Policy: 2016–2016*. Belize: National Institute for Culture and History.

Nettleford, R. (2003). *The Caribbean's Creative Diversity: The Defining Point of the Region's History* 21 March, Distinguished Lecture Series Commemorating the Thirtieth Anniversary of the Caribbean Community]. Available at: http://www.caricom.org/jsp/speeches/30anniversary_lecture_2_nettleford.jsp (Accessed: June 4th)

Onciul, Bryony, (2013). 'Community Engagement, Curatorial Practice, and Museum Ethos in Alberta, Canada', in *Museums and Communities: Curators, Collections and Collaboration,* ed. by Viv Golding and Wayne Modest, London: Bloomsbury, pp. 79–97.

Sandell, R. (2007). 'Museums and the Combating of Social Inequality: Roles, Responsibilities, Resistance'. In: S. Watson (ed.) *Museums and their Communities*. London and New York: Routledge, pp. 95–113.

Shoman, Assad, and Anne S. Macpherson (1995) *Backtalking Belize, Selected Writings*. Belize: Angelus Press, 1995.

Simon, N. (2010). 'Why Participate?' In: N. Simon (ed.) *The Participatory Museum*. Santa Cruz, CA: Museum 2.0.

Smith, Laurajane and Waterton, Emma. (2009). *Heritage, Communities and Archaeology*. London: Duckworth.

Varutti, Marzia (2013). 'Learning to Share Knowledge: Collaborative Projects in Taiwan'. In: Viv Golding and Wayne Modest (eds.) *Museums and Communities: Curators, Collections and Collaboration*. London & New York: Bloomsbury, pp. 59–78, p. 71.

Watson, Sheila (2007). 'Museums and their Communities'. In: *Museums and their Communities*. Oxon: Routledge, p. 6.

5

THE EU-LAC MUSEUMS PROJECT AND COMMUNITY-BASED MUSEUMS

Karen Brown, Marie Claverie and Karin Weil

EU-LAC Museums project beginnings: defining terms

From September 2016 to January 2021, the European Commission Horizon 2020 project titled EU-LAC Museums: Museums and Community: Concepts, Experiences and Sustainability in Europe, Latin America and the Caribbean brought together four institutions from the European Union and four from Latin America and the Caribbean (LAC) (http://www.eulacmuseums.net). By carrying out a comparative analysis of small to medium rural museums and their communities in the EU-LAC region, the project sought to develop the history and theory of museums, communities and their territories. The project achieved this goal by navigating between two impressive historic movements in museum theory and practice—community museums and ecomuseums.

The project partners were tasked with exploring the role of small- to medium-sized local museums in understanding cultural relations between the regions. In designing the project, initially in consultation with ICOM Europe President Luis Raposo and Hugues de Varine, our project theme had its origins in the historic Round Table on the Importance and Development of Museums in the Contemporary World held in Santiago de Chile in 1972 (José do Nascimento Junior, Alan Trampe, Paula Assunção dos Santos, 2012). The subsequent declaration (UNESCO, 1972) and Latin 'New Museology' movement put forward the idea of an 'integral museum,' a concept emphasising the primary responsibility of museums to meet the needs of their communities. This anchoring of the museum in a new 'social function' was integrated into the 1974 ICOM museum definition and has continued to be included to the present day, bringing about a reconsideration of the attitudes, roles and professional practices of museum researchers in relation to their political responsibility (Brown and Mairesse, 2018; De Carli,

DOI: 10.4324/9781003288138-7

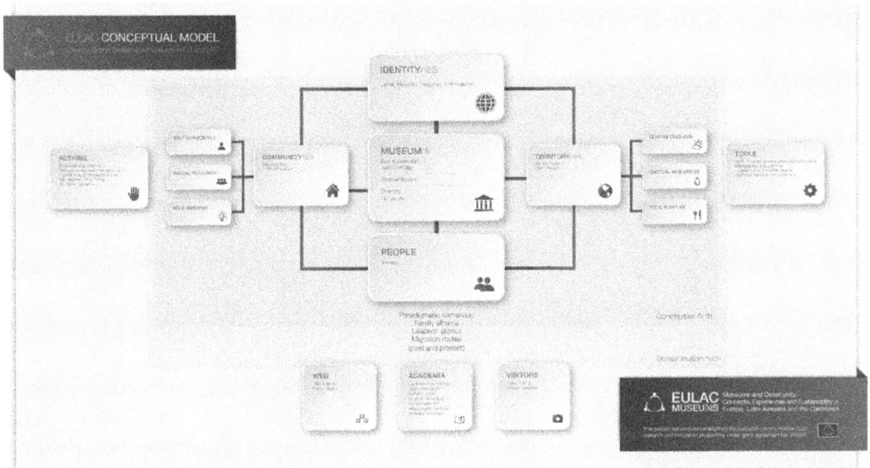

FIGURE 5.1 Initial Conceptual Model for the EU-LAC Museums project. © EU-LAC

2004; De Varine, 2017; De Varine, personal communication, 2000; Mostny et al., 1972; Weil, 2019).

Lisbon, 2016: community-based museums as community-generated initiatives

Our starting point was during the project kick-off meeting held in the National Museum of Archaeology in Lisbon in November 2016. This was the first opportunity we had to meet in person as a consortium, with members from the Caribbean, Chile, Costa Rica, France, Portugal, Peru, Scotland and Spain, to begin discussing what each of us thought a 'community museum' was. On this occasion, with the assistance of our advisors Teresa Morales Lersch (Co-Director, Museos Comunitarios de América network, and co-author of *Memoria: Red de Museos Comunitarios de América. Experiencias de museos comunitarios y redes nacionales*, 2016) and Ann Gunn (University of St Andrews), we asked our group three simple prompts:

- What are three questions you would ask to determine whether a museum is a community museum?
- Think of three things that a community museum is not.
- What is a definition of a community museum that would be useful to you in your work? Include three main aspects.

This collective process threw into sharp relief the fact that a certain lack of information and context existed in each of the regions and that if we navigated between the concepts of 'community museum' and 'ecomuseum,' we had the opportunity

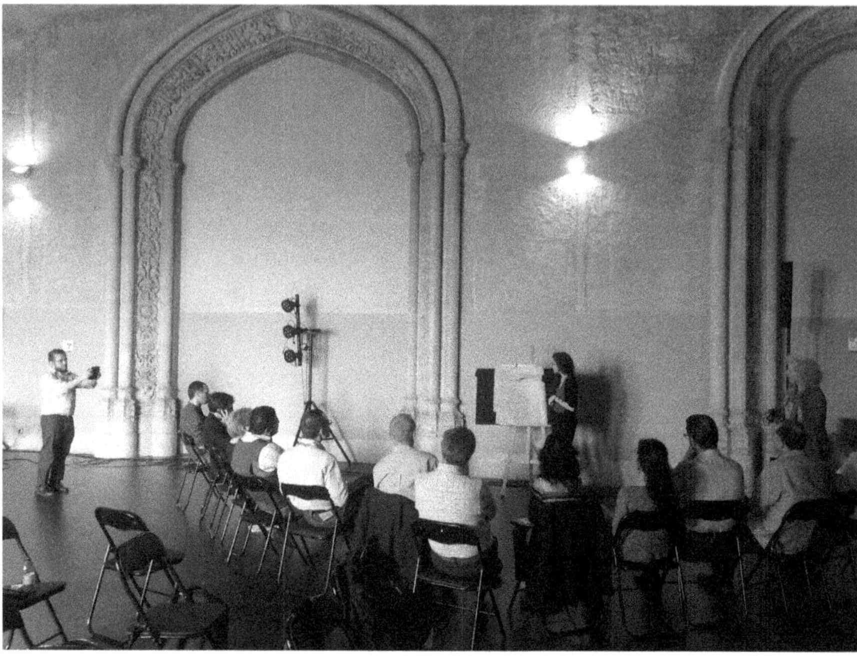

FIGURE 5.2 Workshop on defining a community museum, EU-LAC Museums Kick-Off Meeting, National Museum of Archaeology, Lisbon, November 2016. © Karin Weil

to research meaningful initiatives in museums and community empowerment in Europe and LAC. These gaps were areas where we could grow as researchers, facilitators, teachers and trainers.

During this initial meeting and workshop, we found ourselves encountering, for the first time, the challenge of developing modes of research that took into account our differences, prejudices and diversity. From this workshop and the group reflection in Lisbon, we were able to list areas of agreement around the concept of 'community' (see Table 5.1).

Each of the aspects mentioned in the process of reflection refers especially to the museum as a space created, managed and in direct relationship with the community in which it is located. Despite the differences in social, cultural and environmental contexts and values, these spaces seek to create options for those communities that do not have space or a voice within traditional museums. In community-based museums, the communities create a place to come together to articulate their sense of identity based on their own stories and memory. However, in this same discussion, it became evident that there was a need to deepen and define (a) a frame of reference around the concepts of territory and its multiple connections, (b) what we understand by community and how communities define themselves, (c) the

TABLE 5.1 Key components of a community-based museum: summary of reflections shared during the workshop held at the first consortium meeting, November 2016, Lisbon

Consideraciones para un museo comunitaria	*Considerations for a community museum*
1 Gestionado por y para la comunidad local, en el sentido del desarrollo sustentable	1 Managed for and by the local community, directed to sustainable development
2 Basado en una comunidad específica para preservar, mantener y crear patrimonio comunitario	2 Based in a specific community, to keep alive and to create community heritage
3 Surge como una solución para una necesidad comunitaria	3 Arises as a solution to a community need
4 Creado y gestionado por una comunidad más no exclusivamente	4 Created and managed by a community, but not exclusively
5 Un museo que involucre las actividades de la comunidad y su entorno	5 A museum that involves activities of the community and its surroundings
6 Capaz de integrar varias visiones sobre la comunidad y que incluya un proceso continuo y en transformación de la relación entre las varias generaciones que son parte de esa misma comunidad	6 Capable of integrating several visions of the community, and that includes a continuously transformational process in the relationship between the generations in the community
7 Un espacio de intercambio intergeneracional	7 A space of intergenerational exchange
8 Una herramienta para la acción social y territorial	8 A tool for social action and territorial action
9 Una herramienta organizada directa o indirectamente por la comunidad	9 A tool organised directly or indirectly by the community
10 Un espacio para tratar temas que se relacionan con la comunidad misma	10 A site to address topics that relate to the community itself
11 Comunica la historia desde el punto de vista de la comunidad	11 Communicates the history of a place and its people from the community's point of view
12 Actor principal: la comunidad	12 Community: main actor
13 Definición abierta e inclusiva	13 Open and inclusive definition

relationship between sustainable development and community initiatives and (d) how to address issues of respect, interculturality and social inclusion. The use of the term 'territory' also varied by country. Somewhat like 'the new museology,' the term has been reconsidered not only as a physical and environmental space but also as a form of representation of the social, environmental and cultural fabric. Although in each of the countries, the terms denote the same areas, certainly the ancestral cosmovision of the pre-Hispanic peoples, the sense of belonging, diversity and migration in the Caribbean islands, and the colonial aspects of history with which a number of EU countries are burdened makes each of these countries perceive the term 'territory' differently; however, the concepts of place, meaning, community, social role and processes are used in all of them.

Developing research questions

At the same meeting in Lisbon, it was agreed that each principal investigator would submit ideas based on their own country's context, and from these, a starting framework was to be proposed. The reflection on and development of these concepts allowed the research team to make regional differences explicit and to build the consortium methodology around a clear and more inclusive theoretical framework in the context of regional diversity.

In general terms, the following questions emerged around the historical, environmental and social contexts of each of the regions involved in the project. Attempting to answer these questions allowed us to incorporate regional differences and the implications of these differences into the development of research carried out in the years that followed.

How do we link the concept of territory? What is its relationship with community-based museums?

> Territory—I prefer the word 'place'—is an essential component of the concept of 'belonging', belonging to a landscape, a past; and having a link to the tangible and intangible cultural and natural heritage of a place is an essential component in defining a 'community'. Places are powerful in that they hold memories and help us interpret or understand our community.
>
> *(Researcher from Scotland, Lisbon workshop, November 2016)*

> Territory and community are an indissoluble binomial.
>
> *(Researcher from Peru, Lisbon workshop November 2016)*

The community-based museum is a situated museum. It is located in a territory and is sensitive, and potentially incidental, to the sociocultural dynamics that occur in it. As a public space of a cultural nature, it can be a tool for strengthening community relations over time.

The community-based museum needs to pay attention to the ethical dimension of the processes it works on. Like any discourse, the community-based museum is in a position to make a statement, and it needs to keep its messaging consistent.

The main ideas and generalised reflection on this question can be summarised as follows: in each country, the concept of territory is defined differently, as it is a broad concept that considers tangible and intangible heritage. For many, the community is strongly linked to territory, even though some communities are not defined by their territory.

How do we define 'community'? How does the community define itself?

> A group of people who agree to acknowledge certain characteristics (cultural, political, etc.) and/or interests in common.
>
> *(Researcher from France, Lisbon workshop, November 2016)*

> The community must be defined from within, listening to and respecting its own discourse of self-determination, territoriality, self-government, worldview, historical memory and the relationships that connect it (communality). [Translated from original Spanish.]
>
> *(Researcher from Costa Rica, Lisbon workshop, November 2016)*

A community is defined by relationships of belonging and participation in common codes. It is not necessarily associated with a geographical territory.

In most cases, community-based museums are created with reference to a territory or to a territorialised historical experience. In this sense, they constitute an instance of mediation, a link, between the notions of community and territory (a community that may have a relationship with a territory or its history but does not currently cohabit in it).

The community defines itself by identity and difference. This act of definition contains a dynamic potential; in other words, it can mobilise groups towards the fulfilment of common objectives. It can also be defined 'from the outside' by those who pay attention to common features. This responds to a descriptive register that allows for the distinction of characteristics and limits.

What is the relationship between the community-based museum and the concept of development?

> A community museum can become an articulating element and even an impulse for territorial development. And this is so, because in a community museum, one can differentiate the basic keys to development: social innovation, the existence of territorial resources and the set of relationships between territorial actors, networks that take on various forms.
>
> *(Researcher from Costa Rica, Lisbon workshop, November 2016)*

The community-based museum can be a tool to contribute to the development of a territory or a community, since it contains a story about its past and with it a vision about its present and even certain expectations about its future.

Communities can use the museum to influence political decisions that affect their territories. The project proposed to work with local and regional museums and with the international community of museums, taking into consideration that communities are dynamic, self-defined entities that share common codes, needs and characteristics. Community museums can also be tools that contribute to the development of the community (depending on the objective and activities of the museum), prompting the community to reflect on its own resources, its history and its present situation as it prepares for the challenges of the future. Development can be understood as economic and human (Brown, 2019).

How do we address respect and interculturality?

> [...] the museum must be able to inform about the diverse cultural histories existing in a geographical space through time without hiding the relationships or conflict that may have existed between them, that is, in a historical framework that responsibly and respectfully addresses differences in identity. The museums must be guarantors of human rights, of the convention 169 ILO [International Labour Organization] and of all the legal instruments subscribed by the states for the fulfilment, respect and promotion of cultural and social diversity. They must generate conditions for intercultural dialogue and promote actions under a formal framework of respect for indigenous peoples, children, women and the community in general.
>
> *(Researcher from Chile, Lisbon workshop, November 2016)*

> Clear and respectful communication, ideally face to face at regular intervals. Research across regions will also deepen understanding of other contexts, at first hand.
>
> *(Researcher from Scotland, Lisbon workshop, November 2016)*

The community exists as part of a mosaic of original histories and contexts. To achieve peaceful intercultural relations within and between communities, museums must act in a respectful manner, fostering communication and building bridges between people and communities; it must be a multifaceted space that promotes respect and tolerance of differences.

Next research steps

At various points in the project, we returned to this topic of defining community museums in the process of developing a shared bi-regional research methodology. In 2017, a round table on the topic 'What is a community museum in your region?' was organised during the Carifesta XIII Symposium in Barbados, hosted by the University of the West Indies. The round table garnered perspectives from Jamaica, Barbados and Guyana and included a presentation by the then President of the Museums Association of the Caribbean, Dr Sherene James-Williams.

At the round table, it was explained that community museums take many forms in the Caribbean region, defying a fixed definition and allowing space for adaptation and alternative histories outside the national museum model imported from Europe (see also Ariese-Vandemeulebroucke, 2018).[1] The project compendium *On Sustainable and Community Museums* (Brown, Davis and Raposo, 2019) then aimed to provide reflections and pedagogic tools to suggest how a community might go about conceiving and creating a new community museum or ecomuseum if they so wished. The book contains salient case studies of community-based museums in each partner country and shares the experience and knowledge of our EU-LAC Museums project advisors—Beatriz Espinoza, Hugues de Varine, Peter Davis and Teresa Morales Lersch—concerning the key concepts and features of community and sustainable museums.

Steering Committee meeting of EU-LAC Museums (Antigua Guatemala, 2018)

Having reflected on community and sustainable museums in EU and LAC in the book, and recognising the need to broaden the discussion geographically, the subject of community-based museums was taken forwards into the steering committee meeting of EU-LAC Museums held in Antigua Guatemala in March 2018.[2] On this occasion, we decided to grapple with the question of definition in a more concrete way and to create a survey on the subject, called 'What is a community museum in your region?' This survey was hosted by the University of St Andrews Qualtrics system and attained the university's ethical approval. It was originally designed in Antigua in collaboration with the steering committee members, and it was subsequently edited by the entire consortium and its advisors, notably Peter Davis. The survey results (analysis to follow) have shown that while the community museum is largely about the story of a local community and is a place that caters to that local community, the community welcomes outside expertise in fostering sustainability as long as there are clear terms and parity of esteem between the entities. In contrast to the ICOM museum definition (2007), community museums give prominence to 'people,' 'place,' 'territory' and 'identity,' as well as many definitions reflecting on the intrinsic relationship between the past and the future.[3] In short, community museums are an inclusive and dynamic way of appropriating a community's memory and experiences in the 21st century. They are not only important but also essential for fostering peaceful and resilient societies.

Based on the project research involving community consultations, another milestone was reached when the EU-LAC Museums consortium researchers prepared policy recommendations during two strategic meetings, held first in the University of the West Indies, Barbados (November 2018) and second in Brussels (April 2019). These recommendations were presented to different European Commission and European Union services, including the European External Action Service and the Commission's Directorate-General for International Cooperation and Development, as well as to some invited staff from UNESCO and the United Nations during a policy round table in Brussels in April 2019. These recommendations were subsequently developed into a public report (Brown et al., 2019). This collaborative round table marked a significant moment of recognition that the project could serve as a model for cultural cooperation with LAC.

From theory and practice to policy: EU-LAC Museums policy round table and report (Brussels, 2019)

Reflecting on thematic axes with a balance between the perspectives of Europe and LAC, the report highlights that EU policy can consider and include EU-LAC Museums' project findings. A great deal can be achieved in and through community-based museums, as distinguished from mainstream museums, which are often associated with certain demographics and funding structures that receive state attention and support. Community-based museums offer a lens through which to interrogate both macro and micro, global and local relations. However, community-based museums are often under-represented in policy, while our project findings demonstrate that social engagement and proactive strategies advanced through community-based museums have the potential to challenge and enhance existing EU policy that seeks to contribute to development in LAC, thus making those museums more relevant and sustainable for the future.

Through the vehicle of the policy report, the project consortium aspired to create a bridge between policy and practice, allowing community voices to speak through the project outputs to the policy makers and funders whose priorities are implemented by the project.

The EU-LAC Museums policy report includes, but is not limited to, the following key recommendations:

- 'Community-based museums and heritage initiatives [...] merit more visibility and agency to work through the critical issues affecting human life in different parts of the world.'
- 'Museum activities with social impacts are truly important for cultural diplomacy and should be recognised and supported. In particular, community-based museums have a special role to play as they are, by their very nature, deeply rooted at local level; the engagement of communities is embedded in their core functioning.'
- 'Taking the socio-economic situation of local as well as national areas into account in their strategies, museums and policy makers can contribute to local development. Incorporating museums into local/national development strategies can prove a useful method of medium to long-term partnerships and ensure the meaningful contribution of museums and their policies to local development.'
- 'Policy makers should put the protection of cultural heritage protection (tangible and intangible) at the heart of youth participation and engagement strategies [...].'
- 'Grass-roots understanding should be given the opportunity to inform and affect policy in a bottom-up manner locally, nationally and bi-regionally.'
- 'Museums should be enabled to lay the groundwork for sustainability by recognising the right to self-determination and by making visible the full range of community voices.'

(*Brown et al., 2019.* Report, *p. 19*)

Building on the high-level visibility of the project results and recommendations noted above, on 7 September 2019, the 34th General Assembly of ICOM, held in Kyoto, Japan, adopted Resolution No. 5, entitled 'Museums, Communities and Sustainability,'[4] conceived as part of the EU-LAC Museums project and submitted by the two ICOM Regional Alliances in Europe and LAC. It marked a historic moment for bi-regional and international cultural cooperation concerning the role of community-based museums in today's world.

New ICOM Resolution 'Museums, Communities and Sustainability' (Kyoto, Japan, 2019)

Resolutions adopted by the General Assembly—the decision-making body of ICOM—mark the path of the organisation's actions and missions in the years that follow. Over the past few decades, ICOM has adopted resolutions supporting community museums and highlighting the notion of museums going beyond the formal definition (for example, ICOM 1997, 2010, 2019). Previously, these concepts included key elements such as the recognition of 'the importance of museums in promoting harmony, mutual understanding and exchanges between communities locally, regionally and nationally' (ICOM Resolution 5, 2010, p. 5) and the role of museums as 'fundamental tools for the individual and collective development of critical minds, of self-awareness, of the sense of citizenship and of community's identity' (ICOM Resolution 1, 1995, p. 2). In this same 1995 Resolution, the museum community had already noted that 'some local museums all over the world which are undertaking innovative activities focusing on everyday topics of community life, trying to challenge traditional models and reaching beyond the limits of exhibition spaces, are facing threats of closure and lack of support from their governing bodies,' and it urged 'local and national governments to recognise and support museums as cultural mechanisms in the service of communities, in the valorization of their particular identities, and as unique tools for the collective management of their cultural heritage' (p. 2).

Resolution No. 5 (2019) extends this legacy by acknowledging and recognising the challenges faced by museums as well as their communities in the 21st century. Indeed, it not only advocates for more support and recognition of community-based museums, which today continue to face threats of closure and lack of recognition, but it also takes community-based museums into account in the global reflections on the definition of a museum and the role of museums in sustainable development.

Following the 2016 ICOM General Conference in Milan, ICOM created a new Committee for Museum Definition, Prospects and Potentials (MDPP) to study the current definition of museums. They explored similarities and differences in values and practices of museums in diverse and rapidly changing societies. Combining wide-ranging dialogue across the membership with dedicated expert forums, the committee addressed the ambiguous and often contradictory trends in society and the subsequent new conditions, obligations and possibilities for museums. The EU-LAC Museums project engaged in this process through its symposium, held at the University of St Andrews on 25 November 2017, on the topic of Defining the Museum of the 21st Century.[5] Hosted in association with ICOM's International Committee for Museology, the symposium followed on from one held in La Sorbonne Nouvelle, Paris, and dovetailed similar symposia held in the Universidad Nacional de Avellaneda, Buenos Aires (9–10 November) and the Universidade Federal do Estado do Rio de Janeiro (16–17 November) (Brown and Mairesse, 2018; Brown, Brulon Soares and Nazor, 2018).

Following the processes of active listening and collecting and collating alternative definitions, work done by the MDPP Committee, in July 2019, the Executive Board of ICOM voted for a new, alternative museum definition to be included in the ICOM Statutes instead of the current museum definition. Subsequently, in September 2019, the 26th ICOM General Conference, in Kyoto, Japan, hosted numerous sessions, round tables and debates around this topic, after which the Extraordinary General Assembly, on 7 September 2019, decided to postpone the vote on the new museum definition.

In this context, the EU-LAC Museums project, through the Resolution No. 5, made a significant contribution to this global reflection among museum professionals by highlighting the 'vast number of community-led organisations' with a focus on the fact that the latter 'do not currently fulfil the ICOM Definition of a Museum (2007),' while recommending the museum community to 'remain sensitive to local and regional differences and demonstrate awareness of the geo-political dimension of the concept of the museum, especially relating to the resource needs of community-based museums in lower- to middle-income countries.'[6]

It is worth noting that another pressing theme addressed at the ICOM 2019 General Conference was sustainable development. In addition to opening the conference with a plenary session entitled 'Curating sustainable futures through museums,' another resolution was adopted during the ICOM General Assembly on this specific topic. Resolution No. 1 'On sustainability and the implementation of *Agenda 2030, Transforming our World*' was developed by the ICOM Working Group on Sustainability and submitted by ICOM UK and ICOM Norway. This resolution recognises the role of museums in creating a sustainable future and calls for empowering museums, their visitors and their communities in this regard.

The first recommendation of Resolution No. 5 echoes and complements it by underlining that community-led museums further the 'sustainable use [of natural, cultural and intangible heritages] for environmental, social and economic development of communities, towards achievement of the UN 2030 goals and climate justice.'

The impact of COVID-19 on community-based museums

More recently, and responding directly to COVID-19 and other environmental and political challenges faced by project partners during the years 2016–2020, EU-LAC Museums organised a webinar series, 'Community-based Museums in

Times of Crisis' (12 June 2020, 29 June 2020 and 10 July 2020), which involved 21 invited speakers and reached approximately 905 people from 35 countries via Zoom and Facebook. This global online engagement shows how the research into community-based museums developed by EU-LAC Museums has empowered communities and museum professionals to tell their own stories and address challenges facing the preservation of their heritage and cultural identity.

The COVID-19 crisis has not only exposed us to an epidemiological crisis, but also, above all, it has demonstrated how humanity has abused finite natural resources and voraciously appropriated life-sustaining systems that leave no room for nature. These environmental problems have been mirrored—and arguably exacerbated—by social and economic inequalities that have also been exposed by COVID-19, which has made clear that there is a need for people to reconnect with nature and one another. In this way, the concept of 'community' could lead to progress for the common good of communities and the environment. The need for change is urgent.

The work carried out as part of the EU-LAC Museums project connects in many ways with these contemporary challenges faced by society. Our findings indicate that there is an important role for the global museum community. The project has shown that strong bi-regional relationships and mutual learning are key assets to bring to the global reflection on the roles and definition(s) of museums in different regions. During a second round table in January 2021, we proposed and recommended community-based museums as tools through which many of these needs can be addressed.

Our analyses of community-based museums are based on our own experiences as researchers and professionals working in Europe and LAC, and these analyses have been tested widely through debate and promotion in our ICOM networks and survey. Our project has shown that by working closely with communities and linking with local governments, community-based museums can contribute to sustainable development in different contexts and promote development of peaceful and resilient societies.

Survey: 'What is a community museum in your region?' (2018–2020)

By the census date of 19 April 2021, the survey attained 528 responses from 70 countries written in English, Spanish, Italian, Portuguese and French (mainly in English and Spanish).[7] The responses have proven insightful in qualitative terms for the continuing process of honing our research questions as academics and for taking practical steps to increase international recognition for community museums through our partner, ICOM.

In this survey, 16.4 per cent of the participants reached were from the LAC and 13.4 per cent from the EU, with 70.2 per cent of participants coming from the rest of the world. The largest number of participants were from the UK (largely thanks to the community heritage network in Scotland), followed by Japan (thanks to ICOM Kyoto advertising and dissemination through Japan's network of community museums), then Costa Rica, Peru, Chile, Portugal, Canada, the United States, Italy (thanks to the network of Italian ecomuseums), Spain, Greece and others. Targeted advertising through project networks of community museums and heritage, as well as through social media and our project webinar series, meant that the survey reached an appropriate demographic, with the majority responding that the 'organisation/group interested in local or community development' and that they are part of is a community museum, as shown in Figure 5.3.

In general, respondents engaged well with the survey questions, providing additional comments and reflections where prompted. At times, the specific wording of individual questions or the use of terms such as 'community' or 'sustainability' was

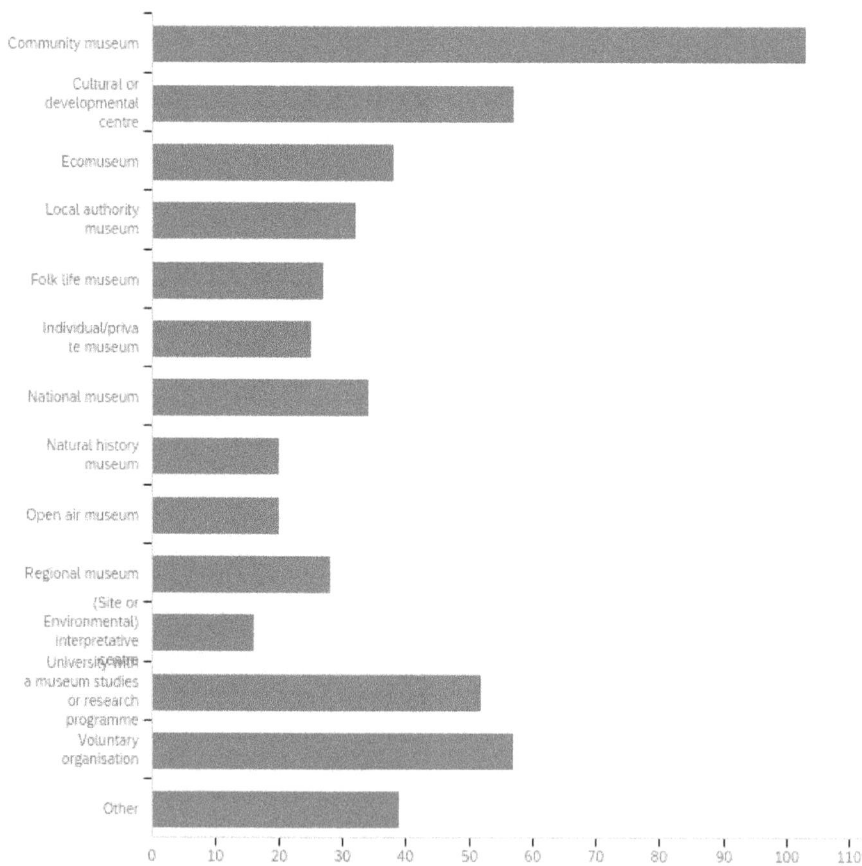

FIGURE 5.3 Survey response to the question, 'If you are part of an organisation/group or individual interested in local or community development, what is its type?'

questioned by one or two respondents, especially those identifying as academics, who helpfully qualified their responses in the comments. The majority of respondents, especially those identifying as working for community museums, engaged generously with the free-text questions, yielding useful data for thinking around the characteristics of community museums on a global scale.

In response to the question 'Which of the following criteria help to define what a community museum is? (Please select any number of answers. Rank your answers in order of priority with 1 being the highest priority),' the option 'a geographical territory' received the most selections, followed by 'a local sense of community,' 'a local sense or spirit of place' and 'a shared local history.' These findings are in harmony with responses to the next question: 'The purpose of a community museum is to provide (please select all relevant fields).' The majority selected in response 'a sense of belonging' (236 selections), followed by 'community participation in heritage matters' (231 selections) and 'heritage preservation' (213). Interestingly, 'collection of objects' received only 124 selections and 'lobbying' had the lowest score, at 50 selections.

Responses to the question, 'In the community museum(s) that you know, whose/what story (past, present or future) is—or should be—being told? (Please select all relevant fields),' are illustrated in Figure 5.4. These responses were supported with free-text comments reinforcing the importance of members of the community telling their own story, especially elders, though occasionally comments suggested

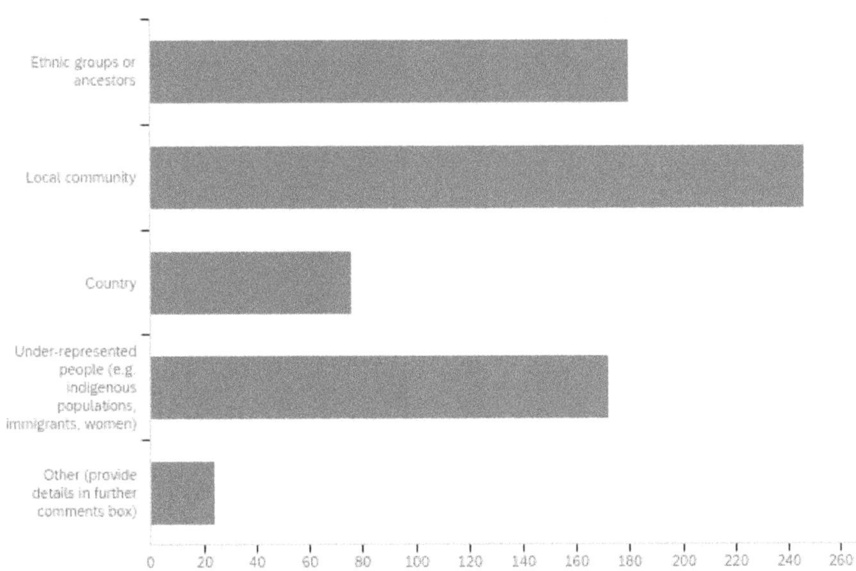

FIGURE 5.4 Survey response to the question, 'In the community museum(s) that you know, whose/what story (past, present or future) is—or should be—being told? (Please select all relevant fields)'

that a local government or ministry of culture should tell the story if the museum were municipal rather than private or independent.

The question 'Who is the main public for the community museum?' had the clearest response of all, building on the evidence that the community museum is characterised as created, governed and used by the local community. The majority, 56.8 per cent, saw the local community as the main public, while 9.8 per cent selected school groups and students, and only 9.5 per cent prioritised tourists. In these free-text comments, many people commented that all options were relevant; the following comment is especially indicative of the answers to this question: 'The community museum serves the community in focus first, and expands outward.' One Scottish respondent commented: 'in [the] Highland region we are pushed more and more to service tourists and I think that this is [to] the detriment of the contemporary relevance and social potential of the museums.'

Answering, 'Do you think that a community museum should best be run by …,' yielded data revealing that respondents thought a combination of community members, local associations, volunteers and professionals should run a community museum, bringing their various expertises together for the local heritage community.

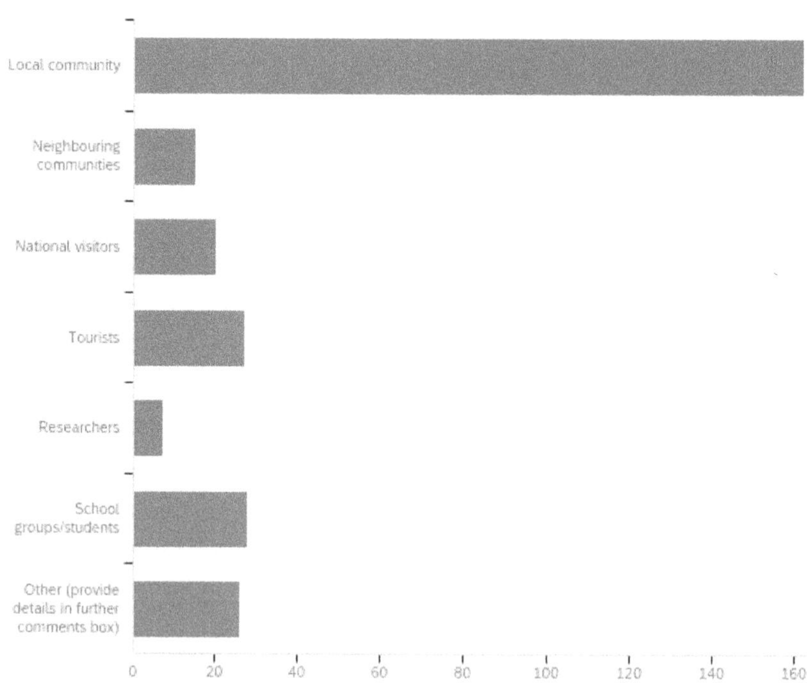

FIGURE 5.5 Survey response to the question, 'Who is the main public for the community museum?'

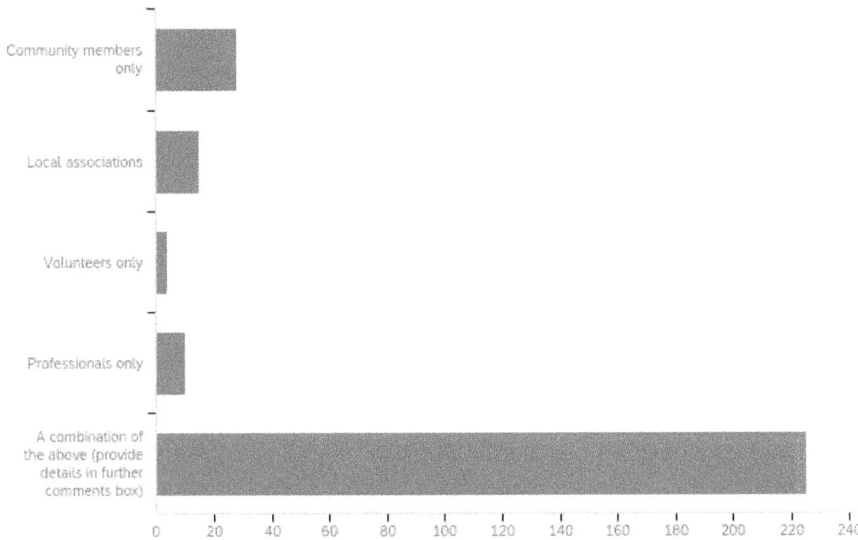

FIGURE 5.6 Survey response to the question, 'Do you think that a community museum should best be run by…?'

For example, one free-text response read: 'Community members provide the substantive material, professionals help to curate and preserve for promotional and archival purposes, volunteers help make it sustainable over time.' Another stated:

> I think it helps to be as professional as possible, but what that means will vary by context. An all-voluntary museum can be more professional than one with paid staff if they have the right expertise on the board and amongst the volunteers. And while its best to have most of the board drawn from the community being represented, some fresh eyes and fresh perspectives can be hugely beneficial.

Several responses in Spanish agreed:

> Comunitarios que estén capacitados como profesionales del patrimonio.
>
> Un museo comunitario es un pacto entre todos los agentes implicados.
>
> Se necesita a los miembros de la comunidad que la conocen, las asociaciones que pueden apoyarlo y los profesionales para darle el contexto y organización.

Respondents made clear through the free-text comments that they saw the need for community agency, supported by professionals where useful, although a minority felt that museum professionals should be in charge, with input from the community. This comment in Italian is indicative of that view: 'la gestione in senso stretto deve essere svolta da soli professionisti, ma appoggiandosi anche ai membri della comunità e/o a volontari'.

The debate was then continued through the next question, 'What makes a community museum sustainable? (Please select any number of answers. Rank your answers in order of priority, with 1 being the highest priority).' 'Capacity building for staff/volunteers/community' ranked highest at 25 per cent, followed by 'substantial start-up funding' at 13 per cent, and 'formal community agreement' and 'competent organisation and financial administration,' both at 12 per cent, as visualised in Figure 5.7.

Clearly, the biggest need for local communities is investment in capacity building on the ground, as well as a defined system of governance made in the community. Once this structure is in place, the participants were very open to external engagement, especially on the international level. For example, 88.9 per cent of respondents thought that 'community museums should engage with international cultural institutions, networks and activities,' with 2.3 per cent responding 'no' and 9.1 per cent indicating 'don't know.' A large number of respondents (222) completed the free-text question 'In what ways should community museums engage with national cultural institutions, networks and activities?', demonstrating an overwhelming positivity towards this type of engagement. In general, respondents saw this type of activity as useful for, among other things,

- the value of networks for strengthening community museums (especially networks between community museums themselves, which national cultural institutions can engage with),
- creating equality in partnership working,
- recuperation or preservation of indigenous community histories,
- sharing between national museum priorities and community voice,

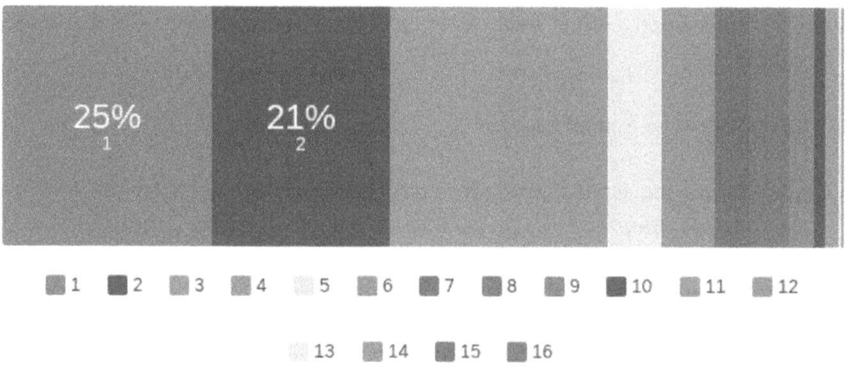

FIGURE 5.7 Survey response to the question, 'What makes a community museum sustainable? (Please select any number of answers. Rank your answers in order of priority, with 1 being the highest priority)'

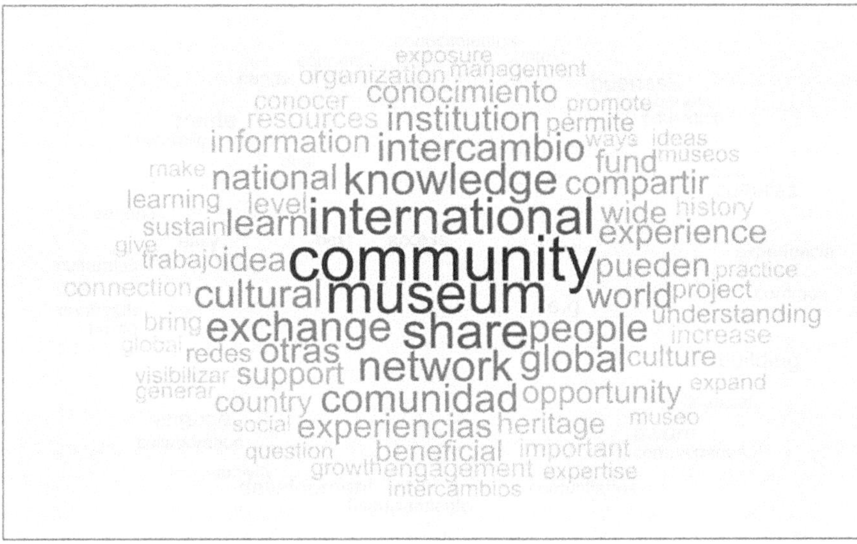

FIGURE 5.8 Word cloud generated from the question, 'Can you offer a definition of a community museum?' © Karen Brown

- validation and recognition of community museum practices,
- creation of seminars and online resources and
- sharing skills, expertise and resources such as preservation, handling and display.

A word cloud (Figure 5.9) visualises the main ideas in all languages received.

Several respondents referred to the link between strengthening the local and connecting community museums at national or regional levels. But they also understood that local distinctiveness is at risk in a global and increasingly interconnected world. One Italian respondent summarises it thus: 'without comparison, it is not possible to recognise the specific characteristics of the community and to progress with a coherent development,' while a Spanish respondent writes that '[t]he exchange of experiences strengthens cultural management.' Looking specifically at national networks, a respondent from Ecuador states that '[t]he network of community museums in Ecuador has allowed the different communities to listen and put into practice the various experiences that each community space has raised.' Another Spanish respondent, commenting on regional networks, notes that '[r]egional networks have real value, since they require permanent collaborative work and transfer of capacities, to establish or project a joint future project.' Another respondent notes in Spanish that such community museum connections 'can teach others their own history and ways of management, creative solutions to everyday problems and an approach to heritage that does not necessarily align with public policy.' These latter values echo strongly those of the Latin American Museos Comunitarios movement through greater awareness of networking and support mechanisms. Several comments also mentioned the role of engagement in the promotion of European citizenship.

These findings on the perceived value of networks and visibility were made even more granular in the light of responses to the next question, which revealed that almost half of the respondents did not benefit from these types of actions, highlighting a clear need for capacity building and networking in the sector. Asked 'If you are involved in a community museum, have you experienced collaboration with other community museums nationally or internationally?', 51.3 per cent answered 'yes,' and 48.7 per cent 'no.' As a follow-on question, participants were then asked if they would consider the possibility of twinning with another community museum (in the way in which cities have twins), and 83.7 per cent said 'yes,' and 16.3 per cent 'no.' However, free-text comments warned about the capacity of time and resources available for other activities in a volunteer-run organisation and also a potential lack of interest among community members to look elsewhere.

The question 'Why do you think that community museums are important? Please comment below on the attributes that they have and what they might achieve that more traditional museums cannot,' attracted answers mainly focused on community voice, storytelling and local history, genealogy, traditions and artefacts, as well as agency over these attributes. In general, the comments demonstrated the belief that, because community museums are close to the local community, they create the best conditions for bringing people together to build community agency and collective endeavours, especially around ideas of identity and among under-represented groups. These attributes are contrasted with 'traditional' museums, which have, according to one response in Spanish, 'high state or private subsidies and are subject to governments of the day and budgets conditioned to the type of discourse that is established.' Another summarised (in Spanish): 'Unlike traditional museums where pieces of materials that were produced by people are exhibited to create a connection with visitors, community museums involve people in the community.' And in English, another responded:

> They are less intimidating and 'authoritarian' than the nationals, and communities often feel connected to them rather than excluded by them. They usually link to local and regional shared places, identities, and even things in living memory, giving a very close and tangible sense of belonging.

And again, 'They closely connect with the grassroots, with the owners and creators of heritage. Community museums also offers [*sic*] a platform for previously marginisalised [*sic*] and excluded voices to be heard in museums and museology.' A respondent in French also pointed out that, while traditional museums may seek to transform the macrocosm into a microcosm, community museums do the opposite, by transforming the unique character of the community into heritage, thus contributing to human diversity. The respondent implored: 'Do not reduce diversity (spirit of place) to an "identity" (fixed, defined and determined).'

Of all the survey questions, 'intangible cultural heritage' was most mentioned in response to the above question; for example, one person responded: 'Sense of community; social bonds; inter-age/gender/class cooperation; preservation of intangible heritage (*not 'freezing' it, using it*) [emphasis mine]' and simply, 'Indoor and outdoor experiences of indigenous practices [...] Intangible has more space.' As summarised in one Spanish comment: 'Being managed from communities in conjunction with other people or organisations, they have the ability to tell a story in a unique way and help preserve their own narrative.' Two others added to this idea: 'above all, they generate bonding strategies with the community and neighbours that traditional museums cannot succeed in exercising,' and 'the community through the museum affirms its material and symbolic identity through its own forms of organisation.' Characteristics apparent from these comments include

- the ability to tell one's own story,
- an understanding of the past at local level in order to plan for the future,
- the freedom to narrate stories on the margins away from state-led narratives found in national museums and
- preservation of local memory, for the next generation and for self-determination.

On the few occasions that tourism was mentioned, it was in connection to ideas of sustainability. For example, one ecomuseum representative commented

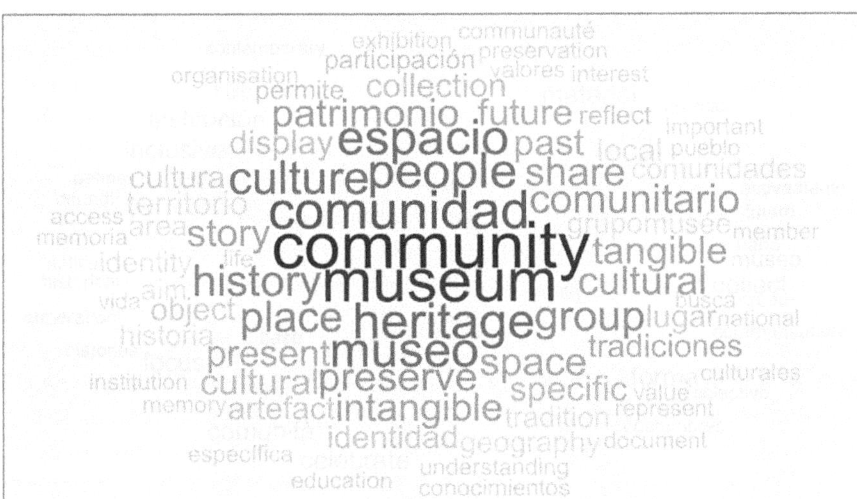

FIGURE 5.9 Word cloud of responses to the question, 'In what ways should community museums engage with national cultural institutions, networks and activities?' © Karen Brown

on how, in contrast to traditional museums whose collections are sited inside a building, an ecomuseum can increase an area's attractiveness as a location to live, visit, work and invest in. These findings affirm the definition of a community museum that our project advisor Peter Davis formulated in 1999 that a community museum is a local museum based in the communities of interest that govern it: 'that is, a small museum with limited collections that serves those people in a defined geographical area' (Davis, 2011, pp. 36–37). Davis explains how such a museum can provide a 'sense of place' for local identities, having affinities with the ecomuseums where local people designate their own territory, encompassing aspects of landscape (geology, scenery), built heritage (architecture), natural heritage and intangible heritage (dialect, songs, stories) (Davis, 2011, p. 81. See also Davis, 2008). The final free-text question of the survey, 'Can you offer a definition of a community museum?', is so topical today in the light of ICOM debates around the museum definition that it merits close analysis of both the geography of the answers and the most frequently used words in the definitions offered.

Notable in the definitions offered is the prominence of the words 'people,' 'place,' 'space,' 'culture' and 'future,' contrasting with, for example, the 2007 ICOM museum definition.[8] Other than the words 'heritage' and 'intangible,' most of the words from the survey responses were absent in the ICOM museum definition.

Conclusion

By creating a space for people to connect with their culture and tangible and intangible heritage, and by finding a balance between understanding the past for building a future, community museums are well placed on the international stage to answer many societal questions from the grassroots. It is for this reason that, in our first EU-LAC Museums Policy Round Table report, the team recommends that 'Community-based museums and heritage initiatives [...] merit more visibility and agency to work through the critical issues affecting human life in different parts of the world' and that 'Museums should be enabled to lay the groundwork for sustainability by recognizing the right to self-determination and by making visible the full range of community museums' (Brown et al., 2019). Resolution No. 5, adopted on 7 September 2019 at the 34th General Assembly of ICOM held in Kyoto, Japan, further reinforced this point, building on past ICOM resolutions (1995, 2013, 2019). The resolution, drafted by EU-LAC Museums supported by ICOM Europe and ICOM LAC, highlighted the 'vast number of community-led organisations' and the fact that these 'do not currently fulfil the ICOM Definition of a Museum (2007),' while recommending that the museum community 'remain sensitive to local and regional differences and demonstrate awareness of the geo-political dimension of the concept of the museum, especially relating to the resource needs of community-based museums in lower to middle income countries.'

Defining may always prove contentious, but in the process of discussing differences in relation to a space for community museums, however they may be defined, has opened to inform discussions around ecological museology in an age of decoloniality through needs-based and co-produced participatory research practices, interviews in the field and the EU-LAC Museums survey.

Appendix

In what follows, a number of the proposals are listed where the respondent named their country and occasionally the community with which they identified. They are all reproduced verbatim, without correcting any spelling or grammar, and English translations are included directly following the original response where relevant. The proposals are listed by region to compare and contrast regional responses.

Europe

Bernera, Isle of Lewis, Scotland: 'A local run museum that collects and preserves local heritage, be that history, archive documents, traditions, stories or artefacts. The collection should be available for the local community and visitors/researchers from further afield to see and access so that they can all learn more about the community and place. Where volunteer run it should always try to use the best practice for preserving and presenting the collection and be able to make use of new techniques where possible. A museum while showing the past must not be stuck in the past itself.'

Isle of North Uist, Scotland: 'Celebrates, affirms and sustains all the languages and modes that make communities distinctive and help them to survive and develop organically and holistically.'

France: 'A community museum is a place that reflects the history, habits, culture and ways of life of a specific area, or a specific group of people, allowing to connect people through stories, memories and objects of a common past, present and towards a future to build.'

Thessaloniki, Greece: 'A community museum exhibits material culture of specific identity groups usually determined of a specific geographic area. They include all kind of collections that depict the community, so they really can be very dynamic. Community museums function as a place where local people can gather.'

Valdostana, Italy: 'Un museo diffuso che preserva e condivide il patrimonio immateriale e materiale della comunità.'

Comunitá di Salbertrand (TO), Italy: 'Condivisione di luoghi, saperi, modalitá di gestione di risorse, conservazione e valorizzazione del proprio patrimonio.'

Comunidad Valenciana, Spain: 'Museo que nacen de la comunidad y no de una administración.' ('Museum born from the community and not from an administration.')

Caribbean

Village Artistique de Noailles, Croix-de-Bouquets, Haiti: 'Musée appartenant à une population et géré par celle-ci avec l'aide de professionnels.'

Diego Martin, Trinidad: 'It is a community-owned place that organises and celebrates the objects, stories, and artwork that is important to the residents of a community.'

Lethem, Guyana: 'A Community Museum is a place of artefacts, physical and virtual, that reflects the local cultures of a socially inclusive demographic and geographical boundary.'

St James, Jamaica: 'Space to share tangible and tangible cultural heritage and stories in the padt [sic] present and enable reflection and discussion on positive chage [sic] for future.'

San Fernando, Trinidad and Tobago: 'A formal or informal museum that reflects the history and heritage of a particular community or community groups whether localised or dispersed.'

Latin America

Bolivia: 'Museo Comunitario. Es un espacio cultural creado por los miembros de una comunidad, en el sentido no restricto de su significado, donde se construye autoconocimiento colectivo, propiciando la reflexión, la crítica y la creatividad, reafirmando los valores materiales y simbólicos de su Patrimonio Cultural, reconocido según sus usos y costumbre.' ('Community Museum. It is a cultural space created by the members of a community, in the unrestricted sense of its meaning, where collective self-knowledge is built, fostering reflection, criticism and creativity, reaffirming the material and symbolic values of its Cultural Heritage, recognised according to its uses and custom.')

Comunidade Vozes de Mestres, Brazil: 'Um lugar comum, com pessoas comuns, a contarem suas histórias comuns, sua cultura, seus modos de fazer, ser em todas as áreas, de forma simples e verdadeira.'

Comunidad vulnerable, Colombia: 'Museo Comunitario, es el arte para y por la Vida.' (Vulnerable community, Colombia: 'Community Museum, is the art to and for Life.')

Colombia: 'Museo comunitario es un espacio de expansión patrimonial material e inmaterial local que permite la participación de la comunidad en temáticas, objetos y acciones.' ('Community museum is a space for the expansion of local tangible and intangible heritage that allows community participation in themes, objects and actions.')

Placilla de Peñuelas, Valparaiso, Chile: 'Museo comunitario es una institución al servicio de su propia comunidad que permite a través de la autogestión, asociatividad y redes de apoyo local, nacional e internacional, desarrollar proyectos y acciones para cuidar, educar y difundir el patrimonio e historia de la propia comunidad con la comunidad. Es un espacio abierto, inclusivo y democrático.' ('Community museum is an institution at the service of its own community that allows, through self-management, associativity and local, national and international support networks, to develop projects and actions to care for, educate and disseminate the heritage and history of the community itself with the community. It is an open, inclusive and democratic space.')

Antigua, Guatemala: 'Es un museo que aportaa comunidad. espacio donde los integrantes de la comunidad construyen un autoconocimiento colectivo, propiciando la reflexión, la crítica y la creatividad. Fortalece la identidad, porque legitima la historia y los valores propios, proyectando la forma de vida de la comunidad hacia adentro y hacia fuera de ella. Fortalece la memoria que alimenta sus aspiraciones de futuro.' ('It is a museum that contributes to the community. Space where community members build a collective self-knowledge, fostering reflection, criticism and creativity. It strengthens identity, because it legitimises history and their own values, projecting the community's way of life inwards and outwards. It strengthens the memory that nourish their aspirations for the future.')

La Merced, Ecuador: 'Son espacios que han sido creados en asambleas de la comunidad en busca de un beneficio de asociación y representación de su pueblo a la sociedad en la cual no se ve representada, con el fin de promover sus costumbres, tradiciones, memoria oral, saberes que han sido transmutados en la historia de la comunidad que busca continuar con su legado.' ('They are spaces that have been created in community assemblies searching for a benefit of association and representation of their people to the society in which they are not represented, in order to promote their customs, traditions, oral memory, knowledge that has been transmuted into the history of the community that seeks to continue its legacy.')

Altiplano, Peru: 'A heritage-focused space that focuses on decentralised groups of people and decentralised cultural narratives and includes a high level of community participation.'

Caracas, Venezuela: 'Aquella institucion organizada en un espacio con una coleccion que representa a su comunidad, un espacio para la educación y el crecimiento de a población que se encuentra mas lejos de las ciudades principals.' Another from San Vicente, Nicoya Costa Rica: es el escenario vivo de un territorio, con tradiciones y diario oficio de sus habitants.' ('That institution organised in a space with a collection that represents its community, a space for education and the growth of the population that is further from the main cities.')

Asia-Pacific

Tuwali tribe of Ifugao, Philippines: 'A community museum is an important tool for the collection, preservation, and exhibition of material and non-material culture of a group of people which can be opened for innovation as to its handling so it can reach out to wider patronage.'

Municipality of Murcia, Negros Occcidental, Philippines: 'Community Museum is a living cultural hub that is being maintaned [sic] by the locals and is being shared to other communities for mutual knowledge and understanding.'

Urban, India: 'Community museums can be defined as the place of conservation and preservation of local traditions, customs, history and heritage involved in documentation, display and dissemination of the same with an aim to generate interest and livelihood enough to support and sustain these communities and their heritage from vanishing through appropriation and assimilation.'

Suwon, South Korea: 'A place that embodies the values of people, life, and places'

Academic museum and museums practitioners, Thailand: 'Museum that runs by community member. Inclusive all members to be engaged.'

Africa

City, Egypt: 'Community museum is the museum in which all local culture, intangible heritage, crafts and hand made could be shown and preserved for purposes of tourism or studying and developing.'

Tirana, Albania: 'Community museums are important because depending on the mission, goals and visionary projects they may have the easiest and most practical way to encourage, educate and inspire people. We need to help the community understand the value of community when they collaborate with each other this way many problems can be solved easily.'

Songhoy, Mali: 'Le musee communauté est le lieu de préservation de la mémoire de l'identité d'une communauté.'

Oko-Anala, Nigeria: 'A community museum is an exhibition center that shows a specific cultural identity of a particular people.'

Nemana, one of the Great Zimbabwe World Heritage Site's local communities, Zimbabwe: 'A museum established by the community, about the community and for the community.'

North America

Ontario, Canada: 'I have worked in community museums for over 35 years. I would not dream of offering a definition.'

Région du Kamouraska au Québec, Canada: 'Une institution culturelle auto-gérée qui est animée par le désir de créer du lien localement sur une base humanitaire.'

Tequesta, US: 'A Museum focussed on local history, tradition, crafts, and or culture. In our case, the preservation of extinct peoples, battles, following growth with inclusion of local celebrites.'

Acknowledgements

This project has received funding from the European Union's Horizon 2020 research and innovation programme under grant agreement No. 693669. For further information, please visit https://www.eulacmuseums.net/.

Notes

1 'What is a Community Museum in your region?' Carifesta, 2017, The University of the West Indies, chaired by Karen Brown.
2 EU-LAC Museums Steering Committee meeting held in Antigua Guatemala, March 2018, consisted of Karen Brown (Project Coordinator), Jamie Brown (Project Youth Programme Worker and Administrator), Lauran Bonilla-Merchev (then President of ICOM Costa Rica), Samuel Franco (then President of ICOM-LAC), Luis Raposo (President of ICOM Europe) and Gustavo San Roman (Professor of Cultural Identity, University of St Andrews).
3 See the updated ICOM Museum Definition (2022) here: https://icom.museum/en/resources/standards-guidelines/museum-definition/ (Accessed 14 September 2022).
4 Find more information about the 34th General Assembly here: https://icom.museum/en/news/resolutions-adopted-by-icoms-34th-general-assembly/
5 Find more about the Defining the Museum of the 21st Century conference here: https://mgcistandrews.wordpress.com (Accessed 12 August 2020).
6 Quote from Resolution No. 5.
7 Participants came from the following countries: Afghanistan, Albania, Algeria, Andorra, Angola, Antigua and Barbuda, Argentina, Armenia, Australia, Austria, Azerbaijan, Bahamas, Bahrain, Bangladesh, Barbados, Belarus, Belgium, Belize, Benin, Bhutan, Bolivia, Bosnia and Herzegovina, Botswana, Brazil, Brunei, Bulgaria, Burkina Faso, Burundi, Cambodia, Cameroon, Canada, Cape Verde, Central African Republic, Chad, Chile, China, Colombia, Comoros, Republic of the Congo, Costa Rica, Côte d'Ivoire, Croatia, Cuba, Cyprus, Czech Republic, Democratic People's Republic of Korea, Democratic Republic of the Congo, Denmark, Djibouti, Dominica, Dominican Republic, Ecuador, Egypt, El Salvador, Equatorial Guinea, Eritrea, Estonia, Ethiopia, Fiji, Finland, France, Gabon, Gambia, Georgia, Germany, Ghana, Greece, Grenada, Guatemala, Guinea, Guinea-Bissau, Guyana, Haiti, Honduras, Hong Kong (S.A.R.), Hungary, Iceland, India, Indonesia, Islamic Republic of Iran, Iraq, Ireland, Israel, Italy, Jamaica, Japan, Jordan, Kazakhstan, Kenya, Kiribati, Kuwait, Kyrgyzstan, Lao People's Republic, Latvia, Lebanon, Lesotho, Liberia, Libyan Arab Jamahiriya, Liechtenstein, Lithuania, Luxembourg, Madagascar, Malawi, Malaysia, Maldives, Mali, Malta, Marshall Islands, Mauritania, Mauritius, Mexico, Federated States of Micronesia, Monaco, Mongolia, Montenegro, Morocco, Mozambique, Myanmar, Namibia, Nauru, Nepal, Netherlands, New Zealand, Nicaragua, Niger, Nigeria, Norway, Oman, Pakistan, Palau, Panama, Papua New Guinea, Paraguay, Peru, Philippines, Poland, Portugal, Qatar, Republic of Moldova, Romania, Russian Federation, Rwanda, Saint Kitts and Nevis, Saint Lucia, Saint Vincent and the Grenadines, Samoa, San Marino, Sao Tome and Principe, Saudi Arabia, Senegal, Serbia, Seychelles, Sierra Leone, Singapore, Slovakia, Slovenia, Solomon Islands, Somalia, South Africa, South Korea, Spain, Sri Lanka, Sudan, Suriname, Swaziland, Sweden, Switzerland, Syrian Arab Republic, Tajikistan, Thailand, The Former Yugoslav Republic of Macedonia, Timor-Leste, Togo, Tonga, Trinidad and Tobago, Tunisia, Turkey, Turkmenistan, Tuvalu, Uganda, Ukraine, United Arab Emirates, United Kingdom of Great Britain and Northern Ireland, United Republic of Tanzania, United States of America, Uruguay, Uzbekistan, Vanuatu, Bolivarian Republic of Venezuela, Vietnam, Yemen, Zambia, Zimbabwe.
8 ICOM museum definition (2007):

> A museum is a non-profit, permanent institution in the service of society and its development, open to the public, which acquires, conserves, researches, communicates and exhibits the tangible and intangible heritage of humanity and its environment for the purposes of education, study and enjoyment.
>
> (ICOM, 2007)

References

Ariese-Vandemeulebroucke, Csilla E. (2018). *The Social Museum in the Caribbean. Grassroots Heritage Initiatives and Community Engagement*. Leiden: Sidestone Press.

Brown, K. E. (2019). 'Museums and local development: An introduction to museums, sustainability and well-being'. *Museum International*, 71(3–4), pp. 1–3.

Brown, K. E. and Mairesse, F. (2018). 'The definition of the museum through its social role'. *Curator: The Museum Journal*, 61(4), pp. 525–39.

Brown, K., Brulon Soares, B. and Nazor, O. (eds.) (2018). *Definir los Museos del Siglo XXI: Experiencias Plurals*. Paris: International Committee for Museology.

Brown, K. E. et al. (2019). *Report on a Policy Round Table by EU-LAC-MUSEUMS held at the European Commission Offices, Brussels, 29 April 2019*. [Online]. Available at: https://zenodo.org/record/3368301#.YyHEaC8w0dU (Accessed 14 September 2022).

Brown, K. E., Davis, P. and Raposo, L. (eds.) (2019). *On Community and Sustainable Museums*. [Online]. Available at: http://www.eulacmuseums.net (Accessed 1 September 2020).

Camarena Ocampo, C. and Morales Lersch, T. (eds.) (2016). *Memoria: Red de Museos Comunitarios de América. Experiencias de museos comunitarios y redes nacionales*. Oaxaca: Museos Comunitarios de América.

Davis, P. (2008). 'New Museologies and the Ecomuseum'. In Graham, B. and Howard, P. (eds.) *The Ashgate Research Companion to Heritage and Identity*. Aldershot: Ashgate, pp. 397–414.

Davis, P. (2011). *Ecomuseums: A Sense of Place*, 2nd ed. London: Continuum.

De Carli, G. (2004). *Un Museo Sostenible: Museo y Comunidad en la Preservacion Activa de su Patrimonio*. Heredia: EUNA; ILAM; UNESCO.

Fúquene Giraldo, Laura, Blanco Wells, Gustavo and Weil G., Karin (2019). 'Redefiniendo la sostenibilidad desde una perspectiva situada:desafíos de museos comunitarios del sur de Chile'. *Polis*, 18(53), pp. 127–144. DOI: 10.32735/S0718-6568/2019-N53-1389 (Accessed 15 September 2022).

ICOM. (2007). Definition of a Museum. Available at: http://icom.museum/the-vision/museum-definition/ (Accessed 17 May 2018).

ICOM Resolutions. (1995). Available at: https://icom.museum/wp-content/uploads/2018/07/ICOMs-Resolutions_1995_En.pdf (Accessed 8 June 2023).

ICOM Resolutions. (1997). Stavanger, Norway. Available at: https://icom.museum/wp-content/uploads/2018/07/ICOMs-Resolutions_1995_En.pdf (Accessed 14 September 2022).

ICOM Resolutions. (2010). Shanghai, China. Available at: https://icom.museum/wp-content/uploads/2018/07/ICOMs-Resolutions_2010_Eng.pdf (Accessed 14 September 2022).

ICOM Resolutions. (2013). Available at: https://icom.museum/wp-content/uploads/2018/07/ICOMs-Resolutions_2013_Eng.pdf (Accessed 8 June 2023).

ICOM Resolutions. (2019). Kyoto, Japan. Available at: https://icom.museum/wp-content/uploads/2019/09/Resolutions_2019_EN.pdf

Lisbon workshop. (2016). EU-LAC Museums Kick-Off meeting at Lisbon Museum of Archaeology, November 2016.

Museos Comunitarios website. (n.d.) [Online]. Available at: https://www.museoscomunitarios.org (Accessed 18 November 2019).

Nascimento Junior, José do, Trampe, Alan, Assunção dos Santos, Paula (eds.) (2012). Mesa Redonda sobre la Importancia y el Desarrollo de los Museos en el Mundo Contemporáneo. *Revista Museum*, 1973. Brasília: IBRAM/MinC; Programa Ibermuseos, 2(235) p. 31.

UNESCO. (1972). Report on the Development and on the Role of Museums in the Contemporary World. Available at: http://www.minom-icom.net/files/023679eb.pdf (Accessed 14 September 2022).

de Varine, H. (2017). Interview with Karen Brown, Paris.

6

MUSEUMS AS TOOLS FOR SUSTAINABLE COMMUNITY DEVELOPMENT

A study of four archaeological museums in northern Peru

Luis Repetto Málaga and Karen Brown

Understanding museums as tools for sustainable community development is one of the priorities of the international research project Museums, Community and Sustainability in Europe, Latin America and the Caribbean (EU-LAC) Museums (2016–2020). This ambitious project has been explicitly designed in response to the European Union Horizon 2020 Work Programme call INT 12 (2015), 'the cultural, scientific and social dimension of EU-LAC relations'. The project seeks to carry out a comparative analysis of small- and medium-sized rural museums and their communities in Europe, Latin America and the Caribbean (EU and LAC) and to develop associated history and theory. By researching state-of-the-art initiatives in museums and community empowerment, and then moving beyond these initiatives to implement actions in partner countries, our aim has been both to transform individual lives within museum communities and to create a method of implementation and evaluation that will be applicable to wider regions.

The basis of the project is that community-based museums allow under-represented communities to stake a place in history, as well as to contribute to environmental sustainability and community empowerment. Our project partners are The University of St Andrews in Scotland (Coordinator), International Council of Museums (ICOM), The Pontifical Catholic University of Peru, The National Museum of Costa Rica, The University of Austral in Chile, The University of the West Indies, The University of Valencia in Spain and The National Archaeology Museum in Lisbon, Portugal. Our research themes are derived from the EU-CELAC Action Plan, namely:

- technology and innovation for bi-regional integration,
- museums for social inclusion and cohesion,

DOI: 10.4324/9781003288138-8

- fostering sustainable community museums,
- exhibiting migration and gender.

We are supported by an eminent Advisory Board and Steering Committee, including the Chairs of the ICOM Regional Alliances ICOM Europe and ICOM LAC (www.eulacmuseums.net).

A number of collective outputs could prove useful to local governments alongside the OECD-ICOM *Guide*, namely, *On Community and Sustainable Museums*, a Report on Policy Recommendations from a Round Table held at the European Commission and a new ICOM Resolution on Museums, Community and Sustainability.[1]

Though many miles lie between us, working together as a bi-regional team has brought about cultural understanding, including: a transformative bi-regional Youth Exchange between Scotland, Costa Rica and Portugal; the launch of an art exhibition, *Arrivants: Art and Migration in the Anglophone Caribbean World*; a bespoke Virtual Museum; a YouTube channel; a Manual on *Integral Museums: Experiences and Recommendations*; researching traditional water heritage practices in southern Spain and northern Peru.

Sustainability has become fundamental to the work of museums. Since the early 1970s, ideas advanced by the 'ecomuseum' movement initiated by Georges-Henri Rivière and Hugues de Varine have questioned the concept of the museum as a repository of collections, challenging institutions to look beyond the confines of their physical buildings.[2] In the Latin American context, authors such as De Carli stress the need for museums to contribute to community development; to achieve this goal, institutions must get to know their communities. They should familiarise themselves with the local people and their beliefs, customs and values, as well as with other aspects of local cultural heritage (De Carli 2004b). One central idea that informs the EU-LAC Museums project in several countries is that of a museum whose core initiatives and actions are anchored in local communities. The purpose of this chapter is to summarise the initiatives and findings of the EU-LAC Museums Peru Case Study—developed within four museums in two regions on the Peruvian north coast—to nurture ties among these museums, their communities and local territorial development.

The discussion is divided into five parts: the first explores four museums where the project was developed, highlighting their shared history and the important relationships they have fostered with surrounding communities. The second part describes the project's approach and the development of two intervention models that it applied, emphasising the challenges and reorientations the project faced due to the impact of the El Niño weather phenomenon. The third part describes the pilots (or 'demos') that were carried out in the countryside of the Chan Chan and Huacas de Moche Site Museums, both in the region of La Libertad. These served to verify the effectiveness of implemented actions. The fourth section describes

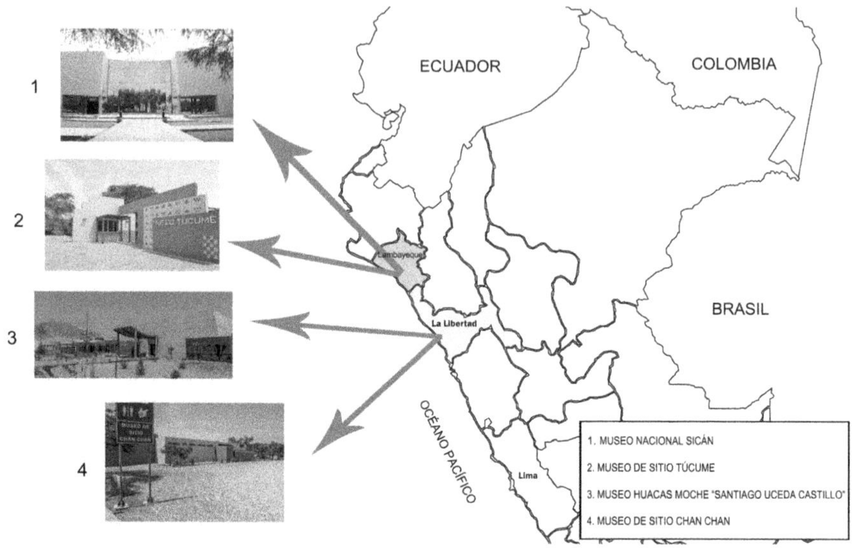

FIGURE 6.1 Location of the four museums that participated to the project, in the regions of Lambayeque and La Libertad (Peru). © Luis Repetto Málaga

the activities carried out to strengthen relationships between museums and communities, particularly with regard to sustainability, regional integration, education, use of new technologies and climate change vulnerability. The fifth and final part highlights the preliminary conclusions of the project (Figure 6.1).

Case studies in the northern coast of Peru

The social vocation of museums was highlighted for the first time in the ICOM Declaration of 1972 in Santiago, Chile.[3] With the advent of New Museology during the 1980s, some questioned the authority of museums as institutions, instead calling for them to become instruments of representation and political power for the communities they serve (Rivière 1985; Fernández 1999; Palomero 2002). The incorporation of this criticism transformed the notion of the traditional and hegemonic museum into one 'at the service of society and its development', as stated in the resolutions of the 22nd General Assembly of ICOM in Vienna (Austria) in 2007. This concept has been especially relevant in Latin America, owing to its colonial history and diverse identities that, as a result, share a common social and political backdrop. In this context, the impact of international tourism has moreover prompted a reassessment of archaeological, cultural, natural, tangible and intangible heritage, which is now widely viewed as a resource for local development.

It is important to note that museums on Peru's northern coast emerged in the aftermath of late 20th century discoveries and were influenced by ideas around the social and educational vocation of museums. During this period, the decentralisation of neoliberal policies by Alberto Fujimori's government led to the strong formation of regional identity, bolstered by the spectacular archaeological discoveries of the Muchik culture (Asensio 2013).[4]

This was coupled with strong investment from regional authorities seeking to develop the tourism potential of this archaeological heritage, resulting in the financing of spectacular excavations and museums. Since their creation, museums on the north coast of Peru have been influenced by the notion that museum institutions should more actively guide the public's understanding of the meaning, use and protection of archaeological remains (Elera and Shimada 2006, p. 217). Moreover, they have become important engines for attracting tourism, which in turn enhances local economies.

Selection criteria and institutional initiatives

The choice of the four museums (Túcume, Sicán, Chan Chan and Huacas de Moche) discussed in this chapter was influenced by our interest in strengthening the ties that had previously been forged between these museums and their communities, and that have had a significant social and economic impact on a local level. The criteria informing the selection of these four museums also align with ICOM Peru's objective of decentralising its discourses and actions within the Peruvian context.

The Sicán National Museum

In the case of the Sicán National Museum, for example, its community-oriented approach began with an initial archaeological excavation in 1978. The Sicán Archaeological Project (SAP) envisioned the Sicán National Museum as a centre for research, conservation and promotion of archaeological heritage, as well as an agent of sustainable development for the communities of the Ferreñafe province. This vision finally came to fruition in 2001 when the museum was inaugurated. Its community-oriented programme has materialised in a wide variety of initiatives.

This included the establishment of a long-term collaboration with the Local Educational Management Unit (*Unidad Local de Gestión Educativa*, ULGE), which helps develop school curricula in the province, promotes school visits and participation in the museum's cultural activities, all with the aim of developing the knowledge and value of the museum's archaeological heritage. One of the activities organised by the museum is the annual festival, 'Ferreñafe sings and dances for Peru', which was created in conjunction with the local community to celebrate the cultural diversity of the province.

The museum has also consolidated its position as a reference for political and community agents in Ferrenafe, who work in the fields of conservation and environmental heritage defence. Sicán National Museum is currently a member of several provincial committees and has taken on a mentoring role with regard to territorial planning, education, the study of heritage and conservation and the development of tourism (Elera 2017). The north coast is an exceptionally fertile landscape, and pressures on local arable land have been constant, consequently affecting the relationship between museums and surrounding communities. This is precisely the case of Sicán National Museum and its archaeological zone in the Pómac Forest, which have both been confronted by occupation and land trafficking.[5] Having fostered a relationship with the communities that live in the buffer zones of the Pómac Forest, the museum has played a fundamental role in addressing this conflict, facilitating the recovery of lands for this now-protected natural and cultural reserve.

The Túcume Site Museum

The Túcume Site Museum opened in 1992 and was renovated in 2014 as part of a community-oriented programme initiated by the Túcume Site archaeological project. The primary objective of the project was to involve the surrounding community in the ongoing preservation of the archaeological site and in revitalising local cultural traditions. This community-oriented approach led the museum to create a dedicated Office of Conservation Education (*Oficina de Educación para la Conservación*); the new office was designed to foster community involvement in museum activities, and in the ongoing preservation of local cultural heritage (Figure 6.2).

This new office has allowed the museum to:

- promote community participation in the planning and activities of the museum,
- organise a series of meetings with local authorities to discuss urban and rural planning and heritage conservation,
- participate in the development of local school programmes to foster an appreciation for archaeological heritage and strengthen a sense of local identity among younger generations.

In addition, the museum's onsite shop functions as a space for showcasing and selling works created by various groups of local artisans and has become an important marketing platform for craftspeople in the community. Initiatives like these, among others, have facilitated the creation of enduring ties among the museum, local archaeological heritage and surrounding communities (Narváez 2017, p. 32).

The Chan Chan Site Museum

The Chan Chan Site Museum, which was created in the 1990s, and the Santiago Uceda Castillo Huacas de Moche Museum, which was associated with a research project at the National University of Trujillo in 2010, have both proven pioneers in integrating technologies for conservation and in disseminating the importance of

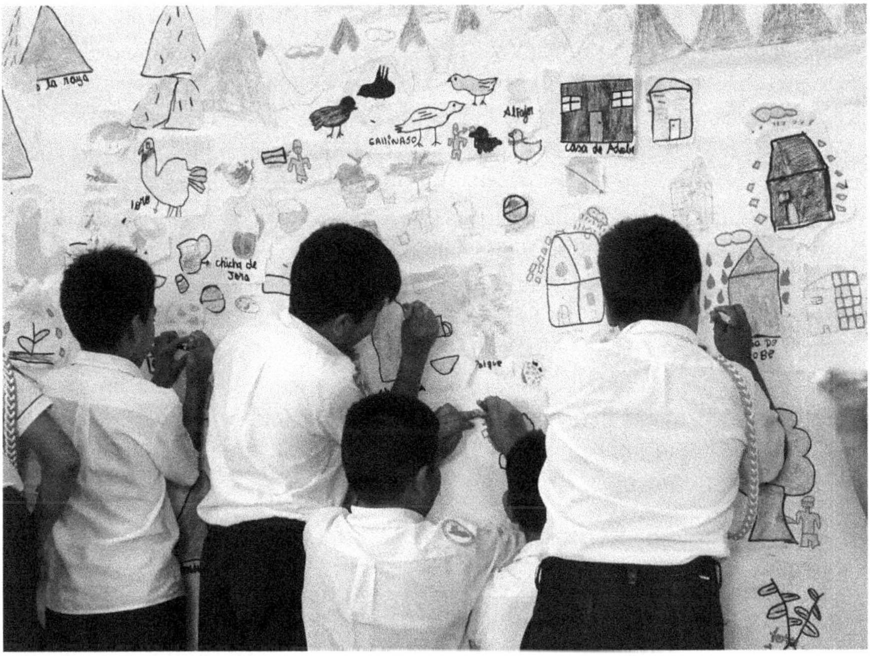

FIGURE 6.2 Heritage Identity Workshop at the Túcume Site Museum. © Luis Repetto Málaga

the heritage they protect. As such, these museums have become genuine emblems of regional identity. The Chan Chan Site Museum is currently undergoing renovations: it is once again incorporating new technologies into both its exhibitions and the onsite restoration and conservation of works.

Within this framework, these museums are notably developing projects and actions aimed at young people in surrounding communities—including school-age children in the capital city of Trujillo and other nearby urban areas—to reassess their link to heritage as a key factor in local development. Carrying out such projects during renovations was, moreover, presented as a unique opportunity for museums to strengthen relations with their communities, and this is reflected in how these institutions approached refurbishment projects.

Promises and challenges

The four museums selected as case studies have been working to strengthen the historically marginalised identities of rural populations in the territory, notably through activities that raise awareness around daily and traditional practices that are key to the cultural heritage of local populations. In this regard, these institutions have the advantage of existing relationships with local communities, working with them to promote the development of the territories or regions in which they are located. However, museums also face challenges arising from these same relationships: these include constant changes in the social, political and economic fabric of local territories, as well as climatic instability on Peru's northern coast.

While such problems and challenges are certainly worth noting, the four museums discussed in this chapter greatly benefit from strong links to surrounding communities. In their respective contexts, these institutions have played key roles in transforming the attitudes of local populations, especially in relation to conservation efforts, encouraging locals to identify with their archaeological heritage, and increasing their capacity to generate economic income through tourism and other associated activities. This experience is also reflected in the four museums' thorough knowledge of the economic, political, socio-cultural and environmental concerns of their communities. This local expertise has allowed the institutions to further strengthen community ties and formulate clear guidelines for their institutional work.

Project methodology: territorial management and the El Niño phenomenon

The first year of the EU-LAC Museums project was dedicated to coordinating and harmonising the proposal within the museums selected, which identified the relationships they had established with surrounding communities and articulated strategic approaches to reinforce these ties in areas of sustainability, regional integration, education and the application of new technologies. This initial planning phase began in September 2016 with the aim of being completed within seven months.

However, it had to be suspended between February and April 2017 due to the impact of the El Niño phenomenon: torrential rain caused landslides and serious flooding, which led to a state of emergency being declared in 12 regions of Peru. The project resumed in May 2017 and concluded in August of the same year. As a result of El Niño, the project design incorporated new elements, taking into account both the vulnerability of north-coast populations to this weather phenomenon and the pressing need to involve museums in territorial management. At this stage, two intervention models were also defined, one to be applied to museums in Lambayeque (Sicán and Túcume) and the other to those in La Libertad (Chan Chan and the 'Santiago Uceda Castillo' Huacas de Moche Museum).

During the El Niño phenomenon, small-scale subsistence activities such as agriculture and artisan fishing were severely affected in rural communities. Even service activities such as micro-commerce or transportation were severely impacted by the destruction of road infrastructure. The number of people affected by this event exceeded one million. A third of the affected population, some 315,000 people, was located on the north coast, mainly in the regions of Piura, Lambayeque and La Libertad. These regions were subject to the catastrophic climatic alterations caused by El Niño until the end of April.

Prior to the El Niño events, it was common knowledge that museums and local communities were inadequately prepared to mitigate the effects of a climatic phenomenon of this magnitude. Any project based on Peru's north coast should always take into account a recurrent and catastrophic phenomenon as part of its sustainability strategy. Accordingly, we consider it essential for museums to participate in coordinating disaster risk management actions, since they are institutions familiar with the characteristics and needs of the local environment.

Two intervention models and new initiatives

Two different coordination models were implemented for each respective region: in Lambayeque, interventions were to be carried out through an external and local actor, the University of Santo Toribio de Mogrovejo (USAT); this corresponds to the civil society actor approach. Meanwhile, in the region of La Libertad, the coordination model drew on the intervention of a more active local authority, the Decentralised Directorate of the Ministry of the Culture (*Dirección Desconcentrada de Cultura del Ministerio de Cultura*). The design of these two intervention models, conceived during the first year of the project, arose from a seven-month research period carried out by the EU-LAC Museums project team from the Pontifical Catholic University of Peru. The result of these processes is discussed in the following two diagrams, which refer to how the Peru team will carry out its approach in each of the selected regions and their corresponding museums for the Peru Case Study (Figure 6.3).

Pilot studies: archaeological sites and nearby rural surroundings

From September 2017 to August 2018, the project initiated activities and workshops to strengthen ties between museums and communities, and two pilots were conducted at museums in the region of La Libertad to assess the feasibility and effectiveness of these actions in compliance with the aforementioned objectives. They pilot studies also aided in identifying opportunities to improve institutional best practices around sustainability, particularly with regard to their relationships with local communities and environments. Pilot 1 was conducted in the nearby rural surroundings of the Chan Chan archaeological site, while Pilot 2 was conducted in the countryside of the Huacas de Moche site, both between March to August and December 2018.

Pilot 1: raising awareness around local ancestral knowledge

Pilot 1 consisted in a series of workshops focused on raising awareness around local ancestral knowledge. These workshops were developed and carried out by the Chan Chan and Huacas de Moche museums within their local communities. They notably discussed the traditional preparation of *chicha de jora* and the

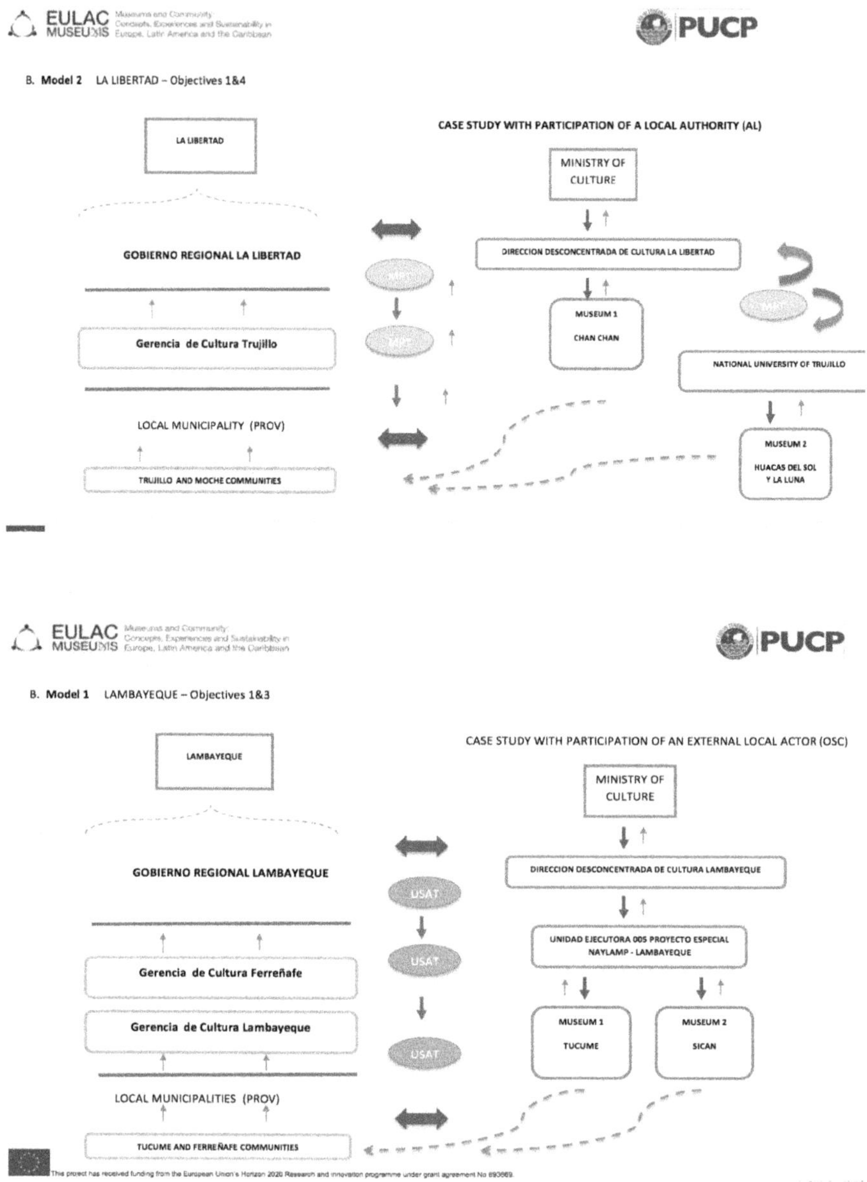

FIGURE 6.3 Model 1: Diagram describing the work scheme of the EU-LAC Museums project for the Lambayeque region in Peru.

cultural knowledge associated with this drink. After the workshop, participants learned about the process of preparing *chicha de jora* and were encouraged to understand how this drink forms part of complex cultural knowledge associated with *picanterías* (restaurants serving spicy lunches), traditional gastronomic spaces and daily life.

Throughout the workshop, it became apparent that most local residents perceive local traditions as disappearing or in danger of extinction. For the participants of this workshop, communities and museums must act in partnership to preserve the local traditional culture, including the preservation of objects and raw materials, by promoting local practices associated with their use. Over the course of evaluating this pilot, we were able to conclude that the goal of achieving project sustainability would depend on our ability to strengthen the museum's role in reviving popular traditions, as well as fostering the knowledge that is fundamental to the construction of local identities—ensuring these are not forgotten.

Pilot 2: heritage preservation and environmental conservation in Moche

Pilot 2 was carried out in the countryside of Moche and consisted of a series of interventions to raise awareness and encourage local communities to participate in the preservation of their environment and local cultural heritage. These interventions involved carrying out a series of meetings, at which both local authorities and members of their communities were invited to discuss and exchange ideas on how to promote the sustainability of museums and their communities. Discussions revolved around how to improve the relationship between archaeological projects and the local community, as well as how to promote water care to ensure the agricultural prosperity of Moche's rural areas.

The pilot yielded a proposal aimed at establishing strategic alliances among the Huacas de Moche Museum, local educational authorities and the cultural, religious and youth associations of the Moche countryside. The proposal's primary aim was to plan and carry out actions designed to improve the environmental, economic and socio-cultural conditions of the local environment, favouring the preservation of local heritage with an emphasis on archaeological heritage.

During the pilot, a TV programme involving local schools was recorded at the Chan Chan archaeological site on 28 April 2018. As the highlight of our second 'demo', our team was in charge of recording a promotional spot at the archaeological site, with the support of site museum supervisors and the Chan Chan archaeological complex. The objective of this spot was to raise awareness among the local population around the conservation of cultural heritage—especially cultural assets built from mud architecture, of which the citadel of Chan Chan is an outstanding example. We toured the archaeological site of Chan Chan with children and young people from local schools, as well as their families and teachers, and stressed the value and validity of oral traditions surrounding this emblematic place; these are still alive in the collective memory of the community (Figure 6.4). The first TV programme was broadcast nationwide on 26 May 2018, at 11:30 a.m. by TV Peru as part of the 'Museums Open Doors' programme.[6]

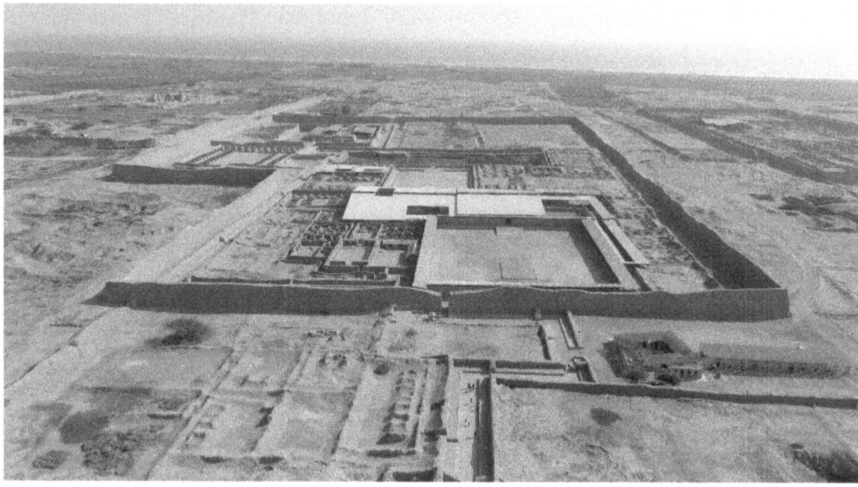

FIGURE 6.4 Aerial view of the Chan Chan citadel. © Dirección Desconcentrada de Cultural La Libertad

Conducting these two pilots proved crucial: they allowed us to evaluate how selected museums and their communities responded to initiatives and actions we implemented to carry out our project. Moreover and above all, they allowed us to assess whether these actions succeeded in achieving the objectives that were initially set and contributed to our goal of promoting more sustainable museum practices in relation to their local communities. Both pilots ultimately allowed us to reflect on how to improve the implementation of our project.

Project development: objectives and activities

The development of the EU-LAC Museums project in Peru is based on the premise that museums have the capacity to significantly contribute to the cultural, educational and economic development of their communities through the recovery, assessment and conservation of their collective heritage and identity. Moreover, this contribution must be carried out through close collaboration between museums and the communities to which they belong. The four museums discussed here all enjoy strong relationships with their respective communities. The activities carried out as part of our project have sought to strengthen these relations in areas pertaining to sustainability, regional integration, education, technology and territory.

New models for sustainable museum practices

In Peru, the lack of a political framework defining the concept of sustainable museum practices from a top-down dynamic has forced local and regional museums to approach sustainability on their own terms. They have adapted the notion of sustainability to local contexts and forged their own set of good practices not only based on trial and error but also based on intimate knowledge of their communities, territories and heritage. Accordingly, we designed activities to ensure that the four museums improved their contributions to the long-term development of their communities, strengthening and promoting the sustainable use of heritage as a resource to jointly address situations of local vulnerabilities and challenges. Existing relations between museums and communities have been strengthened, in part thanks to the participation of local authorities in the development of long-term policies and initiatives that benefit communities.

The only museums that we consider sustainable are those that recognise the economic, political, socio-cultural and environmental concerns of their territory as fundamental considerations in institutional management, and that involve and support community members in actions of preservation, appropriation, capacity building and the responsible use of heritage resources. Accordingly, the actions carried out within the framework of this project are geared towards generating processes that encourage museums to transform the way they work, their relationships with local communities, and their social and institutional contexts.

In the case of Lambayeque, Local Heritage Defence Workshops were conducted at each of the region's museums. These were aimed at local authorities, leaders and public figures and promoted citizen participation in the assessment and defence of local cultural heritage. As a result of these activities, participants committed to forming a System of Heritage Defence Brigades (*Sistema de Brigadas de Defensa del Patrimonio*) in collaboration with regional museums and with supervision from local authorities. The latter are dedicated to protecting archaeological sites through patrol actions, as well as identifying and promoting the most emblematic cultural expressions in their respective area.

In the case of La Libertad, local museums conducted Local Knowledge Assessment Workshops, including one that focused on the use of *totora* reeds (*scirpus californicus*) and *mate* calabash (*lagenaria siceraria*) in the production of handicrafts. Another highlighted the practice of fishing and the traditional preparation and consumption of *chicha de jora*, while a micro-enterprise training workshop aimed to formalise the economic initiatives of participating communities. The overall purpose of the workshops was to highlight the importance of maintaining and continuing these traditional community practices. This empowered the local population to transmit ancestral knowledge in collaboration with the museums, which engaged several generational groups in the process.

Regional integration for social inclusion and cohesion

Actions pertaining to Regional Integration for Social Inclusion and Cohesion (*Integración Regional para la Inclusión y Cohesión Social*) sought to promote regional participation by helping local populations gain confidence around their regional identities and their ability to assess ancestral knowledge. The influence of the work of museums on their territories typically generates diverse impacts that reach beyond local contexts and may affect regional dynamics around identity and heritage in both positive and negative terms.

We emphasise here the process of transforming relations among the four selected museums and their communities to establish a sense of belonging within the population. The overall purpose of this project is to foster greater cohesion between regional populations through the work of museums. It also aims to promote local cultures by creating more integrated regional societies in the long term, with individuals finding common ground by assessing their past and present heritage.

In the Lambayeque region, for example, the Túcume Site Museum is notable for its particular management model: one that involves the community surrounding the archaeological site in the institution's decision-making process. However, it still struggles to involve sectors of the population mostly associated with the district's urban centre, who seem less interested in preserving the cultural values that the museum promotes and protects. Moreover, while the Sicán National Museum is recognised by the communities of the La Leche river basin as an unconditional ally in the defence of local culture—thanks to its long history of working on multiple research, educational, tourism and heritage promotion initiatives—the museum still finds it difficult to fully commit to the specific needs of the multiple territories and communities under its jurisdiction.

The situation is similar in the region of La Libertad. In the case of the Chan Chan Site Museum, its efforts to preserve and disseminate information about the World Heritage Site have succeeded in elevating the mud citadel as an undeniable symbol of local cultural identity for all inhabitants in the region. Nevertheless, the museum still struggles to change the mindset of communities living nearby the archaeological site, so that they have an interest in preventing plundering at the site.

Meanwhile, by working closely with certain local artisans, the Huacas de Moche Museum has managed to position a certain type of handicraft—inspired by the results of archaeological investigations at the site—as a hallmark of the region's own quality. However, this has not lessened tensions between site archaeologists and the inhabitants of rural Moche: the latter continue to perceive the museum as an entity that has restricted access to and use of resources in many areas of the territory.

The activities developed at this above-mentioned museum aimed to bolster positive ties between museums and the local population by encouraging them to highlight the importance of their own territorial references, natural resources, cultural landscapes and artistic expressions inherited from the pre-colonial past, thereby consolidating their own local identities. These identities can, in turn, be grouped under the same regional identity component that consolidates the common past of all these communities, and that contributes to their unification as a collective people inhabiting the same territory: one that possesses common values and traditions based on cultural heritage, and that also faces common problems.

Systematising local knowledge and heritage

To that end, several Local Heritage Diagnostic Assessment Workshops were conducted by UNESCO expert Ciro Caraballo Pericci, who sought to identify, systematise and evaluate local knowledge and heritage values recognised by the population within the territorial scope of our four selected regional museums. The workshops led to a recognition that the local population plays an important role as transmitters of knowledge, ensuring the continuity of certain traditional local practices such as cooking, sewing or handicrafts production. They also spurred local authorities to commit to developing a training programme on participatory citizenship for younger generations: one that seeks to transmit the idea of the permanence of their community identities and cultural traditions. We firmly believe that these workshops, coupled with the other activities considered in our project, will contribute to achieving our objective of regional integration for social inclusion and cohesion. They do so by promoting unity among the population and instilling confidence in their cultural institutions, particularly through the recognition of heritage values common to inhabitants of the same territory.

To conclude this process, each of the regions staged a fair in which the experiences and results of the EU-LAC Museums project were showcased. The region of Lambayeque held the EU-LAC Museums Intercultural Fair, which involved the Túcume and Sicán museums, together with members of their communities: these included artisans, artists, producers, local authorities, managers and inhabitants. The fair showcased products that were created during the various workshops and project activities: new artisan designs, organic horticultural goods. It also staged video art presentations based on the local heritage of communities surrounding the Túcume and Sicán museums.

Meanwhile, the La Libertad region organised the first Cultural Identity Renaissance Week in the Moche countryside, for which the Huacas de Moche Museum created a programme to showcase rural Moche's various culinary, artisanal, artistic and cultural expressions, developed within the framework of the project activities through fairs, parades, guided visits to the archaeological sites and festivals.

Collaborations with the Chile Case Study

To complement these efforts, the EU-LAC Museums project has shared insights from the Peru Case Study with the Chile Case Study, allowing the project's Chilean colleagues to apply their sustainability initiatives and documentation, which were developed for museums in the Chilean region of Los Rios to counterparts of the Peruvian regions of Lambayeque and La Libertad. This provided the latter institutions with in-depth knowledge of the quotidian impacts of the programme's proposed interventions on museums. The Chilean team's methodology was implemented by the Peruvian team at the end of July 2018, allowing them to use the methodology developed by their colleagues in the Local

Heritage Diagnostic Assessment Workshops once these were completed in the four selected museums.

Based on the information collected on Peru's northern coast, the Chilean team found that knowledge about water management is one of the most valuable cultural resources of the communities on Peru's northern coast: it allowed them to adapt to hostile desert conditions and transform the territory into a fertile valley, ensuring the survival and prosperity of local culture up to the present. In addition, carrying on from the bi-regional integration goal discussed in the working meetings of the EU-LAC Museums project in May 2018, water was acknowledged as an essential element of cultural and heritage contexts, both in the case of Valencia, Spain (also an EU-LAC Museums partner) and in the cases of Peru and Chile.

Accordingly, the consolidation and organisation of a joint programme among teams in Spain, Peru and Scotland was proposed, with the aim of sharing the experiences, knowledge, best practices and weaknesses of the organisations and relevant players involved in water management. Moreover, a proposal was put forward to discuss how the contribution of interdisciplinary knowledge from universities associated with the project can promote actions to link territories in bi-regional contexts. The overall purpose of this is to improve the sustainability of museum practices at all institutions participating in the EU-LAC Museums project.

On World Water Day in Moche and following the initiative, judges from local utility companies Water of Valencia (Spain) and Water of Corongo (Peru), which are both recognised as intangible heritage of humanity, met with the irrigation board of the Moche countryside. This resulted in an exchange of enlightening knowledge and prompted self-reflection on the part of members of both countries.

Education as a key social function of museums

Education is one of the key social functions of museums and is central to their commitment to local communities. One key objective of this project concerns disseminating the research and work carried out by museums in their communities. However, it also inspires promoting dialogue between locals and experts regarding the interpretation of heritage in the territory they share. Through their educational work, museums effectively represent and transmit the cultural values of their communities, which serves to both raise the confidence of local groups around their own identity and inform the rest of society of the importance of preserving this heritage.

For the Peru Case Study, our four selected museums maintain a commitment to their local communities by continuously developing initiatives designed to strengthen learning and knowledge about local traditions: ones based on the practices of the territory's pre-colonial culture. This commitment translates into a clear intention to promote the creation and strengthening of regional identities, as well as a desire to promote local development through education and the appreciation of local heritage. By implementing actions that focus on education and disseminating knowledge about local culture, regional museums can help make their communities and territories more resilient in the face of any economic, political, socio-political or environmental challenges.

In the region of Lambayeque, the Túcume and Sicán museums worked to establish themselves as true centres of local culture by keeping their doors open to the community, and particularly through continually offering training workshops on a variety of topics. The central theme of the first workshops focused on the performing arts and self-expression through movement, as well as the use of audiovisual media and visual anthropology. These workshops were conducted in response to the local population's need to increase their abilities to represent and therefore contribute to preserve various expressions of local culture. These include dances, stories and performances linked to the oral memory of their community and the region's pre-colonial past.

The central objective of the second set of workshops was to support local ventures and traditional practices, such as producing handicrafts, gastronomy and organic horticulture geared towards the tourism market. These workshops provided locals with a space for learning and sharing knowledge, with a view to improving their craft production processes—basketry, weaving and embroidery techniques—as well as updating their knowledge of agricultural production, much of which is passed down ancestrally. The workshop also addressed ways to specifically adapt to the local tourism market targeted by regional museums (Figure 6.5).

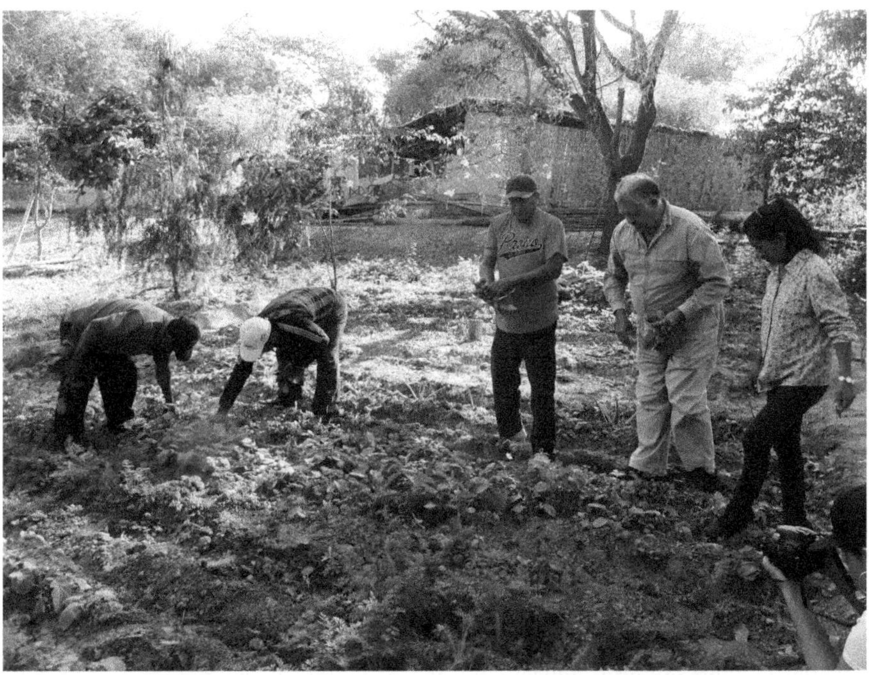

FIGURE 6.5 Horticulture workshop at the Sicán National Museum. © Luis Repetto Málaga

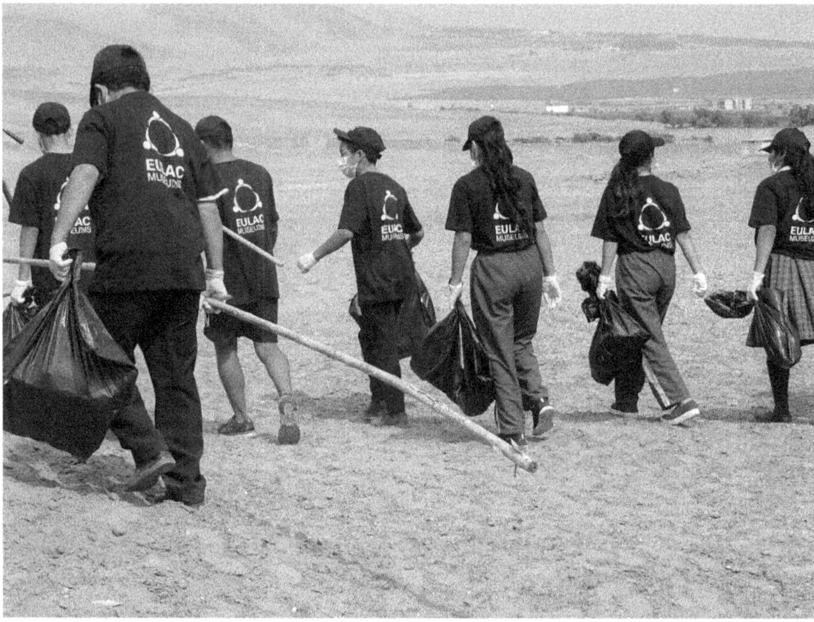

FIGURE 6.6 Cleaning campaign in the countryside of Moche, La Libertad. © Luis Repetto Málaga

In the region of La Libertad, the Chan Chan and Huacas de Moche museums carried out activities aimed at strengthening the identity of communities living around nearby archaeological sites. These activities mainly centred around heritage identity workshops, in which students from local schools participated by visiting the archaeological sites associated with museums, learning more about the daily work of these institutions and the latest research on local heritage. As complementary activities, community cleaning days of the buffer zones near these archaeological sites were also carried out, in order to preserve the environment and the landscape setting of the Chan Chan and Moche countryside; recreational activities were also conducted such as competitions, guided tours and painting murals, in order to increase the population's knowledge of the work of local museums and the dissemination of main cultural and iconographic references at the archaeological sites of Chan Chan and Moche (Figure 6.6).

Memory preservation and the use of technologies

The educational work carried out by our four selected museums contributes to collecting, safeguarding and treasuring regional and territorial memory, which is essential to their cultural, educational and economic development. But to fulfil this mission, it is also necessary to ensure that communities have the necessary means to safeguard these memories. This is the only way that communities can put cultural knowledge to use, allowing them to understand who they are, remember where they come from and consider where they want to go in the future.

This is precisely how the use of technology can greatly contribute to the work of museums in relation to communities and their heritage. Ensuring the sustainability of museums' educational work implies strengthening their means of disseminating knowledge by incorporating new and modern technologies. Through these, all community members can strengthen their confidence in their local identity, cultural values and regional heritage as well as develop a critical and reflective mindset to help them face daily challenges.

For this reason, the Peru Case Study chose to use the web portal of the EU-LAC Museums project, as well as mobile and 3D technologies, to improve the educational experiences provided by the four selected museums. These experiences were complemented with technological tools allowing users to access information about objects, oral histories and audiovisual material about their territorial cultures.

In terms of activities, the Túcume and Sicán museums in Lambayeque participated in the 3D Digitisation and Spherical Technology Workshops as part of an initiative spearheaded by the team at the University of St Andrews (Scotland). This initiative led to the publication of an open-source manual that can be used by any community museum in the world. As a result of this collaboration, part of these two museums' collection of cultural assets were scanned using this technology and their 3D models can be seen on the EU-LAC Museums project website. The previously mentioned workshops on audiovisual media and visual anthropology were also part of the project activities, designed to link the educational work of museums to the use of modern technologies.

In the region of La Libertad, the Chan Chan Site Museum developed an app and a new website to promote this World Heritage Site. It also allows users to take a virtual tour of the museum's facilities, with updated information on the latest research with respect to this heritage and the ancestral knowledge still present in the practices of local communities.

The Huacas de Moche Museum, in collaboration with the National University of Trujillo, created a video to promote the activities of the project among the community of rural Moche, disseminating the results to the citizens of the region.

Additionally, two TV programmes were produced on the subject of community museums—one for the region of Lambayeque and the other for the region of La Libertad—which were broadcast on free-to-air TV nationwide. These programmes detailed the current progress of the project, with a focus on its participatory approach and the importance of collaboration between museums and communities in achieving regional sustainability.

Natural disaster risk prevention

In addition to the objectives initially set, our project necessarily had to take into account a feature that characterises the two regions in which our four selected museums are located, and which affects the entire northern Peruvian coast in general: the El Niño phenomenon. The impact of this phenomenon in 2017 led to the interruption of project activities during part of the design phase. It also served as a stark reminder of the vulnerability of the northern coastal territories, whose social-change processes have been shaped by natural disaster events of this magnitude since pre-colonial times. Accordingly, we committed to developing activities related to natural disaster risk prevention as part of the project. The aim was to make the relationship between museums and their communities more sustainable by reducing the vulnerability of local territories, human groups and heritage that are impacted by natural phenomena.

In view of these objectives, each of our four selected museums conducted natural disaster risk prevention workshops with local authorities, in order to raise awareness around the importance of preventing risk in the face of recurring natural phenomena, such as El Niño. Another aim was to prepare citizens to respond in case of emergency through mobilisation efforts, allowing them to protect their heritage and safeguard the integrity of their communities and livelihoods. As a result of these workshops, participating authorities signed a memorandum on integration with the local communities, committing to arrange future meetings to jointly develop community risk maps.

Conclusion

The museums and communities discussed have demonstrated their potential to promote actions and consolidate working models that strengthen museum institutions and their relationships with local communities and environments. More sustainable spaces can be achieved by improving relations between museums and their local communities, schools, families and the visiting public. Museums can only become true platforms that bring about sustainable development in their local environments by fostering in-depth knowledge of local realities, as well as recognising the history and heritage of local populations. Our project involved creating a series of activities, including joint learning workshops with specific objectives, methodologies and proposals; however, the overall goal of these was from the outset to strengthen ties between museums and their communities. Cultural heritage and local development represent the starting point and fundamental strategy at work in this endeavour.

Acknowledgement

This research received funding from the European Union's Horizon 2020 Research and Innovation programme under grant agreement No. 693669.

Notes

1 Available at: https://zenodo.org/communities/eu-lac-museums [Accessed 29 November 2019].

2 Founding museologists of the ecomuseum movement and directors of the ICOM from 1948 to 1965 and from 1965 to 1974, respectively.

3 See http://www.ibermuseus.org/wp-content/uploads/2014/07/copy_of_declaracao-da-mesa-redonda-de-santiago-do-chile-1972.pdf.

4 This denomination was first popularised by the anthropologist and activist Richard Schaedel (1920–2005) in response to ethnographic work carried out by Enrique Brüning (1848–1928) on the indigenous people of Lambayeque. The term *Muchik* was used to refer to the mestizo-peasant communities descended from the indigenous people that inhabited the north coast of Peru since pre-colonial times, and to underline the continuity of their cultural traditions to date: approximately 2000 years. Although the use of their native language (the Yunga language or Muchik) was already extinct, Schaedel (1996) argued that the 'essence' of this people's ethnic identity would remain associated with their continuous use of technology (including the management of traditional crops such as native cotton) and with the cognitive processes underlying their customs and beliefs (as in traditional medicine and *curanderismo*). Currently, this hypothesis is being explored further by researchers of the Sicán National Museum, who have also pointed out the importance of the influence of the Quechua people in the history of the north coast civilisations, having introduced the concept of a 'Muchik and Quechua ethno-cultural matrix' (Elera 2014, 2017).

5 Land trafficking can be defined as 'the usurpation, illegal appropriation, and commerce of lands. It is closely linked with rural–rural and urban–rural migration and can be seen as an activity that organizes and facilitates migration' (Shanee and Shanee 2016). See https://journals.sagepub.com/doi/10.1177/1940082916682957 [Accessed 25 November 2019].

6 The programme's official trailer is available on YouTube under the video entitled 'Museos Puertas Abiertas (TV Perú) – Ciudadela Chan Chan y Huacas Moche'. While the official version of the programme has not yet been published on YouTube, an unofficial upload can be found at the following link: https://www.youtube.com/watch?v=jTPxr1rbzjg&t=2s [Accessed 25 November 2019].

Bibliography

Asensio, Raúl. (2010). *Arqueología, museos y desarrollo territorial rural en la costa norte del Perú. Proyecto Desarrollo Territorial Rural con Identidad Cultural (DTR-IC)*. Santiago, Chile: Rimisp, IEP.

Asensio, Raúl. (2013). '¿De qué hablamos cuando hablamos de participación comunitaria en la gestión del patrimonio cultural?' *Argumentos*, Vol. 7, No. 3. [Online]. Available at: http://www.revistargumentos.org.pe/participacion_patrimonio.html [Accessed 3 September 2019].

Asensio, Raúl. (2014). 'Entre lo regional y lo étnico: el redescubrimiento de la cultura mochica y los nuevos discursos de identidad colectiva en la costa norte (1987–2010)', in *Etnicidades en construcción. Identidad y acción social en contextos de desigualdad*, edited by R. Cuenca. Lima, Peru: IEP, pp. 85–123.

Borea, Giuliana. (2003). 'Hablando de nosotros en un país pluricultural: identidad, poder y exotismo en los museos peruanos', *Arqueológicas*, No. 26, pp. 267–275.

Burón DIaz, Manuel. (2012). 'Los museos comunitarios mexicanos en el proceso de renovación museológica', *Revista de Indias*, Vol. LXXII, No. 254, pp. 177–212.

De Carli, Georgina. (2004a). 'Vigencia de la Nueva Museología en América Latina: conceptos y modelos', in *ABRA Magazine*, Faculty of Social Sciences of the National University of Costa Rica.

De Carli, Georgina. (2004b). *Un Museo sostenible: museo y comunidad en la preservación activa de su patrimonio* (1st ed.). San José C.R.: Oficina de la UNESCO para América Central.

Elera, Carlos. (2017). 'El quéhacer institucional del Museo Nacional de Sicán en la cuenca de La Leche, Lambayeque', *Quingnam*, Vol. 3, pp. 35–60.

Elera, Carlos and Izumi Shimada. (2006). 'The Sicán Museum: Guardian, Promoter, and Investigator of the Sicán Culture and the Muchik Identity', in *Archaeological Site Museums in Latin America*, edited by H. Silverman. Gainesville: University Press of Florida.

EULAC MUSEUMS Project Peru Case Study. (n.d.) *Interim Report Nr.1.* Lima: Pontifical Catholic University of Peru.

EULAC MUSEUMS Project Peru Case Study. (n.d.) *Interim Report Nr.2.* Lima: Pontifical Catholic University of Peru.

EULAC MUSEUMS Project Peru Case Study. (n.d.) *Interim Report Nr.3.* Lima: Pontifical Catholic University of Peru.

Fernández, Luis A. (1999). *Introducción a la Nueva Museología.* Madrid: Alianza editorial.

Narváez, Luis A. (2017). 'Patrimonio, territorio y comunidad. El ecomuseo Túcume', *Quingnam*, Vol. 3, pp. 19–34.

Palomero, Santiago. (2002). '¿Hay museos para el público?' *Museo*, No. 6, pp. 141–157.

Rivière, G. H. (1985). 'Definición evolutiva del ecomuseo [Evolutionary definition of the ecomuseum]', *Museum*, XXXVII, Vol. 37, No. 4, pp. 182–184.

Schaedel, Richard. (1996). *Recolección histórica-etnográfica sobre la problemática Muchik.* [Historical-ethnographic collection on the Muchik problem]. Lima: Manuscript.

Shanee, Noga and Shanee Sam. (2016). [Online]. 'Land Trafficking, Migration, and Conservation in the 'No-Man's Land' of Northeastern Peru', in *Tropical Conservation Science*, October-December, pp. 1–16. Available at: https://journals.sagepub.com/doi/pdf/10.1177/1940082916682957 [Accessed 18 November 2019]

7

CONNECTING MUSEUMS THROUGH CITIZEN SCIENCE

Jamaica/US partnership in environmental preservation

Tracy Commock and Dionne Newell

Introduction

The term 'citizen scientist' encompasses the collaboration between members of the public and science professionals in tasks such as species monitoring, collecting and transcribing data, with the aim to increase understanding and management of the natural world (Ballard et al., 2017). The combination of specimen collections, scientific and public education expertise and wide audience reach make natural history museums ideal vehicles for driving conservation science and education using the citizen science concept (Ballard et al., 2017). Natural history museums have the potential to educate the general public about science and environmental issues. Newmark and Rickart (2007) proposed several ways in which natural history museums can build on this potential. Natural history museums have historically embarked on collaborations with community naturalists, ranging from amateurs to experts in the field, with many museums founded by naturalists who support their development and maintenance (Andrea Sforzi et al., 2018). This is supported by Dorfman (2019), who documents the relevance of museums in the past and outlines how their roles have changed over the years. Recently, museums have changed drastically from merely displaying artefacts and specimens to allowing visitors to relate to issues beyond the objects and focus on matters of universal importance. The project discussed in this chapter allowed both museums to work with the general public to address an environmental concern.

While a plethora of literature explores natural history museums and citizen science in the context of European museums, the experiences of Caribbean museums have rarely been documented. Notwithstanding, the challenges of maintaining relevance in changing times, limited financial resources and support and the need to continue the museum's traditional role of specimen conservation and collections

DOI: 10.4324/9781003288138-9

access are not unique to Jamaica and other Small Island Developing States (SIDS). SIDS claim to contribute the least to climate change but are the most vulnerable to its effects (Wong, 2011). SIDS have high percentages of endemic species and are recorded as biodiversity-rich areas, with their biodiversity being highly threatened. As repositories of specimens representing a country's biodiversity, natural history museums play a critical role in raising awareness about biodiversity and documenting species that may be affected or driven to extinction by climate change and other threats.

We at the Natural History Museum of Jamaica (NHMJ) collaborated with the Patricia and Phillip Frost Museum of Science (Frost Science) based in Miami, Florida (US) to engage young citizen scientists between 2014 and 2015 in restoring degraded habitats. Before the project, we were not familiar with the concept of citizen science and its potential for successful conservation programmes. However, through public engagement, research and education, staff of the NHMJ recognised the inherent value of the museum and its offerings. The challenge of biodiversity loss and the associated environmental degradation added to the museum's recognition of the need to engage the wider society in building awareness of the importance of nature and the need to conserve its valuable components. The project offered wide-ranging opportunities for students to gain knowledge on scientific concepts such as biological diversity, taxonomy and environmental conservation. The knowledge gained helped them to better understand the interrelationships among species in two types of ecosystems and the value of conservation. The participants were also exposed to various types of museum research and conservation activities relating to wildlife observation and identification and ecosystem restoration, including planting trees and the removal of solid waste. Many participants were introduced to the museums for the first time, and others worked towards travelling out of the country to visit the partner museum.

The project was considered a success and its results were subsequently shared at the Institute of Jamaica's Research Symposium in November 2017. Prior to this, our early experiences during the project were shared at the 2014 symposium of the Museums Association of the Caribbean. Both the Frost Science Museum and the NHMJ used the project to show how museums could use the concept of citizen science to address environmental issues over a short period of time. This project highlighted the creativity of a museum in a SID in dealing with limited resources versus the assumptions of a museum with much 'at its fingertips'. In addition to our similar practices in museum displays, using various technologies and environmental stewardship, participants from Jamaica and the US, particularly the citizen scientists, were exposed to each other's cultures, through social media and exchange trips where they met in person. The cultural differences and similarities were enlightening to both parties.

All participants significantly contributed to the success of this project, but we wish to especially highlight and acknowledge the citizen scientists. The project was conceptualised and implemented without prior knowledge of the principles of citizen science. We support and recommend other museums to follow the 'Ten principles of citizen science' (Robinson et al., 2018, pp. 29–30). In addition, we would recommend reading Davis and Klein's (2015) 'Investigating high school students' perceptions of digital badges in afterschool learning' for an outline of the opportunities and challenges of using digital badges as incentives.

The Natural History Museum of Jamaica

The Institute of Jamaica, established in 1879, founded the NHMJ as one of its first divisions. The museum fulfils the scientific component of the organisation's mandate 'For the Encouragement of Literature, Science and Art'. The NHMJ has, for many decades, been involved in educating the Jamaican people on the island's unique biological diversity. Programmes range from teaching sessions in-house and outreach activities to the use of displays utilising specimens representing Jamaica's plants and animals. The science museum has been a centre of attraction for many Jamaicans, from its days of displaying live fauna in the 1960s to the more recent exhibition gallery that showcases a variety of specimens depicted in their unique habitats. The museum is located in the heart of downtown Kingston, where many inner-city communities are found. In spite of the challenges associated with limited funding, insufficient parking for visitors, the aversion by many potential visitors to venture downtown, and keeping up with technological advances and societal demands, the museum has been able to successfully attract and engage the surrounding community and the wider public through its innovative programmes and activities. The museum uses programmes, such as Afternoon with a Scientist, to bring large numbers of students to the museum, where they are introduced to professions in science by in-house and external scientists and given tours of the facilities. Environmental expositions also provide opportunities for the museum to showcase its collections while sharing current global messages about protecting species and the environment. In more recent times, the museum embarked on a renovation exercise to upgrade the displays. Real-life models and state-of-the-art interactive displays were developed to attract a technology-savvy audience.

The Patricia and Phillip Frost Museum of Science

The Frost Science Museum located in Miami, Florida was formerly known as the Miami Science Museum. Established in 1949 as the Junior Museum of Miami inside a local residence, the museum mushroomed into a 4,500 m^2 facility by 1960, located on a 12,000 m^2 site in Coconut Grove. In 1966, the Space Transit Planetarium was added to the property, and by 2017, the new 23,000 m^2 facility in downtown Miami added the 1,900,000 L aquarium housing fish, rays and sharks. The museum's mission is to inspire people of all ages and cultures to enjoy science and technology and to better understand ourselves and our world. Recognising science as a pathway to understand the wonders and challenges of the universe and to navigate a sustainable future, the museum has a vision to create compelling STEM (science, technology, engineering and math) education that inspires learning and innovation and enables people to explore real, rigorous science (Phillip and Patricia Frost Museum of Science, n.d.).

Frost Science has embarked on several activities to achieve one of its goals, which is to effectively share the power of science with as many people as possible. These activities include research partnerships, youth programmes and projects that focus on expanding access to science learning. Partnerships include schools, universities, research institutions and other museums. Additionally, Frost Science builds partnerships with community-based organisations, local government and private businesses throughout Miami-Dade County and beyond.

Museums Connect

Museums Connect was an initiative of the US Department of State's Bureau of Educational and Cultural Affairs and administered by the American Alliance of Museums (AAM) from 2007 to 2017 (American Alliance of Museums, 2018). This initiative focused on strengthening connections and cultural understanding between people of the United States and people abroad through innovative projects facilitated by museums and executed by local communities. Nine projects were funded in 2014 and Jamaica was the only recipient from the Caribbean. The programme encouraged museums to have a significant impact on local communities and their citizens, outside the physical walls of the museum. In less than a month after posting its profile on the programme website, the NHMJ was approached by Frost Science, and following a series of online exchanges through emails and Skype meetings, the project entitled Citizen-Led Urban Environmental Restoration was conceptualised. The project was crafted in alignment with the AAM themes 'Investing in Green Practices' and 'Developing Amateur Experts'. Project activities served the main aim, which was to create communities of informed, environmentally active citizens, with the understanding that each individual can effect change in their environment. It was proposed that the citizen scientists would participate in urban habitat restoration and outdoor discovery, learn about conservation issues critical to both Miami and Kingston and ultimately take action to improve their environments and contribute meaningfully to their communities.

The NHMJ and Frost Science partnership

Jamaica and South Florida share similar environmental issues, including environmental degradation, pollution, sea level rise and urban sprawl. Both museums are located in urbanised communities with limited access to green spaces. The communities are mainly mixtures of businesses and low-income families. The NHMJ and Frost Science recognised the similarities of the institutions' goals and related activities. Both shared a common goal to educate the public while fostering environmental stewardship, especially among the youth in local communities. The museums' outreach programmes focused on youth and young adults through many programmes that involved some level of community participation.

The goals of the project were crafted by the partner organisations after careful examination of the mandates of the museums and the guidelines provided by AAM. The overall goals included

- developing protocols for environmental restoration involving citizen science,
- establishing a relationship between museums and individuals,
- building a network of environmentally knowledgeable young citizens,
- restoring different yet similar ecosystems and spaces in Jamaica and the United States through the participation of citizens and
- fostering cultural awareness and exchange.

The expected project outcomes included

- well-rounded citizen scientists with increased knowledge of local and international environmental issues,
- citizens with a greater understanding of cultural differences in terms of attitudes and perspectives on environmental issues,
- young people with more knowledge about access to environmental services and opportunities,
- citizen scientists with an understanding of scientific protocols and a developed sense of environmental stewardship for their own communities,
- project scientists with an understanding of environmental issues and restoration strategies relevant to both project locations,
- project scientists with improved skills on strategies for communicating environmental science to the general public, using traditional and current technological methods, including social media platforms,
- partner museums and project scientists with increased knowledge of the flora and fauna important to both sites, in the context of broader conservation issues unique to and shared by each partner site (i.e. invasive species, urban sprawl) and
- communities that maintain and enjoy their restored natural spaces and develop an increased sense of pride and appreciation for their environment.

The sites

The NHMJ identified a space facing the building of the IOJ Programmes Coordination Division (IOJ Junior Centre), located in the Portmore municipality, as an ideal location in need of conservation. The Junior Centre engages the youth (aged 6–18 years) from surrounding communities in afternoon and weekend programmes in the visual and performing arts. The Junior Centre served as an ideal source for potential project participants. The space was a dust bowl functioning as a major thoroughfare for pedestrians, who used it to access the nearby major bus park and transportation centre. The space was littered with solid waste and had sparse vegetation, including a few palm trees and small shrubs. Portmore is an area originally populated by mangroves, coastal ponds, marshes and some dry-forest

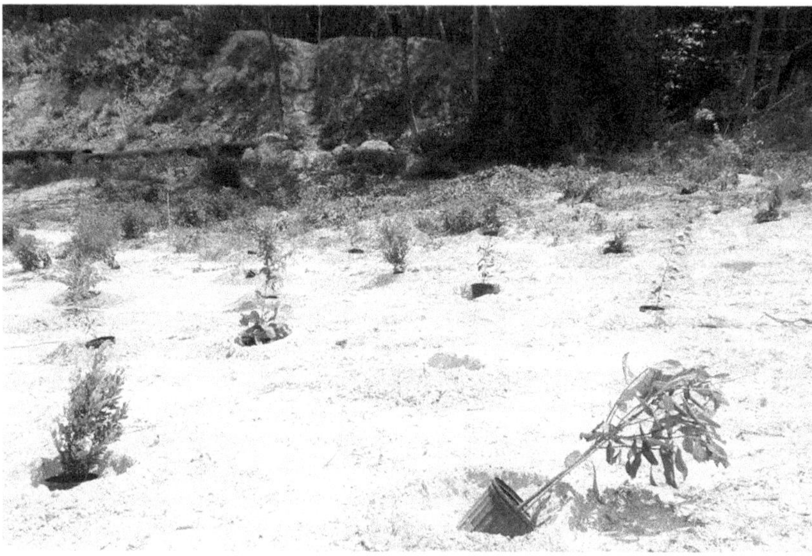

FIGURE 7.1 Site preparation at the Virginia Key North Point, Miami, Florida.

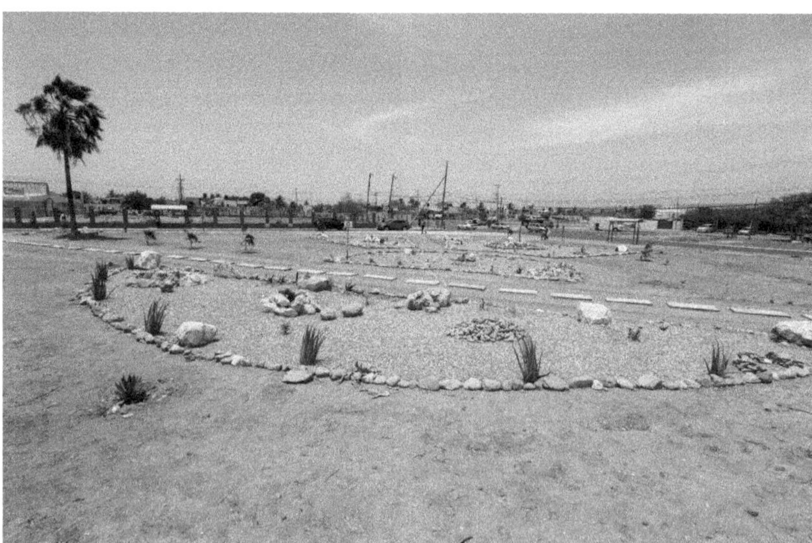

FIGURE 7.2 Rock Garden developed during restoration of the Greater Portmore site, Jamaica.

vegetation. Since the 1950s, the area has been increasingly transformed by urban housing schemes. The original vegetation of the site was considered during the restoration activities.

Frost Science, through its Museum Volunteers for the Environment (MUVE) programme, was already working on restoring 17 acres of coastal habitat in Virginia Key North Point (VKNP)[1] and decided to expand their efforts at that site. VKNP is an uninhabited, 1,200-acre barrier island located a few miles from downtown Miami; it has a diverse matrix of native habitats, including mangrove wetlands, tropical hardwood forests, dunes, active sea turtle nesting beaches and a freshwater wetland, and it is surrounded by healthy seagrass beds and isolated coral reefs.

Project activities

Scientists and their training

Three scientists were recruited by each museum to serve as mentors for the citizen scientists during the life of the project (see details under 'Recruitment'). The six scientists (and their affiliations during the project) were Dr Kathleen Sullivan-Sealey (Associate Professor of Biology, University of Miami); Benjamin Wilson (PhD candidate, Marine Science, Florida International University); Danielle Ogurcak (PhD student in Earth Systems Science, Florida International University); Keron Campbell (Botanist, NHMJ); Elizabeth Morrison (Zoologist, NHMJ) and Damion Whyte (Terrestrial Biologist and Environmental Officer, Urban Development Corporation).

Frost Science led a series of three 4-hour workshops using the research-based Portal to the Public approach. Portal to the Public is a professional learning community that facilitates work in the fields of science communication and public engagement (Advancing Research Impact in Society, n.d.). Organisations can use the framework to design programmes that fit their goals and prepare scientists for face-to-face interactions with the general public. Portal to the Public, used by several museums, is an innovation of the Institute for Learning Innovation, with support from the National Science Foundation and the Institute of Museum and Library Services.

Skype was used to facilitate meetings between the US-based scientists and partners and their counterparts in Jamaica. The training sessions for the scientists were aimed at arming the scientists with the skills to communicate efficiently and effectively with young participants and other non-scientists. The training focused on creating interactive and hands-on experiences using strategies that included sharing personal stories, the pleasure of discovery and the avoidance of jargon.

As part of their deliverables for the workshop, scientists were required to document the various research techniques in developing the project's field guide. The field guides were then used by the citizen scientists recruited for the project restoration events, during field observations, data collection and monitoring activities.

Stakeholder engagement

The engagement of community members, including organisations and businesses, was integral to the success of the project.[2] The Portmore Municipal Council (PMC) played an integral role in ensuring that the project adhered to the Government of Jamaica guidelines relative to protocols concerning the engagement of citizens, the erection of signs and restoration activities. PMC staff provided information on existing water facilities for the maintenance of the green space and the engagement of approved businesses for the purchase of benches, signs and other project-related materials. Notable participants included the acting mayor, PMC staff, the counsellor for the area and associates, who were actively involved in the project activities, including the recruitment of student participants, site restoration and the official opening of the restored space. The Public Affairs Section of the US Embassy also provided support in exposing the project participants to their on-site facilities, which include a research library with computer stations, and in issuing temporary visas to all the participants who qualified for travel to the United States. Successful students and citizen scientists who were selected to participate in the travel exchange programme included some individuals who were receiving a passport and visa for the first time. Members of the diaspora community based in Florida were also involved in the cultural exchange component of the project. Persons of note included Jamaica's Consul General to Miami, who participated in an official tour of the new Frost Science facility in downtown Miami, which

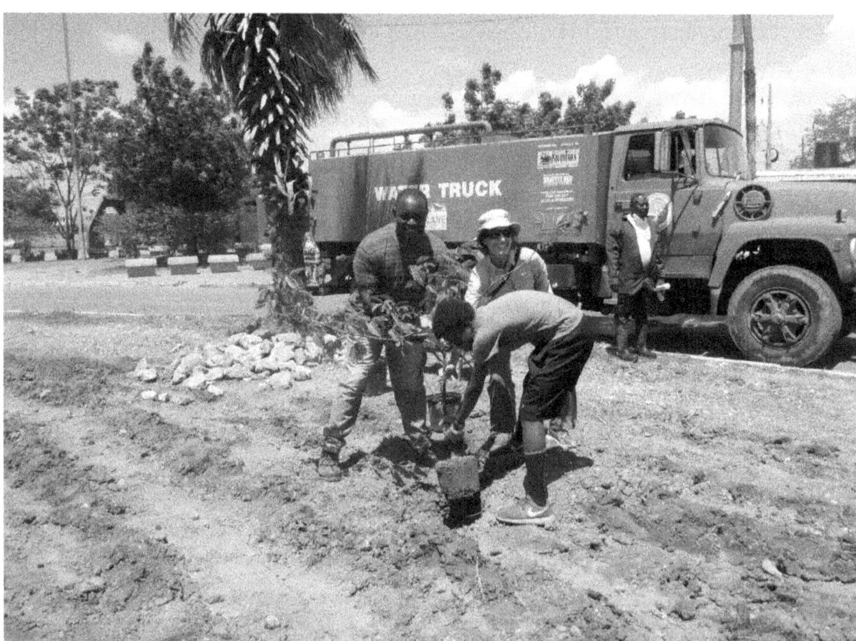

FIGURE 7.3 Mayor Leon Thomas and project participants from Miami planting trees at the Greater Portmore site.

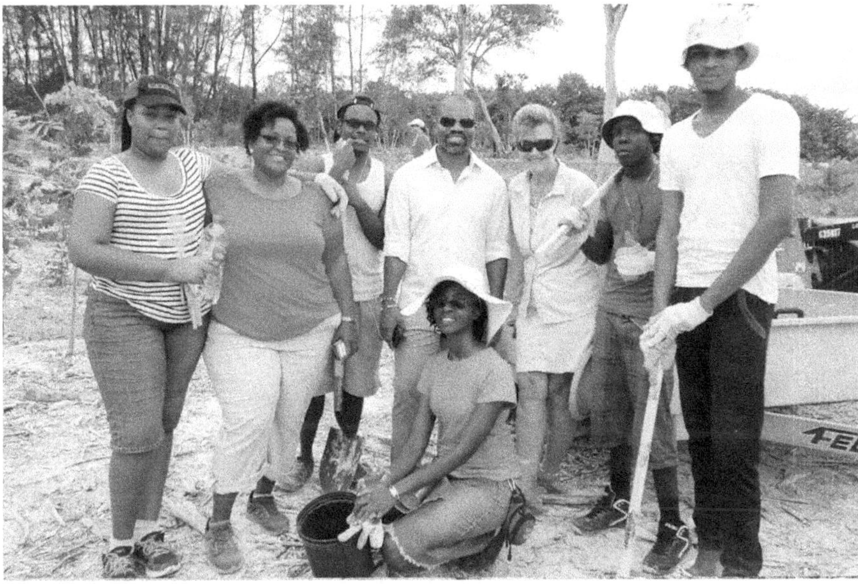

FIGURE 7.4 Jamaica's Consul General to Miami, Franz Hall, with project participants from Jamaica at the Virginia Key North Point site, Miami, Florida.

was undergoing construction. The Consul General, along with influential members of the diaspora in Miami, also participated in an official event hosted by the Frost Science Museum. The event was a fundraiser for NHMJ gallery renovation and served also to recognise the partnership of the NHMJ and the Frost Science Museum. Students from Jamaica who travelled to Miami had the opportunity to share their experiences with a wide cross-section of audience members, including the heads of both museums and other leaders of the Miami community.

Recruitment

The NHMJ partnered with the Programmes Coordination Division (IOJ Junior Centres) to recruit committed candidates for the project. Since the focus was on the space in front of the Greater Portmore Junior Centre, high school students (citizen scientists) from the immediate location were targeted. Citizen scientists would be challenged with the task of identifying suitable plants for greening the space while ensuring that they would attract fauna such as butterflies and birds. Specific walkways would be established and benches for pedestrians would be added so that people would have the opportunity to enjoy their surroundings.

The NHMJ used its experience with its Afternoon with a Scientist programme to engage in-house and external scientists in training secondary school students (citizen scientists) in species identification, conservation and environmental restoration. Secondary school students were sourced from the IOJ Junior Centre programmes, which focus on engaging children and teenagers in the visual and performing arts.

The Frost Science Museum was engaged in MUVE, a volunteer-based habitat restoration project that used social media, eco art and science exhibits to engage local residents in restoring coastal environments that once thrived in Miami. To ensure a healthy environment, local volunteers would learn about environmental stresses and actively participate in projects and activities. The student volunteers would become the project citizen scientists engaged by education and project staff of the Frost Science Museum.

The recruitment process aimed to have students who would participate in all aspects of the project for its duration of 8–12 months. In the letters to schools and community organisations, the NHMJ requested individuals who were passionate about the environment, were pursuing science subjects and had an interest in environmental restoration and preservation. The NHMJ developed an application form for students and used this to help in the selection process. Students were asked why they were interested in participating in the programme, what they expected to gain from the experience, how they thought they might use the experience in the future and what their planned majors were for college. Thirty student citizen scientists (aged 14–18) from each country were recruited. Participants were invited to complete a questionnaire developed by the partners and secure the consent of parents or guardians.

Additionally, pre-surveys were issued to the students to gather data about their levels of understanding of biodiversity and environmental conservation. The pre-survey also aided to assess content and cultural knowledge and included asking students to list ways that humans can destroy or restore natural environments, name any flora or fauna they knew to be native or invasive to either country, describe their impressions or knowledge of the partner country and rate their use of and comfort with social media.

The scientists recruited for the project also went through an application process but using a designated application form. Their forms included similar questions to those included in the student application, but applicants also had to indicate what level of education they had received, their research areas, their philosophies on science communication and their experience, if any, in informal science education activities with the public. As part of their application process, they committed to attend restoration events and actively engage with students at the events and via project social media. A pre-survey was also administered to project scientists, which included similar questions to those in the student survey, as well as questions about their experience and comfort with engaging with the public about their research, in person or via social media. The latter response assisted with structuring the science-communication training.

Restoration activities and virtual meetings

Restoration of the Greater Portmore site was the main focus for the NHMJ and Jamaican participants, and the challenges were faced with energy and determination.

The aim was to reform the space into a useful, aesthetically pleasing oasis in the middle of a bustling community.

The service of a landscape architect was retained to redesign the site, bearing in mind the main issues of concern which included improper disposal of garbage by pedestrians and the dry, dusty landscape overrun by weeds. The resulting layout included designated areas for plants, walkways, rest stops and garbage receptacles. Approval from the mayor of Portmore and the councillor for the area were signals for work to commence.

The project followed the six-step model for environmental restoration as outlined in the proposal:

1 identify site,
2 prepare logistics,
3 conduct assessment,
4 prepare site,
5 restore site and
6 monitor site.

Heavy machinery and workers removed dangerous objects and vegetation before the citizen scientists took charge and contributed to restoration activities.

Activities in both countries relied heavily on the involvement of volunteers. The core participants for the events were the citizen scientists engaged under the project, but other volunteers also participated. The monthly activities had scientists working closely with the young participants to share knowledge in practical and hands-on activities.

Site preparation included the removal of garbage and identification of plant and animal species. During these activities, invasive species were removed and beneficial species retained. The project scientists guided the citizen scientists to collect and identify specimens of plants and animals from the site. The museum staff also encouraged the collection of specimens and gave practical demonstrations on preservation techniques. Early in the process, the soil quality was assessed and topsoil was acquired and spread over the site.

The field notebook designed for the project was mandated to be used during these activities as participants documented all activities, including introducing plants, removing invasive plants, surveying bird and planting native flora. The Frost Science and the NHMJ worked together to design and organise the structure of the field guide, which included

- an overview of the project,
- a description and photo of each of the sites to be restored in the US (VKNP coast) and Jamaica (Greater Portmore urban green space),
- images and descriptions of some of the most common native and invasive flora and fauna in both areas,

- science process skills[3] in a step-by-step format (e.g. plant collecting, sweep net sampling, bird surveys, biomass estimation, sieving sand and plant growth and abundance),
- blank grid pages with instructions on mapping the restoration site and documenting observations,
- blank pages for notes and
- one page for each of the six project scientists, which included their education histories, descriptions of their research and goals for their experience in the project.

Project scientists were instrumental in creating the field guide, as they contributed lists of key flora and fauna, outlined science process skills important to their research and wrote their own scientist page. These field guides were printed on all-weather, waterproof paper in professional-style field journal booklets, and each student had their own field guide for the project. The project exposed students to several other skills that they would not have learned in the classroom, such as bird watching and identification, laying transects and water-quality monitoring.

The virtual meetings were the most exciting parts of the monthly restoration events during the life of the project. These virtual meetings were facilitated through Skype and allowed for participants from both countries to meet to share their experiences and ask questions related to culture, science and the environment.

The comment below was posted on the project's blog by one of the citizen scientists following a restoration activity:

> When we had our first event on Virginia Key this January, I learned and saw so many new things. With my group, we were picking up sea beans that washed up on shore from other islands. There were so many different colours, shapes, and sizes along the shoreline. I believe it is imperative to have projects like these to save the environment and keep it intact. I encourage others to try and join to participate in many other projects similar to this one.
>
> *(Lakayla Moody, US citizen scientist)*

Another citizen scientist remarked during a virtual meeting:

> I totally enjoyed myself at the event, and based on the expressions on the faces of my peers, they seemed to enjoy it too. It was extraordinary. We had a great and superb time learning, but most of all, I made new friends, too.
>
> *(Andrew Henry, Jamaican citizen scientist)*

Exchange visits

Each museum conducted three sets of exchange trips to the partner country. The first exchange trip involved two project staff from Miami who travelled to Kingston and two project staff from Kingston to Miami. These trips were reconnaissance activities where project 'kick-off' meetings were held to outline project plans in more detail. Project staff from each location met project partners and scientists, and they toured the restoration sites to gain an understanding of the partner environments and develop defined plans of action for each site. Local project scientists were present at each meeting to discuss the conditions and needs of each site. Frost Science project staff Lindsay Bartholomew and Fernando Bretos travelled to Jamaica, and NHMJ project staff Tracy Commock and Dionne Newell travelled to the US. In addition to working together on details of event timelines and participant recruitment strategies, it was essential that each trip involved visits to each partner's identified restoration site. This resulted in creating an outline of goals for each site as well as specific activities in which citizen scientists could participate. Project staff were able to identify and acknowledge the similarities and differences between the two sites and this helped to determine successive project plans. These trips also provided an opportunity to meet community partners and officials in both locations, including the Acting Ambassador of the US Embassy in Kingston, Lee Martinez, and Jamaica's Consul General in Miami, Franz Hall.

Midway through the project, one Frost Science staff member (Chelle King), one project scientist (Danielle Ogurcak) and four citizen scientists (Dayna Richardson, Lakayla Moody, Kenyartha Clark and Khalif Muhammad) travelled from the US to Jamaica. This was followed by the exchange trip, where one NHMJ staff member (Dionne Newell), one project scientist (Keron Campbell) and four citizen scientists (Delano Ellis, Shemar Spence, Joelle Vidal and Oshane Somers) travelled from Jamaica to the US.

As the project progressed, a second set of exchange trips took place: two Frost Science staff members (Lindsay Bartholomew and Chelle King) and four citizen scientists (Minerva Olazabal, Brianna Cineus, Wayne Holmes and Robinson Wagnac) travelled from the US to Jamaica. Conversely, one IOJ staff member (Kerri-Ann Palmer), one project scientist (Elizabeth Morrison) and four citizen scientists (Andrew Henry, Cesar Buelto, Tara-Chin Benloss and Tajh Reynolds) travelled from Jamaica to the US.

These exchange trips were reported as enormously successful and exciting. They represented the result of efforts of citizen scientists, as well as scientist mentors who actively participated in project activities. Participants learned first-hand about the partner's restoration activities and country, met new friends as well as community leaders, officials and the media and actively participated in their restoration efforts.

During trips to Jamaica, the visiting scientists explored the Blue Mountains, where many of them saw mountains for the first time. The visitors also met then serving US Ambassador to Jamaica, Luis G. Moreno. They described the Jamaican cuisine as delicious, and the students worked and socialised with new friends through planting trees at the restoration sites and socialising at the nearby beach. During one of the trips to Jamaica, the visitors were able to see the local frenzy associated with the annual staging of the Boys and Girls Athletic Championships. They were awed by the show of school spirit at every level of the society. The culminating event in Jamaica involved an official ceremony where government representatives, community leaders and project participants opened the restored site in Portmore.

The trips to the US were firsts for a few of the citizen scientists. Many were awed by the Miami cityscape, defined by its high-rise structures. The differences in landscape were highlighted in a blog written by Tracy Commock and Dionne Newell following the November trip to Miami:

> We finally got a chance to see the project site, Virginia Key and crossed the highest elevation in Florida, the bridge of maybe 2 metres high, WOW. We could swap some of these fantastic views of the skyline with some of the mountains in Jamaica.

The Jamaican students were exposed to the many fascinations of the science museum, including the numerous interactive displays and the planetarium. One highlight was the fundraiser event hosted by the Frost Science Museum, which was held to benefit NHMJ during their trip to Miami. The Jamaican contingent contributed to a final event, during which they and other volunteers, joined by Jamaica's Consul General in Miami Franz Hall, collectively planted over 900 trees on Virginia Key.

The participants' experiences underscore the role of museums in exposing them not only to more than just scientific principles but also to social and economic activities and their linkages to the environment. The citizen scientists were truly fascinated by the exchange in both countries and expressed their experiences as follows:

> The fun, adventure and everything—every single detail of this trip, has been fantastic! Miami is beautiful and diverse. The people are just as colourful and grand as their city. I have made friends with the scientists and students in the MUVE programme who will not be forgotten. I definitely look forward to returning to see my legacy grow and change an environment for the better.
>
> *(Cesar Buelto, Jamaican citizen scientist)*

> There I learned that one of the endangered butterflies in Jamaica was the fascinating Giant Swallowtail Butterfly. I also learned that one of the endemic butterflies of Jamaica was the breathtaking Clear Winged or Glass Winged Butterfly. While being in Jamaica and also doing our restoration project in Miami on Virginia Key, I learned that the Periwinkle Plant grows in both areas, and in Jamaica it is called Jamaican Vinca. I one day hope to return to learn

more and reunite with my Jamaican friends, because this was an experience that I will never forget, and forever it will remain in my heart.

(Dayna Richardson, US citizen scientist)

On the second night in Miami there was a function hosted to observe the museum connection between Frost Museum of Science and the Natural History Museum of Jamaica. The JahMaians [Jamaicans + Miamians] were all asked to do a briefing about the programme and what we have learned. We all gave varied comments concerning the operations and progress of the site and group as a whole. When it was my time to speak, I was nervous but I held my composure and nailed it. The applause came raining like raindrops from a category five hurricane when I remarked that 'the project is a team effort and I strongly support the saying that "Teamwork Makes the Dream Work"'.

(Shemar Spence, Jamaican citizen scientist)

Social media... #jamuve

NHMJ and Frost Science used basic 'tools' to implement the project but have since become aware of researchers who think that citizen science projects need to adopt new technologies to allow participants and organisers to communicate, participate and interact effectively (Bonney et al., 2014; Newman et al., 2012, p. 1; Sturm et al., 2017).

The use of social media was one of the means by which citizen scientists were engaged, which proved highly successful throughout the project. The project team coined the term #jamuve for all online platforms. Frost Science has an existing programme called Muve, and it was agreed that 'jamuve' would highlight activities related to the Museums Connect project. Social media was the means by which many of the participants were introduced to each other, maintained contact and eventually became friends. Through social media outlets, the project created a network of concerned active constituents in both countries. Citizen scientists were encouraged to document activities, accomplishments and challenges by posting pictures, videos, blog posts, questions and comments for scientists and their colleagues. They were further encouraged to act by a digital badge incentive system that awarded them for accomplishments and participation. Citizens would earn digital badges by attending restoration events, submitting to photo and essay contests, writing blog entries, posting questions to scientists or collecting the most data. These incentives were used to help determine which citizen scientists would earn a place on exchange trips to the partner country. Recognising potential limitations for some participants, citizen scientists who did not have easy computer or mobile access were allowed to submit media to project staff at restoration events so that they could receive hard copy digital badges and become eligible to participate in the trips. Computers were set up at NHMJ to assist these constituents in participating virtually. Data collection sheets and field guides for flora and fauna were provided in hard copy form as well as in digital form on the project's web portal. A hard copy pen pal exchange was also initiated between those citizen scientists with limited access to the internet or mobile technology.

Through the use of platforms such as Instagram and WhatsApp, all project participants from both locations were encouraged to communicate as well as post live updates on activities, including restoration events as they occurred. This was a means to facilitate real-time interaction. Although some of the citizen scientists used platforms such as Twitter, for most people, especially the Jamaican contingent, WhatsApp was the preferred platform. There was a constant stream of posts daily: citizen scientists communicated with each other as well as with project scientists, posting photos and seeking help in identification of plants and animals. Interactive training sessions for project scientists in both locations were held using Skype.

> My absolute most favourite part of this whole restoration project was talking to the Jamaican students via Skype! I loved asking them questions and talking about what I was learning. They are so intelligent and kind, which is why I hope to go there and personally meet them.
>
> (*Minerva Olazabal, US citizen scientist)*

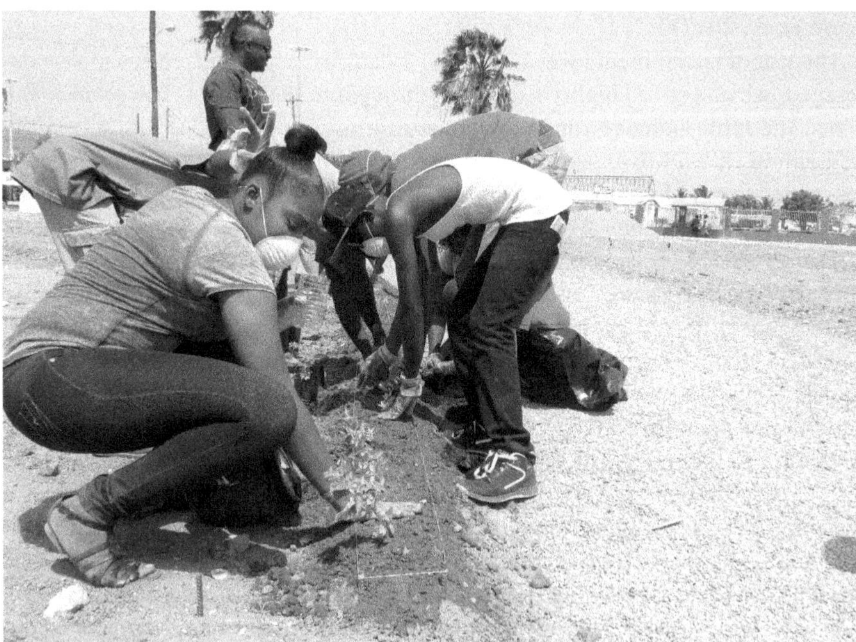

FIGURE 7.5 Citizen scientists conducting restoration activities at site in Greater Portmore, Jamaica.

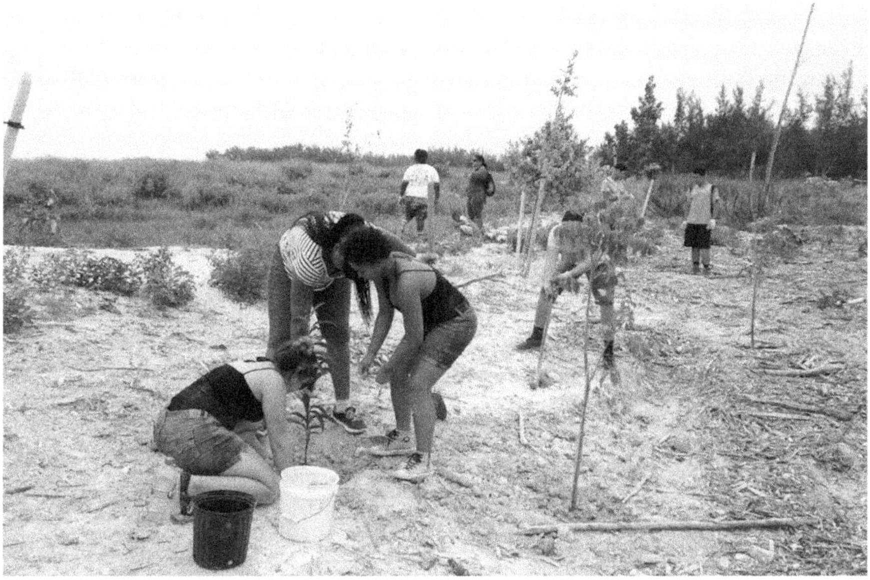

FIGURE 7.6 Citizen scientists planting sea oats and other native plants at Virginia Key North Point, Miami, Florida.

Challenges

In spite of the overall success of the project, challenges faced during implementation became learning experiences that strengthened efforts to succeed. Both museums shared common experiences in collaborative projects where institutions have varying roles and responsibilities in project implementation. However, there were delays in the transfer of funds as well as the signing of related documents due to limitations in communication. The delay in the signing of the project charter actually caused the delay in the start of the project, giving it an official start date of 24 September 2014 and end date of 31 July 2015. The museums therefore had 10 months to implement a project originally designed to be implemented over 12 months. Both institutions were unaware of each other's operational cultures and procedures, and this created initial barriers to its smooth implementation. Bureaucratic requirements for importation of equipment resulted in delays in acquisitions; however, the museums were able to overcome this by finding more suitable (lawful) methods. Many of the local citizen scientists had limited access to computers or reliable internet service, which prevented them from participating in social media activities. The project, however, provided computer equipment, which was used during training and other exercises. We were mindful of the time commitments by our citizen scientists, who were all full-time students who also had to balance regular co-curricular activities and family commitments with the requirements of the project. During the restoration phase, Jamaica was in the midst of a drought, which resulted in significant delays in planting native species in the area.

This project benefitted logistically by Miami and Kingston being in the same time zone (or just one hour difference, depending on daylight saving time), having English as an official language and being on a similar academic school calendar. However, a few challenges have been identified and were addressed throughout the development and implementation of this project. Although no translation was required, some citizen scientist students had difficulty understanding Jamaican dialect, but project staff were present to assist. One challenge was that NHMJ citizen scientists did not have easy access to computers or mobile technology. Miami students had access to computers at school, at Frost Science's Best Buy Teen Tech Center and personal mobile devices, while computer facilities at NHMJ and Jamaican schools existed but were limited. In Jamaica, the Junior Centre, through their Saturday opening hours, facilitated the execution of the project when activities were held during the school year. Advanced Training Opportunities Programme (ATOP) students met on Tuesdays and Thursdays, and commitment was sought from parents regarding students' participation on days when project activities competed with family time.

Impact and outcomes

The project documented the following outcomes:

- Citizen scientists gained increased knowledge of and involvement in their environments, both locally and regionally, including flora and fauna before and after restoration and scientific protocols.
- Scientist mentors gained skills and strategies for communicating environmental science to the public, in person and via social media.
- Project staff from both partner museums learned how to better engage citizens in environmental restoration in urban spaces.
- The target audience learned about flora and fauna in both locations.
- Participants increased their appreciation of restored natural spaces.
- The target audiences in the US and Jamaica developed a broader knowledge about and understanding of one another's cultures.

Lessons learned and other observations

Several important lessons were learned during the implementation of the project.

Sensitivity to cultural differences was high on the list. Both museums are located in urban communities, with participants from various ethnic groups and social standing. The Jamaican participants interacted mainly with persons of the same ethnicity. That was not the case for the students in Miami. The students from Miami who were of Jamaican descent were happy to share their cultural similarities. Through the WhatsApp group, the students shared jargon and the use of Jamaican dialect. Experience with understanding flexibility in implementing projects was one of the main lessons for both partners. As a museum from a SIDS with the ongoing challenge of insufficient resources, it was easier for us to adapt to the situation and quickly develop solutions.

Visits to the Frost Museum highlighted the possibilities and reach of museums and museum professionals when resources are available. The Frost Museum was also in the process of building a new museum space, which members of NHMJ also had the opportunity to tour during the project. It was heart-wrenching but encouraging nevertheless to see what could be achieved locally if even a fraction of the resources observed in that context were made available. Museums in Jamaica could benefit from more financial support and access to learning resources, including interactive exhibits.

One of the most important lessons learned was to be adaptive in our approach to project implementation, and this was vital to the success of our project. At the Greater Portmore restoration site, for example, we encountered logistical problems related to how we would deliver water to the plants after planting. This was exacerbated by an ongoing drought affecting Jamaica at the time. Our collaboration's efforts would be futile if the plants planted by eager volunteers failed to thrive without water. NHMJ took quick action by contacting the mayor of Greater Portmore to request his support. In Miami, this approach would be more complex, given that politicians in such a large metropolitan area are more detached from this type of issue. But the Greater Portmore officials kindly offered their assistance and assured water delivery throughout the project. The lesson is to try every potential solution when building a community project and take into account every factor, whether it is varying access to technology, involvement of community officials, changing physical conditions or new ideas.

Sustainability

The NHMJ recognised the need to know what aspects of the project could be sustained after the life of the project. The continued monitoring of the restored site in Jamaica was one of the key priorities. During the implementation of the project, soil quality and access to water were major challenges. Although new soil and drought-resistant plants were introduced to the site, there has been a major problem with stray animals and insufficient water. The NHMJ received written commitment from the PMC for their assistance with the site throughout the project and has maintained good working relations to date.

As outlined in the project document, the museum seeks opportunities to use the site in other IOJ programmes, especially if there is an opportunity related to climate change, biodiversity and native versus invasive species monitoring.

Current projects

The NHMJ has been implementing the Biodiversity Awareness Project with TransJamaican Highway (TJH) 2000 East–West since 2011. This environmental education project is an example of partnership with a private sector organisation to achieve awareness of science and environmental issues.

The project involves the participation of five schools that are situated in the parishes and communities traversed by the highway built and managed by TJH 2000 East–West. The students are exposed to lessons on important plants and animals in their communities, and they conduct practical exercises that include the development of vegetable gardens. Project outcomes to date include the construction of a biodiversity centre, trips to the museum and its field stations as well as participation in special events hosted by the NHMJ.

Reflections and conclusions

The project provided students with the opportunity to interact with their peers locally and internationally on environmental stewardship. This experience created a baseline for the development of future leaders in environmental conservation. Activities such as bird watching and identification, preparation of insects and plants for permanent, preserved collections, and restoration activities expanded their minds and opened up a new world of possibilities for them.

The Museums Connect project allowed the scientists and other project participants to have closer interactions with students and the general public, which, for many persons, was a new experience outside of their routine job functions. The project also allowed museum professionals to dispel the myth that natural history museums are cemeteries with old preserved specimens in drawers and jars. Museum professionals were challenged to think outside the box and develop creative ways of engaging the public, especially the young citizen scientists.

The activities undertaken throughout the project worked towards national and international goals relevant to environmental conservation and sustainable development. Jamaica's Vision 2030 was built on seven guiding principles that put people at the centre of national development and four national goals towards achieving first-world status (Planning Institute of Jamaica, 2012). Partnership and sustainability were two of the guiding principles that were central to the success of the project, which placed people at the centre of fulfilling its overarching objectives. Jamaica and the United States, through science-based museums, embarked on a cultural exchange programme facilitated by sharing financial and technical resources. This partnership resulted in the creation of environmental conservation stewards empowered with increased knowledge of environmental issues and possible solutions. Jamaica recognises that economic, social and environmental problems and solutions are all interconnected and that an integrated approach, underpinned by good governance, in addressing these issues will ensure that the country's development is sustainable.

Goal one of Vision 2030 speaks to the empowerment of Jamaicans to achieve their fullest potential. The positive response and enthusiasm displayed by the citizen scientists throughout the restoration activities and training exercises were good indicators of their heightened interest in the environment and the need for its protection. Additionally, the interactions between the citizen scientists and the project scientists and staff exposed the students to the various careers that could be undertaken in science, environmental conservation and museums. This increased exposure fostered the ability of the citizen scientists to recognise that they can achieve their fullest potential using their talents and abilities to elevate their standard of living and quality of life (Planning Institute of Jamaica, 2012). Goal four of Vision 2030 outlines Jamaica's plan for a healthy natural environment. Several challenges were outlined in the Vision document, including poor management of solid waste and loss of biodiversity. The activities undertaken during the project, such as the replanting of indigenous plants and trees to encourage birds and flying insects as well as the removal of solid waste, fulfilled the project's aim of addressing these issues. The restoration of the space in Greater Portmore supports this goal by way of creating a green space that creates harmony among development activities and fosters environmental conservation and sustainability (Planning Institute of Jamaica, 2012).

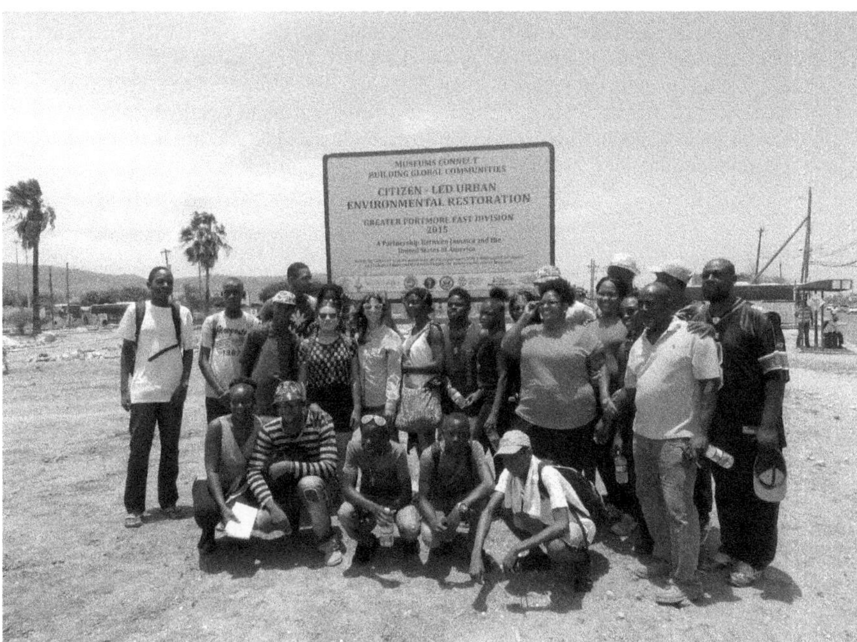

FIGURE 7.7 Jamaican and American project participants at site in Greater Portmore, Jamaica.

Internationally, several nations have been consciously working since 2015 to achieve the 17 Sustainable Development Goals (SDGs) by the year 2030. Jamaica is among the committed nations. The experiences from the project contributed significantly to attaining three of these goals, namely 4, 11 and 17. Goal 4 addresses better education, and the project provided the opportunity for all participants to gain experiences in such a way that they developed increased knowledge and were equipped to educate others. Goal 11 focuses on building sustainable cities, which is in line with the component of sustainable urban and rural development, a guiding principle of Jamaica's Vision 2030. The efforts to create an oasis in an urban 'desert' to improve the natural environment, improve aesthetics and create a usable space for the residents were addressed in the project. The project also allowed the community to see that they could contribute significantly to sustaining their own city. Goal 17 speaks to partnerships focused on attaining the SDGs. Building and sustaining partnerships to achieve common goals was definitely an achievement of the project.

Notes

1 Read more about the Virginia Key restoration project here: https://www.scientificamerican.com/citizen-science/virginia-key-restoration/
2 Community partner organisations in Miami included Miami-Dade County, the City of Miami, Miami-Dade College, Florida International University, the University of Miami, Miami-area high schools, Council for Opportunities in Education and the Stokes Institute, The Mission Continues, Wounded Warrior Project and HandsOn Miami. Notable individuals who partnered with the project included Dr Gregory R. Frederick. In Jamaica, community participants included the PMC, Greater Portmore Joint Council, Jamaica Urban Transit Company, Urban Development Corporation, Mico University College, high schools in Greater Portmore and neighbourhood youth clubs.
3 Skills used by scientists when conducting research that include observing, measuring and communicating.

References

Advancing Research Impact in Society (n.d.). 'Portal to the Public'. Available at: https://www.researchinsociety.org/portal-to-the-public (Accessed: 25 April 2020).

American Alliance of Museums (2018). 'Interested in International Museum Partnerships? Download the Museums Connect Evaluation Report'. Available at: https://www.aam-us.org/2018/05/30/museums-connect-evaluation-report/ (Accessed: 29 May 2021).

Andrea Sforzi et al. (2018). 'Citizen Science and the Role of Natural History Museums'. In Hecker, S., Haklay, M., Bowser, A., Makuch, Z., Vogel, J. and Bonn, A. and foreword by Carlos Moedas, Commisioner (2015-19) Research, Science and Innovation, European Commission (eds.) *Citizen Science Innovation in Open Science, Society and Policy.* London: University College London Press, pp. 429–44. London: UCL. Available at: https://doi.org/10.14324/111.9781787352339 (Accessed: 26 April 2020).

Ballard, H.L. et al. (2017). 'Contributions to Conservation Outcomes by Natural History Museum-led Citizen Science: Examining Evidence and Next Steps'. *Biological Conservation*, 208, pp. 87–97. Available at: https://doi.org/10.1016/j.biocon.2016.08.040 (Accessed: 29 June 2021).

Bonney, R. et al. (2014). 'Next Steps for Citizen Science'. *Science*, 343(6178), pp. 1436–7. Available at: https://doi.org/10.1126/science.1251554 (Accessed: 29 May 2021).

Davis, K. and Klein, E. (2015). 'Investigating High School Students' Perceptions of Digital Badges in Afterschool Learning'. *CHI 2015: Proceedings of the 33rd Annual CHI Conference on Human Factors in Computing Systems*. New York: Association for Computing Machinery, pp. 4043–6.

Dorfman, E. (2019). 'Changing Epistemologies in the Museum: An Evolving Relationship with Nature'. *Museum International*, 71(1–2), pp. 30–7. Available at: https://doi.org/10.1080/13500775.2019.1638024 (Accessed: 26 April 2020).

Newman, G. et al. (2012). 'The Future of Citizen Science: Emerging Technologies and Shifting Paradigms'. *Frontiers in Ecology and the Environment*, 10(6), pp. 298–304. Available at: https://doi.org/10.1890/110294 (Accessed: 29 June 2021).

Newmark, W.D. and Rickart, E.A. (2007). 'Are Natural History Museums Telling the Right Story?' *BioScience*, 57(5), p. 390. Available at: https://doi.org/10.1641/B570502 (Accessed: 29 June 2021).

Phillip and Patricia Frost Museum of Science. (n.d.). 'About Us'. *Frost Science*. Available at: https://www.frostscience.org/about-us/ (Accessed: 17 May 2021).

Planning Institute of Jamaica (2012). *Vision 2030 Jamaica National Development Plan*. Available at: https://jis.gov.jm/features/vision-2030-jamaica-national-development-plan (Accessed: 24 April 2020).

Robinson, L. et al. (2018). 'Ten Principles of Citizen Science'. In Hecker, S. et al. (eds.) *Citizen Science – Innovation in Open Science, Society and Policy*. London: University College London Press, pp. 27–40. Available at: https://doi.org/10.14324/111.9781787352339.

Sturm, U. et al. (2017). 'Defining Principles for Mobile Apps and Platforms Development in Citizen Science'. *Research Ideas and Outcomes*, 3, p. e21283. Available at: https://doi.org/10.3897/rio.3.e21283 (Accessed: 24 April 2020).

Wong, P.P. (2011). 'Small Island Developing States'. *WIREs Climate Change*, 2(1), pp. 1–6. Available at: https://doi.org.rproxy.uwimona.edu.jm/10.1002/wcc.84 (Accessed: 24 April 2020).

8

EVOKING WONDER TO INSPIRE ACTION AROUND CLIMATE CHANGE—A COLLABORATIVE EXHIBITION PROJECT IN THE ISLANDS

Natalie Urquhart[1]

In a world plagued by fake news, museums are considered to be one of the most trustworthy sources of information. According to recent American Alliance of Museums (2018) statistics, museums are more highly rated by the public than local papers, non-profit researchers, the government or academic researchers. Millions of people visit US museums annually, and millions more use their online portals for research and information. In the Caribbean, our visitor numbers (and operational budgets) may be smaller, but we are well positioned to inform and inspire our diverse audiences and to revitalise public pedagogy around pressing local and global challenges.

A question that we must now ask ourselves as Caribbean museum professionals is: in light of the intensifying socio-environmental pressures we face, how can we evolve beyond our traditional position as keepers of memory to become curators of future stories, or agents of change, to provide both the spark and the tools? There is no doubt that the role of museums has transformed over the past few decades, with our core functions and social responsibilities having become increasingly interconnected. Today's museums can no longer function simply as windows to the past but, as Jette Sandahl (2019, p. v) proposes, they must 'become inextricably part of and active agents in society'. They must, that is, 'engag[e] visitors in dialogue surrounding contemporary social issues, and in shaping the way we see, think about and act towards others and the world around us' (Janes and Sandell, 2018, p. xxvii). To succeed within this new framework, we must also adapt as professionals and engage our public through inspiring exhibitions and programming, revisit how we provide access to our collections and strategically build genuine and lasting collaborations with our communities.

DOI: 10.4324/9781003288138-10

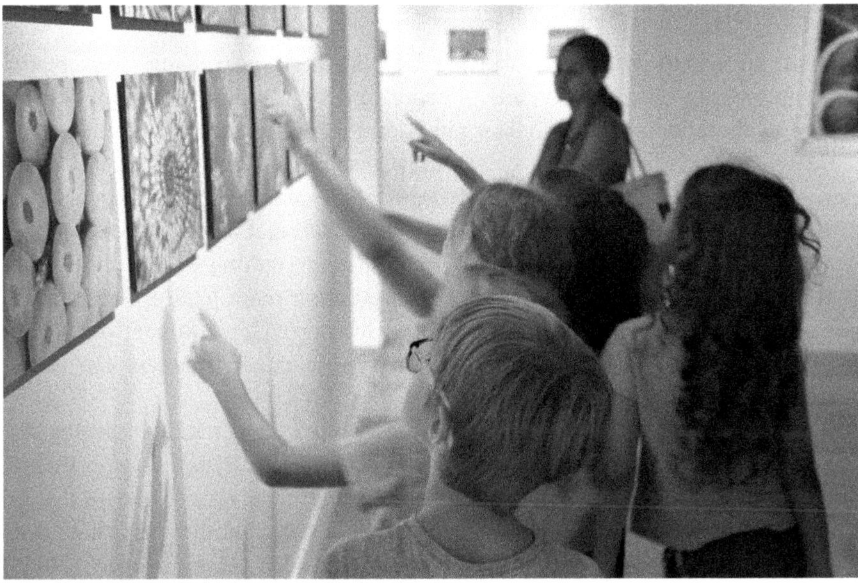

FIGURE 8.1 National Gallery of the Cayman Islands. © NGCI, 2018.

Each of our institutions has different priorities around what constitutes an 'urgent issue' within our communities. At the National Gallery of the Cayman Islands (NGCI), as a regional art museum on a small island in the Caribbean, our programming is increasingly concerned with climate change. Grand Cayman sits at an average height of 7 feet, making it particularly vulnerable to rising sea levels. As islanders, we are all deeply aware of how shifts in climate, even subtle ones, affect us. From rising seas and coral bleaching to the loss of fresh water, islands are among the nations most vulnerable to climate change impact (Cameron, Hodge and Salazar, 2013) with many Small Island Developing States (SIDS) already experiencing the drastic impacts of climate change (United Nations Office of the High Representative for the Least Developed Countries, Landlocked Developing Countries and SIDS, 2015). Following a decade of increasing storm surges and hurricanes—the damage and economic losses associated with Hurricanes Irma and Maria in 2017, alone were larger than the annual gross domestic product of several of the smaller islands that were impacted (Nurse and Edwards, 2019, p. 37)—there can be no doubt that the violent hurricane seasons we have recently experienced are directly related to raising global temperatures. For all of us in the Caribbean, this climate conversation will be a defining issue of our times; as trusted public spaces and sites, Caribbean museums have a unique opportunity to emerge as key players in the shaping of local narratives around this topic.

Museums, art and climate change

Museums the world over are actively engaging with the climate change conversation and reinterpreting older permanent collections or creating new programming and artistic collaborations to engage their public in this conversation. There are many innovative projects from which we can draw inspiration. In Washington, DC, for example, the Smithsonian National Museum of Natural History has installed an impactful digital gateway to its traditional Ocean Hall that urges viewers 'To dive in… to discover to make a difference'. In addition to cutting-edge technical displays, it features a remarkable coral reef installation made entirely from found plastic objects, entitled *Turtle Ocean* (2016), which provides a wonderful example of how artistic interventions in the traditional museum environment can inspire wonder and reinforce messaging.

Plastic Ocean (2017), an art installation created by Tan Zi Xi and displayed at the Sassoon Docks Art Project, addresses similar issues to *Turtle Ocean*. Visitors find themselves in a situation unnervingly like being underwater—except that the surface is completely covered in a layer of plastic waste. Another inspiring example is the work being carried out as part of the artist-led *Cape Farewell* (2012–) project. The project has invited artists and creatives to join expeditions to explore arctic science, sustainable island communities, urban regeneration and the clean-tech industry. The results are presented in innovative exhibition collaborations with museums around the world. At the National Building Museum in Washington, DC, the interactive installation *ICEBERGS* (2016) by James Corner Field Operations similarly addresses concerns about melting polar ice caps and material waste by constructing the work out of reusable materials such as scaffolding and polycarbonate panelling.

In the contemporary art sphere, we find equally compelling public art and exhibition projects that seek to inspire audiences to engage in conversations on climate change. For example, UK artist Chris Bodle's *Watermarks* (2009) involved projecting future water levels onto buildings, thereby transforming abstract scientific data into a stark visual reminder about potential flood levels. *Rethink – Contemporary Art & Climate Change* (2009–) is a joint project of the National Gallery of Denmark, Nikolaj Copenhagen Contemporary Art Center and the Alexandra Institute, which travelled to several European museums. It invited 30 artists to use the subjective and immediate qualities of art to engage the public in climate discussions in a variety of ways. *Don't/Panic* (2012), an exhibition in Durban, South Africa, curated by Gabi Ngcobo, sought to create a dialogue around culture and climate change from a uniquely African perspective. Meanwhile, *Weather Report – About Weather Culture and Climate Science* (2017), developed by the Deutsches Museum, Munich, and supported by UN Climate Change, used historical European artworks to create a sense of understanding around our complex climate systems and the increase in extreme weather conditions.

The thread that joins these projects is the notion that artistic interventions can create positive reinforcement. This builds upon Gary Braasch's (2013, p. 38) observation that 'a sense of connection with the causes and consequences of climate change in a positive manner [...] tend[s] to create more possibilities for engagement and action'. For example, speaking about the *Weather Report* exhibit, Nick Nuttall (spokesperson for the exhibition patron, UN Climate Change Executive Secretary Patricia Espinosa) noted the importance of going beyond economic arguments in communicating about climate change. He emphasised how such exhibitions can inspire people to take climate action:

> Some people need to be moved by art to think differently and see their perspective change. We need a broad movement of different perspectives to shape a broad response globally, nationally in our homes, in our schools.
>
> *(United Nations Climate Change, 2017)*

Much research has been done on how to connect people to climate communication (see Braasch, 2013; Cozen, 2013; Guy, Henshaw and Heidrich, 2015; Lyons and Bosworth, 2018). As a topic, climate change is viewed by many as a very abstract problem that is hard to visualise in a way that relates to everyday life, especially in regions not yet being affected by extreme weather. This can result in individual disconnection and a failure to act (Weber, 2006). In their work on visual art as a way to communicating about climate change, Roosen, Klöckner and Swim (2018, p. 91) write that much of the climate communication we engage with fails to inspire due to a lack of relatable narratives and metaphors. Further compounding this issue is negative imagery—which fails to create an emotional response or, worse, promotes a sense of despondency—along with a lack of the kind of storytelling that makes information memorable (Roosen, Klöckner and Swim, 2018). Several studies have shown that emotional reactions to environmental risks such as climate change can trigger a tendency to act (Böhm and Pfister, 2000; Klöckner, Beisenkamp and Hallmann, 2009; Roosen, Klöckner and Swim, 2018). If powerful experiences are created for people to experience a narrative about climate change personally, these experiences can be more moving and ultimately may trigger a process of changed behaviours and actions on an individual level (Bullot, 2014). Inspiring experiences, such as those referenced above, enable visitors both to access information and to draw connections between the work's content and their own lives, encouraging an engagement in the climate conversation long after the museum visit ends. It remains for us as museum professionals to create more of these types of experiences for our local audiences.

Developing *Coral Encounters* at the National Gallery

In 2017, the NGCI began developing a project based on this notion of positive reinforcement as a way of communicating about climate change. An art exhibition and community programme entitled *Coral Encounters* was designed around the concept

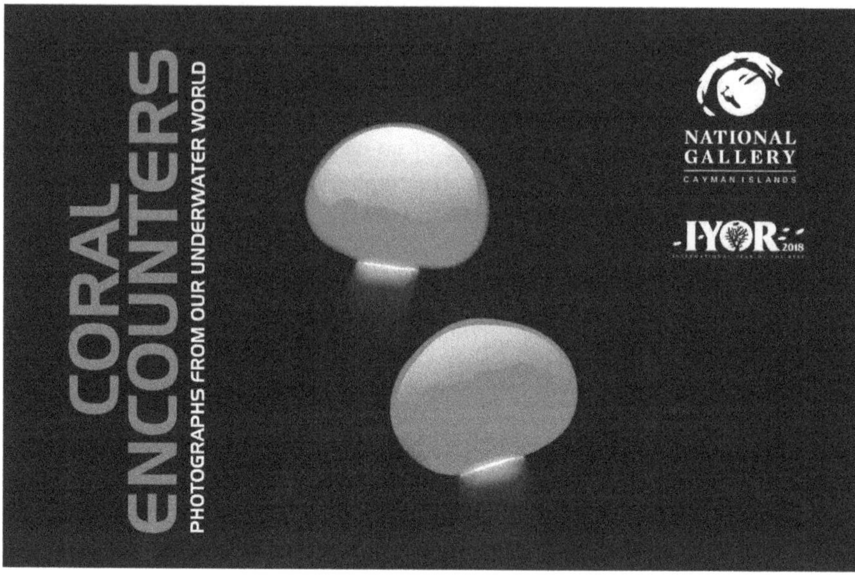

FIGURE 8.2 *Coral Encounters* promotional poster. © NGCI, 2018.

of 'evoking wonder to inspire action', encouraging viewers 'to discover, and dive in, to make a difference'. The following description of this exhibition, from design to programming and partnerships, serves as an example of how smaller Caribbean museums and galleries can have a large impact on local and regional conversations and actions on climate change.

Coral Encounters, a collaborative project by NGCI and the Central Caribbean Marine Institute (CCMI), in conjunction with the International Year of the Reef (IYOR) initiative, explored coral reef health and its relation to climate change through the work of 30 photographers and six partner organisations. Positioned at the crossroads between art and science, it brought together artists, marine professionals, scientists, curators and educators to inspire dialogue, support ongoing research and management efforts and strengthen long-term collaborations for coral reef conservation—a critical conversation for low-lying island nations.

This project was designed around a collaborative programming model, which at its core was built on our partnership with CCMI, a leading global marine research centre located on the smallest of the three Cayman Islands. Based on an island steeped in maritime heritage, the National Gallery's education programme is heavily focused on this aspect of our history, including maritime heritage, ocean conservation, coral health, plastics and recycling. Sustainability as a theme, however, fell largely outside the traditional expertise of our art museum staff (Hebda, 2007, p. 335). The initial partnership was designed to enable NGCI to draw on CCMI research and programming support, with NGCI in turn providing lecture facilities for the CCMI schedule in Grand Cayman. By 2017, discussions were underway to launch the CCMI's *Reefs Go Live* project (discussed further below) at the National Gallery in early 2018 to coincide with the IYOR. This launch quickly developed into a wider exhibition project which would draw upon local partners while benefiting from the IYOR scope and resources.

The vision for the project was to use the subjective and immediate qualities of art to engage the Caymanian population in climate discussions (Hulme, 2010, p. 22). These would be centred around Cayman's world-renowned coral reef system, which is at the heart of the local water-based tourism industry. From the start, NGCI's curatorial team were keen to avoid a purely documentary-style exhibition, given the organisation's primary mission as an art museum. We sought to create an experience that would engage visitors in a fresh way rather than offer up traditional marine images that they had experienced before. This was achieved by focusing on the natural design elements found in the corals themselves. As land-based beings, most of us struggle to understand the unique processes of the ocean. To many islanders, the underwater environment is a distant place beyond their reach, brimming with alien lifeforms: stinging corals, glowing anemones and giant carnivorous clams that are bizarre, beautiful and otherworldly. The idea that began to germinate was to create an entry point into this 'weird and wonderful world beneath the waves' (Urquhart, 2017), and to inspire students to care enough to get involved in the efforts to save it. Building upon the premise that climate-related messages should aim to be inspiring (Roosen, Klöckner and Swim, 2018), and that

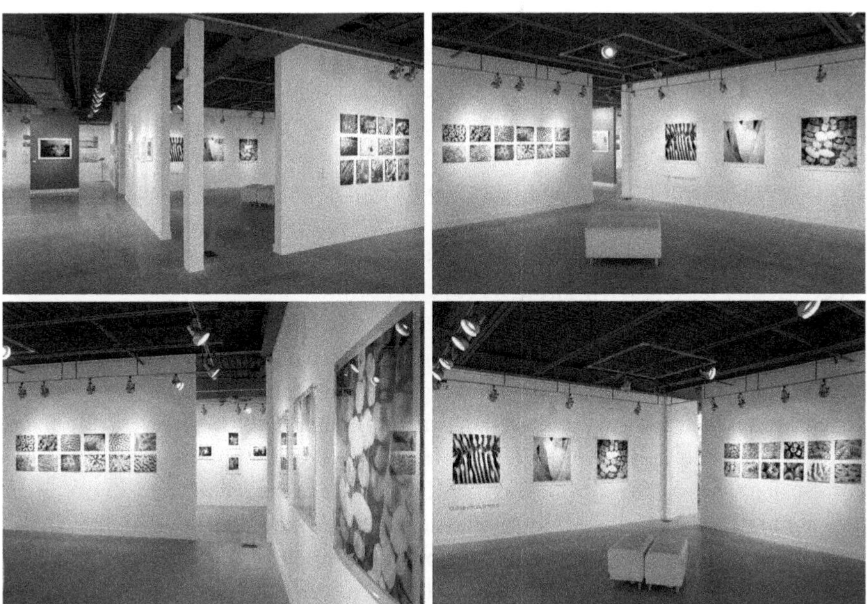

FIGURE 8.3a-d *Coral Encounters* installation views. © NGCI, 2018.

inspiration may be a motivating factor in the process of change (Thrash and Elliot, 2003, p. 96), we settled on science fiction to tell our very serious story. The concept of 'a galaxy far, far away yet ten feet from the shore' thus formed the basis of the exhibition brief.

Once our core project partners had been established, an open call was developed and shared with local and international photographers through the IYOR, NGCI and CCMI websites. Of the multiple entries, 30 photographers were selected from 7 countries, and a total of 72 images featured in the exhibition. Importantly, the decision to feature predominantly macro photography enabled us to focus attention on the remarkable patterns and colours found in the natural environment, which helped us avoid presenting a more typical documentary-style exhibition of larger reef systems. This stunning collection of images offered a unique chance to glimpse one of the world's most diverse environments and to get up close with some of the species that inhabit it.

In an effort to create an environment that would evoke a sense of wonder, the exhibition design drew heavily upon the popular science fiction film *Close Encounters of the Third Kind*. Sci-fi fonts, wayfinding graphics, bold neon colours and large vibrant grid systems were used to create visual impact, while luminous lighting created an otherworldly feel. Sectional headings extended this with titles that included other science fiction references, such as *Stranger Things* and *Between Two Worlds*. Ultimately these well-loved popular cultural references and the vibrant design helped to provide an access point for visitors of all ages and a chance to engage them in a more serious conversation around coral reef health (Gough, Berigny and Dunn, 2016, p. 16). For example, the activity sheets created for school tours focused on exploring pattern and design and included conversation prompts through which wider themes of coral bleaching, sustainability and recycling could be discussed. In addition, an important part of the exhibition design also included a pop-up education space that sought to create space for reflection. Labels provided information and facts about coral growth, healthy reef ecosystems, threats to the reef and so on, concluding with suggestions about ways in which visitors could begin protecting our reefs both at the community level, through advocacy, and individually, through recycling efforts.

Once the backdrop for the wider conversation had been established through the exhibition itself, a larger programme of workshops, lectures and events was developed. Again, we drew upon several long-standing partnerships to help provide content, ranging from government departments, other NGOs, small marine-based businesses as well as individual celebrities who were championing the conversation locally. They included the Cayman Islands Department of the Environment, the Cayman Islands National Trust, Plastic Free Cayman, Eco Divers, Sustainable Cayman as well as individual environmentalists. Each of these groups and individuals came forward with ideas for lectures, screenings, panel discussions and even underwater workshops. The scientists provided educational facts and ensured that exhibition labelling was correct, and the divers helped us identify mysterious

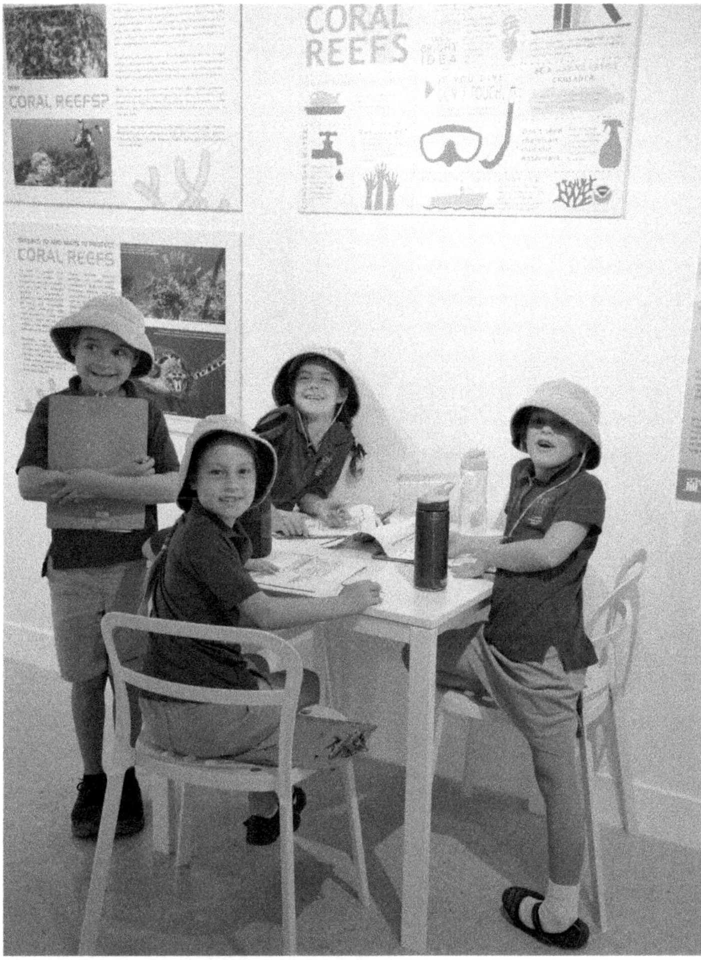

FIGURE 8.4 Students participating in a *Coral Encounters* tour exercise in the pop-up 'Exploration Station'. © NGCI, 2018.

species unknown to our museum staff, as well as where these creatures were located in Caymanian waters.

All of this information fed into our 'Minds Inspired' schools programme, which sought to maximise the context of the exhibition and optimise the opportunity for young visitors to engage in the topic beyond the museum field trip. Schools started their visit to the exhibition with a guided tour that introduced the science behind coral reefs, including a scavenger hunt worksheet that encouraged students to find certain corals based on their unique shape, pattern and colour. This was followed by time in the pop-up lab, which reinforced the fact that coral reefs were at risk from climate change and explained how students could each make a difference.

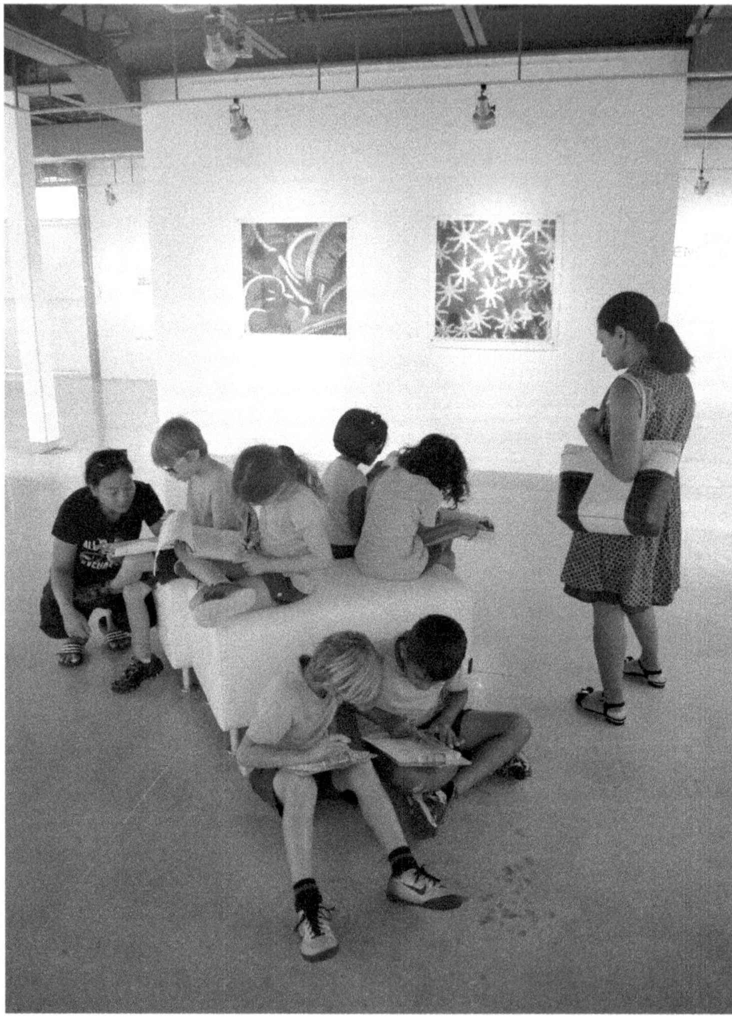

FIGURE 8.5 Students completing activity sheets during a guided tour of *Coral Encounters*. © NGCI, 2018.

Students were then invited to participate in an activity that illustrated how coral bleaching occurs. Finally, classroom resources were provided to all visiting schools that included information from the exhibition, several suggested classroom art activities and a bibliography for further research.

A second component for schools was CCMI's *Reefs Go Live* project, which was piloted during the exhibition. *Reefs Go Live* uses Virtual Live Experiences (VLEs) to connect students and the public to real-time coral reef activity. Students were able to watch CCMI scientists deliver live lessons from the underwater

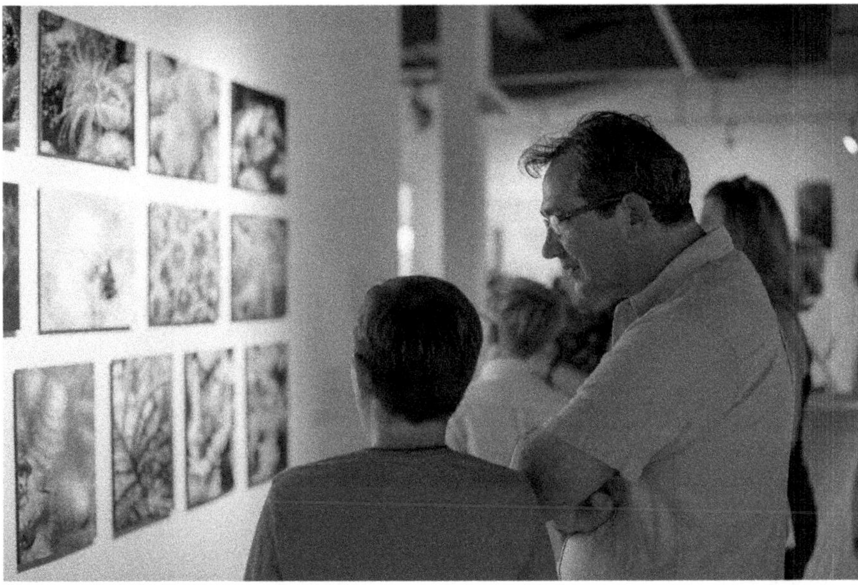

FIGURE 8.6 A father and son discuss *Coral Encounters* at a National Gallery Family Fun Day. Photo by Carol Lee, © NGCI 2018.

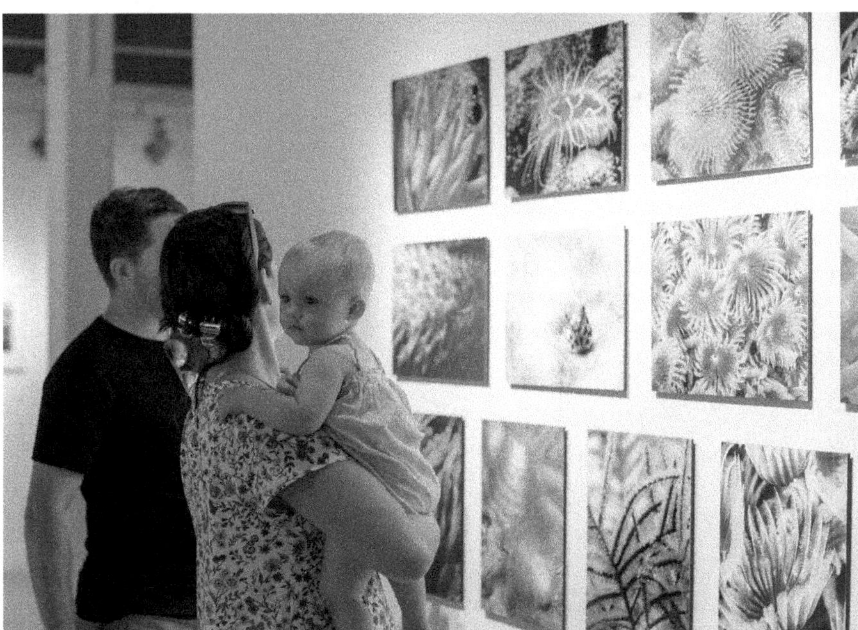

FIGURE 8.7 Visitors enjoy the *Coral Encounters* opening reception. © NGCI, 2018.

environment and to ask questions in real time as the underwater camera moved over the living reef. This groundbreaking work opened up the underwater world to students and helped to reinforce the exhibition narrative that they had encountered at the museum.

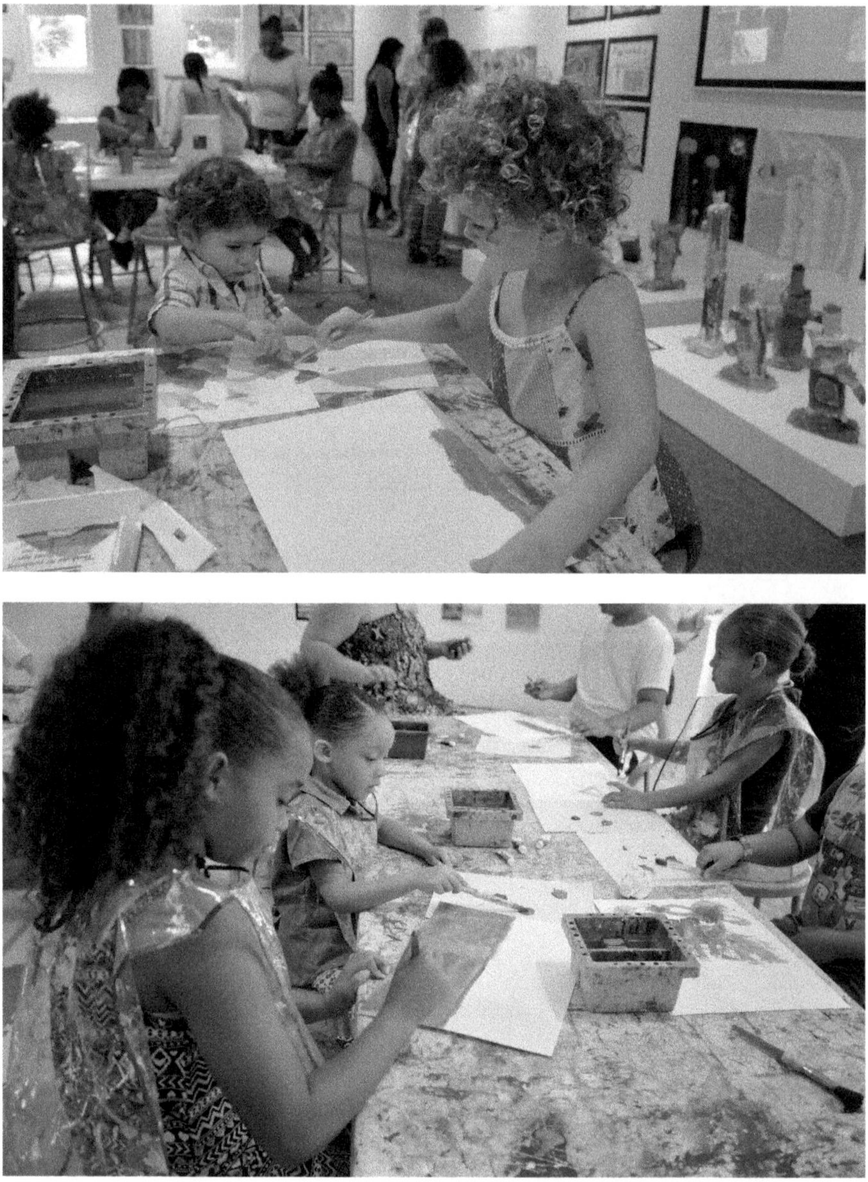

FIGURE 8.8a&b Young students interpreting their *Coral Encounters* experience onsite and back in the classroom. © NGCI, 2018.

The wider community programme also featured screenings of *Reefs Go Live* in addition to a wide range of events around coral reef health and related topics. These included 'Recycled Art Workshops', 'Under the Sea Family Fun Days', lectures on 'Coral Reefs 101', 'Photographing & Protecting Coral Reefs', 'The Value of Restoring Cayman's Reef Fish', 'The Repopulation of the Nassau Grouper', 'Coral Reef Farms' and others. In addition, we held special teacher appreciation evenings with resources on how to teach students about the importance of coral reef health, and panel discussions around wider sustainable living approaches and single-use plastics. Importantly, a 'Call to Action' was included during each event. This featured ways in which visitors could connect directly with local organisations working in this area and provided a pledge that visitors could make to protect our coral reefs.

Conclusion

With a record number of schools engaging with the exhibition and strong uptake of the related community programming, *Coral Encounters* was considered by the core partners to be an overwhelming success. As Semmel (2019, p. xix) notes in her introduction to *Partnership Power*, 'today's challenges are complex and multidimensional requiring different organizations to come together [...] in order to

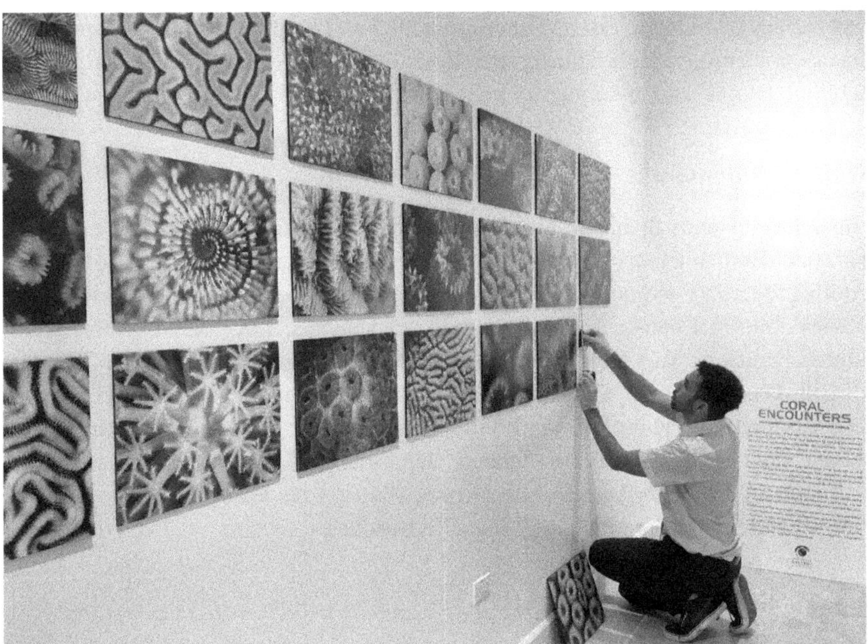

FIGURE 8.9 Little Cayman Museum *Coral Encounters* Travelling Exhibition Installation. © NGCI, 2019.

"move the needle" on effective change'. Despite our limited budget, the project made a big impact locally due to the collaborative exhibition model with multidisciplinary partners (museum staff, designers, artists, marine biologists and educators), and the open sharing of knowledge and resources. This enabled our message to have a greater impact and attract new audiences. Moreover, due to the wide scope of the community programming, the project drew extensive local press coverage, in addition to international exposure through our strategic alignment with IYOR. This in turn has helped draw attention to local advocacy efforts around reef health, recycling initiatives and single-use plastics and has also contributed in part to the growth of youth climate protests that we are currently experiencing in the Cayman Islands. The project has subsequently travelled to Grand Cayman's Sister Islands, with plans to develop the exhibition for regional travel. Talks are also underway to establish a formal artist residency programme focused on climate change and ocean health, which will be co-hosted by NGCI and CCMI.

Coral Encounters illustrates that, given our unique position as a trusted information source (Cameron, Hodge and Salazar, 2013, p. 9), even the smallest museums can become 'ideally placed to foster individual and community participation in the quest for greater awareness and workable solutions to our global problems' (Janes, cited in Cameron, Hodge and Salazar, 2013, p. 380). By joining with partners outside our sector and through the creative and innovative use of available resources—whether photographs on foam board or big-budget installations—we can create engaging experiences that inspire our audiences to action and help them find solutions to the issues our communities are facing. Perhaps collectively, we might eventually inspire 'a billion people' and, by doing so, help enact urgent and effective change.

Acknowledgements

The author extends thanks to all of the participating photographers and exhibition partners: Emma Camp, Tracy Candish, Eleanor Chalfen, Cathy Church, Tim Codling, Martin Colognoli, Julie Corsetti, Ellen Cuylaerts, Chase Darnell, Sharon Davies, Darvin Ebanks, Laura L. Fall, Carl Hawkes, Lucy Janes, Aubri Keith, Lauren Knuckey, Yuri Korchynski, Jim MacCallum, Elena McDonough, Lindsay McGill, Dusty Norman, Courtney Platt, Elizabeth Riley, Peggy Sinclaire, Brooke Sipple, Monte Lee Thornton, Mark Tilley, Jason Washington, Dale Williams, Bryan Winter, the CCMI, the Cayman Islands Department of Environment, the National Trust for the Cayman Islands, Eco Divers, Plastic Free Cayman, Cathy Church's Photo Centre, Courtney Platt and Stacie Sybersma.

Notes

1 This chapter was originally developed for the opening plenary session at the 2018 Museums Association of the Caribbean Conference, Barbados, held in conjunction with the EU-LAC Museums project and the University of the West Indies. Having recently returned from Washington, DC, where I had been privileged to attend a Global

Leadership Forum for museums hosted by the Smithsonian Conservation Commons' Earth Optimism initiative in conjunction with the Earth Day Network. Those gathered spent the day brainstorming ideas and plans to activate programming around Earth Day 2020 and some of us left believing in the impossible. Perhaps it is our institutions—big and small—that are collectively best poised to make a difference in this very urgent conversation. For more information, see Smithsonian Conservation Commons (2021) and Earthday.org (2020), respectively.

References

American Alliance of Museums (2021). *Museum facts and data.* [Online]. Available at: https://www.aam-us.org/programs/about-museums/museum-facts-data/(Accessed 27 May 2021).

Böhm, G. and Pfister, H.-R. (2000). Action tendencies and characteristics of environmental risks. *Acta Psychologica*, 104(3), pp. 317–37.

Braasch, G. (2013). 'Climate change: is seeing believing?' *Bulletin of the Atomic Scientists*, 69(6), pp. 33–41.

Bullot, N.J. (2014). 'The functions of environmental art', *Leonardo*, 47(5), pp. 511–12.

Cameron, F., Hodge, B. and Salazar, J.F. (2013). 'Representing climate change in museum space and places', *WIREs Climate Change*, 4(1), pp. 9–21.

Cozen, B. (2013). 'Mobilizing artists: green patriot posters, visual metaphors, and climate change activism', *Environmental Communications*, 7(2), pp. 297–314.

Earthday.org (2020). *Earth Day 2020.* [Online]. Available at: https://www.earthday.org/earth-day-2020/ (Accessed 27 May 2021).

Gough, P., Berigny, C. and Dunn, K. (2016). 'Climate change education through art and science collaborations', in Wilson, L. and Stevenson, C. (eds.) *Promoting climate change awareness through environmental education.* Hershey: Information Science Reference, pp. 16–36.

Guy, S., Henshaw, V. and Heidrich, O. (2015). 'Climate change, adaptation and eco-art in Singapore', *Journal of Environmental Planning and Management*, 58(1), pp. 39–54.

Hebda, R.J. (2007). 'Museums, climate change and sustainability', *Museum Management and Curatorship*, 22(4), pp. 329–36.

Hulme, M. (2010). *Why we disagree about climate change*. Cambridge: Cambridge University Press.

Janes, R. and Sandell, R. (2018). *Museum activism*. Abingdon: Routledge.

Klöckner, Christian, Beisenkamp, A. and Hallmann, S. (2009). 'Emotional reaction of children aged 9–14 years to climate change and how emotions are linked to action', *IOP Conference Series Earth and Environmental Science*, 6, p. 572002

Lyons, S. and Bosworth, K. (2018). 'Museums in the climate emergency', in James, R.R. and Sandell, R. (eds.) *Museum activism*. Abingdon: Routledge, pp. 174–185.

Nurse, K. and Edwards, D. (2019). 'Climate impact and action: the travel and tourism industry in the Caribbean and Small Islands Developing States,' *Sustainable Futures Policy Brief*, 3, pp. 1–8.

Roosen, L.J., Klöckner, C.A. and Swim, J.K. (2018). 'Visual art as a way to communicate climate change: a psychological perspective on climate change-related art', *World Art*, 8(1), pp. 85–110.

Sandahl, J. (2019). 'Addressing societal responsibilities through core museum functions and methods: the museum definition, prospects and potentials', *Museum International*, 71(2), pp. iv–v.

Semmel, M. (2019). *Partnership power: essential museum strategies for today's networked world*. Lanham: Rowman & Littlefield.

Smithsonian Conservation Commons (2021). *About earth optimism*. [Online]. Available at: https://earthoptimism.si.edu/about/ (Accessed 27 May 2021).

Thrash, T.M. and Elliot, A.J. (2003). 'Inspiration as a psychological construct', *Journal of Personality and Social Psychology*, 84(4), pp. 871–89.

United Nations Climate Change (2017). *Creating an understanding for climate science through art*. [Online]. Available at: https://unfccc.int/news/creating-an-understanding-for-climate-science-through-art (Accessed 27 May 2021).

United Nations Office of the High Representative for the Least Developed Countries, Landlocked Developing Countries and Small Island Developing States (2015). *Small Islands Developing States in numbers: climate change edition*. New York: United Nations.

Urquhart, N. (2017). *Coral encounters exhibition brief*. National Gallery of the Cayman Islands.

Weber, E.U. (2006). 'Experience-based and description-based perceptions of long-term risk: why global warming does not scare us (yet)', *Climatic Change*, 77(1–2), pp. 103–20.

PART II

Connecting regions

Communities and museums co-curating heritage and memory

9

THE CASE FOR A RHIZOMATIC RESEARCH APPROACH IN CARIBBEAN MUSEOLOGY

Natalie McGuire

Introduction

Community inclusiveness in museum theory and practice can take many forms, from considering different physical and learning abilities to ensuring equal representation and openness across racial and socio-economic barriers, and having publics lead cultural projects in museum spaces. Yet, the underlying connection between these outputs is arguably recognition of the value of putting community voices and participation at the forefront of museum design and practice. This is particularly significant in a Caribbean context. In *Caribbean Discourse* (1989), Édouard Glissant disrupts Western hierarchies of national unity in identity through a series of interventions into Westernised philosophy. He challenges the confinements of postcolonial meaning-making by suggesting that identities in the Caribbean are not fixed to a singular essence, unity or place but rather involve multiplicities and continual cultural encounters (Ostrander, 2015).

A concept with which to navigate this comes in the form of the *rhizome*. For his concept of a rhizomatic approach to identity, Glissant draws primarily on Deleuze and Guattari's (1987) work in *A Thousand Plateaus: Capitalism and Schizophrenia*. They construct frameworks of meaning-making that are analogous to root structures of plants, suggesting that there has been a conditioned perception that knowledge is passed from the root upwards in a linear fashion (Deleuze and Guattari, 1987, p. 5). They then argue that meaning-making is in fact *rhizomatic*, in that meaning is created through multiple points of communication, each with a multitude of multiplicities; ways of knowing are therefore produced through a process of assembling, moving away from the hierarchical notion of interpretation as

DOI: 10.4324/9781003288138-12

a linear act (p. 11). Glissant (1989, p. 11) summarises his use of Deleuze and Guattari's *rhizome* as follows:

> The root is unique, a stock taking all upon itself and killing all around it. In opposition to this [Deleuze and Guattari] propose the rhizome, an enmeshed root system, a network spreading either in the ground or in the air, with no predatory rootstock, taking over permanently. The notion of the rhizome maintains, therefore, the idea of rootedness but challenges that of a totalitarian root.

Glissant's (1997, p. 142) writings in *Caribbean Discourse* and *Poetics of Relation* posit creolisation as a rhizomatic process, as relational, and involving multiple and constant processes of transformation. Museology is also arguably relational, given that it is essentially an interdisciplinary field that studies the relationships among people, objects and scenarios (Rússio, 2010). Waldisa Rússio, writing in 1981, described the museum as 'comprised of [people] and by life, which allows the museological process and its method to be substantially interdisciplinary, since the studies of man, nature and life are related to different branches of knowledge' (Soares, Valentino and Limoeiro, 2019, p. 103).

This chapter assembles inquiries into community inclusiveness within museological practices in the Caribbean. It does so by assessing current museology within the Anglophone Caribbean countries of Barbados and Trinidad and Tobago, terminology used in practice, and research outputs in museum spaces, as well as by investigating a new tool for museological research based on Glissant's concept of the rhizome.

Towards community-focused museology

What is understood as museology, and how has that been interpreted in the Caribbean? In 1989, Vergo (p. 1) began his introduction to *The New Museology* with the line: 'What is Museology?' At the time, academics and practitioners had already been struggling for decades to agree on a concise and universal definition. Vergo presented a set of critical essays charting a 'new' direction for what museology could be, stressing that his inquiry sought a definition not only for a theoretical or philosophical function, but also to determine *who* should be concerned with matters of museums. Would museology just be a concern for those working within a physical museum site? Or is the impact of a museum such that it intertwines with other aspects of the social, academic, economic and environmental fabric of society?

This question would be addressed again almost three decades later in the most recent International Committee for Museology (ICOFOM) monograph, *A History of Museology: Key Authors of Museological Theory*. Like Vergo, editor Bruno Brulon Soares (2019) began his introduction by questioning the current state of museology and went on to classify three distinct types of museology. These are *normative museology*, which was marked by a '"dictionnarization" of museum knowledge' and aimed to shape the field in its early years, when it was solely connected to museum practice; *theoretical museology*, which refers to the establishment of ICOFOM as a committee to pursue the theory of museology with a view to aligning it with the social sciences; and *reflexive museology*, which marked the shift from museology as a science to museology as an interdisciplinary field (Soares, 2019, p. 26). In addition, Soares (2019, p. 33) recognised that museology is being developed through different schools of thought in different geographical regions: for example, Latin America, France and Germany are more inclined to consider museology a science, while the United Kingdom and North America tend to explore the humanities aspect of museology, which has led to the emergence of museum studies as a discipline.

Although the field of Anglophone Caribbean museology has been growing, modes of representation within Caribbean museums have often fixated on either retaining colonial stories or constructing postcolonial national identities (Cummins, 2004; McFarlane, 2012; Farmer, 2013). Museums as modern institutions have a policy history tied to colonialism. They emerged in late 18th- and early 19th-century Europe as a means of showcasing objects that had been plundered—an analogy, perhaps, with the nations that were themselves plundered (Bennett, 1995). With movements of independence sweeping across Anglophone Caribbean countries in the mid to late 20th century, the postcolonial museum in these spaces developed, in part, to become an institution used to construct national identity, which strove to integrate the changing cultural context while serving communities in order to remain relevant (Bennett, 1995; Farmer, 2013). But how have narratives around community agency in museology been reflected within Caribbean museum research?

The increased interest in community museum research, as well as significant events such as the 1972 Round Table of Santiago de Chile sparked a shift in museum praxis, where the emphasis on internal aspects, such as collection size, diminished and the consideration of external aspects, such as how they function to serve the communities they represent, increased (Beans, 1994). This prompted museums not only to demonstrate community inclusiveness through constructions of an arguably still colonial sense of national or cultural identity but also to incorporate the voices of both indigenous and migrant communities in interpretative materials within national museums (Witcomb, 2003).

The contemporary museum as an institution, however, can still be seen as a space with a significant degree of social power, having responsibilities as the gateway to the cultures of its communities (Sandell, 2002). Within new museology, one aspect of the perceived relationship between museum staff and communities is the emphasis on curators who have the authority to define people and space:

> Their role is a more demanding one which involves responsibility for actually defining the community being represented [...] Such an understanding of the

> curatorial role does not assume that there is a community 'out there' that the museum can represent.
>
> *(Witcomb, 2003, pp. 153–4)*

Here, the knowledge hierarchy is disguised as 'responsibility', with the suggestion that the curator can be the only figure to define the community, as opposed to facilitating communities' self-determination. This argument also avoids addressing the issue that the institutions creating guidelines for representing these communities often perpetuate a colonial ideology that generates a sense of universal internationalism without accountability of provenance, thus masking the continued imperialistic agenda of retaining the cultural property of 'others' (Busse, 2009). As Szekeres (2002) points out, striving to create 'belonging' actually risks contestation around ideas of cultural ownership.

Definitions of community

How, then, are definitions of the term 'community' articulated within contemporary museology? In *Ecomuseums*, Davis (1999) states that these tend to include: shared geographical location, shared religious practices, a common political system, shared culture (tangible and intangible heritage) and a notion of community identity. Watson (2007, pp. 3–4) recognises the multiplicities and self-determination involved in the museum space around the concept of community but nonetheless prefers to apply neat categorisations of community museum approaches to public engagement. Classifying the experiences of individuals from the outside in this way arguably reduces people's agency, as the individuals are not always aware of, and may not agree with, the categorisations of community used. In a survey text of museum–community practice by Golding and Modest (2013), an attempt to further stratify the concept of community is evident, whether it is described as a 'community of communities' (Golding, cited in Golding and Modest, 2013, p. 20), a 'nation not a neighbourhood' (Gable, cited in Golding and Modest, 2013, p. 41) or 'porous, multifaceted, ever-shifting loosely connected groups of people' (Onciul, cited in Golding and Modest, 2013, p. 81).

When extending the definition of the community museum, Morales Lersch and Camarena Ocampo (2018, p. 224) draw on the Latin American context to argue that:

> The key component of the concept in this context is the decisive action of the community, developing a collective initiative to strengthen its identity and capacity for self-determination. Community actors, expressing themselves through a variety of consensus-building processes, create the community museum. In this sense the community is not one of many local actors – it is the determinant force, the protagonist and the creative motor of the museum.

In this understanding, the notion of community identity is directly correlated to indigenous identity. Within the Caribbean, however, identity is multifaceted, and with histories of migration (whether forced or voluntary) into and out of the region, a more relational approach to defining community and the community museum is better suited. Brown (2018, p. 115) has spoken specifically of the challenges for the regions involved in the EU-LAC Museums Project (European Union, Latin America and the Caribbean) when seeking an agreed-upon definition for use within the project:

> As is the case for the ecomuseum, defining the community museum in this bi-regional forum is proving a contentious task, mostly because of stark differences between our lived realities. In grassroots community museum contexts, it is more a matter of principles and the role of the community in decision-making within the museum than a concrete definition, just as for ecomuseums it is more a question of a *process*, rather than defining a fixed and transportable model.

When Brown talks about 'grassroots community museum contexts', she may be alluding to the persistent issue that museological terms are still often defined from the top down, where people are described by outsiders, rather than being offered the opportunity to describe themselves. As discussed in the scholarship arising out of the EU-LAC Museums Project, this may not be a relevant approach to community-run museum spaces.

The case for a rhizomatic research approach

Recently, the recognition that new museology still carries a hegemonic metanarrative, despite claiming community inclusivity, has shifted the theoretical discourse into examining relational turns. Dewdney, Dibosa and Walsh (2013) developed the term 'post-critical museology' to describe an alternative framework that addresses the concern about metanarratives or agendas. Post-critical museology couples researchers with practitioners, bridging the theoretical gap between traditional museology and museum studies. It is collaborative, transdisciplinary and reflexive and works further to decentralise authority and create multiple avenues of meaning-making. As such, it is a framework that could lay foundations for a *rhizomatic* research approach within museology.

Post-critical museology takes a democratic approach to collaboration with communities in the research and development of museum projects. The notion of knowledge-*sharing* as opposed to knowledge-*collecting* is of interest in conducting this research in the Anglophone Caribbean to counter the top-down approach usually adopted. In the Caribbean particularly, 'education' isn't always considered an inclusive tool, as it reinforces hegemonic ideologies, and has historically erased Afro-Caribbean experiences (see Freeman, 2005). In her 2004 article, Hickling-Hudson (2004, p. 294) called for the Caribbean to decolonise education and move towards 'knowledge-societies' facilitated by 'activist educators' who contribute 'to the breaking down of neo-colonial barriers which kept the Caribbean societies insular and trapped in the language and isolationist education traditions of the former colonizing powers'. This aligns with Freire's (1973, pp. 66–8) concept of 'critical pedagogy' as a learning technique to dismantle colonial authority and empower people through ownership of their own histories. My research explores how humanising research in Caribbean museology, by challenging the way data is collected and interpreted, can allow for more inclusion and agency in community participation.

According to Dewdney, Dibosa and Walsh (2013, p. 226), post-critical museology 'locates itself in the everyday and in spaces "outside", "between", and "beyond" those of the foundational boundaries of knowledge disciplines'. This suggests that museology should exist in museum practice and work to democratise the power structures inherent in curatorial approaches. George Abungu reflected on a similar ethos during the plenary session at the ICOM General Conference, held in Kyoto in 2019:

> It brings us back also to [...] the role of the curator, and I think especially in the western world. The curator knows it all, the curator interprets it, the curator [...] prepares it and has a big day [...] On the other side, we are saying the museum should be a hub, it should be a place of dialogue. Which means the communities and everybody of concern, everybody should actually be able to play a role.
>
> *(ICOM, 2019)*

With these issues in mind, I now turn to how museological scholars have responded to community-focused approaches adopted in Caribbean museums.

Current mappings of Caribbean museology

There has not been much museological theory produced from the Anglophone Caribbean, as quite often the museum spaces themselves have been implicated in the colonial project as instruments for racist, oppressive power structures and nation building. This has meant that current Eurocentric models of museology are often imposed onto museums in the region, rather than the region generating its own models out of localised practice. Thus, there is a profound absence of Caribbean voices influencing the development of international museological policy. This status quo has also resulted in several centuries of people of African descent in the Anglophone Caribbean being denied opportunities to represent their own stories in local museums, silencing shared heritages that span a far longer timeframe than the 400 years with which museums often begin Caribbean social histories and also extend far beyond the geography of the Caribbean basin.

For the most part, museums have been consciously tackling what Vergo (1989, pp. 2–3) describes as the "subtext of voice" that derives from conscious or subconscious political, social and economic agendas and biases of the museum board, which, via directors, curators, scholars, designers and sponsors, goes out to society and the education systems that influence these cultural practitioners.

McFarlane (2012, p. 14) comments on the Eurocentric bias in museum exhibition display in postcolonial Jamaica that:

> One can only surmise that, either visitors themselves have been the subjects of colonial educational paradigms which have been accepted as truth; visitors lack critical analytic tools with which to question historical representation; or, visitors to museums see no need to question the veracity of exhibitions.

Similarly, Cummins (2013, p. 3) points to the need to address Caribbean-based practices within new museology:

> Museums and collections that had their origins in the colonialist era [...] existed and still exist in a number of Caribbean island nation states, as they do elsewhere in the world. Coupled with this tendency in the Caribbean context is the legacy of slavery which, even now, is in danger of being submerged and obscured in museums and heritage sites by narratives that occlude or diminish the histories of the enslaved.

Both Modest (2012) and Cummins (2013) draw attention to imperial collecting practices that focused more on nature than culture and ultimately informed the growth of museums in that manner. Cummins (2013, p. 33), for example, notes that 'for colonial institutions in the Caribbean, natural history continued to be equated with national history in the first half of the twentieth century'. Favouring the collection of objects related to natural landscapes can be interpreted as active oppression of contested histories in order to uphold colonial identities tied to the landscape. Therefore, museums in the Anglophone Caribbean, in their display of 'natural history as national history', have perpetuated a version of Caribbean history that has submerged and erased human stories, particularly narratives of enslavement, African heritages prior to enslavement and modern indigenous cultures. This aligns with Haitian writer Michel-Rolph Trouillot's (1995, p. 22) notion that there are silences of history imposed by Eurocentrism.

Within Anglophone Caribbean historical discourse, there have been epochal transitions as the society shifted from being based on ideologies of enslavement to those of the postcolonial moment; this was particularly marked in the early period of independence from Britain in the 1960s. The social, economic and cultural changes brought about by the independence movements from the mid-1950s onwards affected both the museum as institution and the establishment and sustainability of grassroots initiatives. Hume and Kamugisha (2013, p. xxiii) suggest that, alongside 'discourse on the dynamic process of cultural change and identity formation, emerged a related and somewhat localised preoccupation with articulating a Caribbean aesthetic, and by extension an independent, non-alienated subjectivity'. Tancons (2012, p. 39) proposes a shared Caribbean cultural practice or aesthetic—Carnival—under the institutional framework of the museum, suggesting that it could be pivotal in shifting Caribbean practitioners from an 'exhibitionary complex' that hinders curatorial potential. Farmer (2013, pp. 172–3) explains that, post-independence,

> existing museums in the region were co-opted by postcolonial governments to become agents of identity creation. This saw existing institutions, such as the Barbados Museum and Historical Society (BMHS), change focus through government intervention to facilitate the creation of a new national identity.

However, Freire (1974, pp. 3–18) indicates that this type of project, which essentially asserts a linear progression from the colonial to the postcolonial moment, is marked by contradictions between a perpetuation of past ideologies and the demands of a choice-based future. Museums as spaces have arguably encouraged these epochal reflections through displays that rely heavily on chronological narratives and timelines of human history that clearly separate the 'then' from the 'now', reinforcing centres of colonial power and creating peripheries of marginalised communities. This approach has been described as a 'contact zone' in which museums are 'an asymmetric space where the periphery comes to gain some small, momentary and strategic advantage, but where the centre ultimately gains' (Boast, 2011, p. 66). Therefore, museums have, for the most part, actively contributed to the absence of critical discourse on race and colonialism in the Anglophone Caribbean and perpetuated the narrative of the colonial ideological centre. Recently, there have been calls for museums to dismantle their ostensible neutrality and acknowledge their role in shaping social ideologies and discussions on contested histories.

Museology is not separate from these discussions, and the dominance of museum theory from European countries needs to be addressed. How, then, can museology in the Anglophone Caribbean be developed to allow for deeper analyses of historical and contemporary events, particularly of the Afro-Caribbean experience? And how can this analysis provide the surrounding communities who encounter these spaces with opportunities for engagement?

The Caribbean is in some ways still working through structural racism in institutions, as we see with the retention of colonial politics in the representations and display of culture in the region. McFarlane (2012, p. 40) proposes that one way to counter these inherent biases in meaning-making within the museum display is to apply Critical Race Theory to exhibition design to 'facilitate examination of the veracity of underlying historical premises of exhibition narratives; explore whose interests is served in utilising Eurocentric narratives; and open the door to the work of Afrocentric historians who present oppositional perspectives'.

The forging of connections across the Caribbean region has also had an impact on the way that museology is articulated. Jean-Philippe Maréchal (1998) called for an 'island museology': that is, a specific framework to link Caribbean museum practice and help foster awareness of cultural heritage protection. The notion of connecting the archipelago is significant within Caribbean museology, and from the 1980s, there have been important developments in sharing scholarship. These have mainly taken the form of ICOM-affiliated regional meetings, such as those held by the Museums Association of the Caribbean (established 1989), ICOFOM LAC (established 1989), and the International Association of Art Critics—Southern Caribbean (established 1997). Regional exchange has been a way to share museological developments as well as to strengthen contributions at international meetings. This has extended into the diaspora, particularly through the Black Diaspora Visual Arts programme (established in 2007). Recently, regional exchange has been brought online through the EU-LAC Museums Project and the establishment of a Virtual Museum of Caribbean Migration and Memory, which includes capacity-building through the digitisation of artefacts, firmly situating Caribbean intellectual practice in the development of digital museology (Brown, 2018; Cassidy et al., 2018).

Support for, and interest in, museological developments within the Anglophone Caribbean is therefore evident. However, there is a risk that the museological narratives of Caribbean museums might be co-opted for international projects. There has been a trend of external museum study and museological surveys done on the region from international tertiary institutions who have Museum Studies or Museology departments or programmes, which has framed, in part, the Caribbean within international museum theory development. Quite often, these take the form of European researchers or institutions creating maps or databases of museums in the Caribbean, 'mining' intellect from local practitioners with very little benefit to those in the Caribbean itself. For agency of these sites to exist within these projects, elements of co-curation and demonstrations of sustainable relationships with local partners should arguably be evident.

My research offers a museological study that, while in conversation with larger international bodies, is grounded in a Caribbean-based tertiary programme and created out of local voices, using a multi-vocal approach. It draws on the work of Glissant to explore the possibilities of museology as rhizomatic. In doing so, this work eschews claims to authority or epistemological superiority over work done within museology but hopes to broaden the possibilities of how museum work in the Anglophone Caribbean can be theorised beyond already existing Eurocentric models or frameworks.

A rhizomatic approach to museology

> Rhizomatic thought is the principle behind what I call the Poetics of Relation, in which each and every identity is extended through a relationship with the 'other'.
>
> *(Glissant, 1997, p. 11)*

There is a case in post-critical museology for employing transformative action research in implementing qualitative methodologies to shift paradigms and approaches to practice. How can a rhizomatic research tool be used within a transformative action research approach to put community voices at the forefront of practice? Britton (1999, p. 18) describes how the work of relation (*la relation*) by Glissant has enabled resistance through language for communities living in creolised societies, and how the 'rhizome becomes a figure for identity in creolized culture'. Yet, the questions raised by creolisation processes are complex beyond Glissant's reflection. Indeed, Kamugisha (2019, p. 101) argues that theories of creolisation as Caribbean cultural identity have often negated its complicated role in ongoing racism: 'haunting creoleness still is its association with antiblack racism, elite domination, and its inability to provide a true—and as important—*continuous*—reassessment of the place of "blackness"'.[1]

The basis for using a rhizomatic research approach to museology is the recognition that knowledge production is not apolitical. Actively developing a methodology to decentralise authority within the creation and archiving of museum theory therefore presents alternate possibilities for knowledge-sharing within the case studies of the Anglophone Caribbean. Academics in the (broader) Caribbean museological field have recognised the importance of Glissant's influence on museums (Golding, 2009; Ostrander, 2015). Golding (2009, p. 82) applies Glissant's concept of relations in her interrogation of poetics in museology because 'Relation is understood as a space where the poetic and the political are intertwined'. As an example, in 2015, the Perez Art Museum in Miami held an exhibition, including Caribbean artists, entitled *Poetics of Relation*. The museum stated that Glissant's theories on relation 'can be used to understand the investigations of each of the artists exhibited, how through their works they address their own centered identities as responsive to various sites simultaneously, across the diverse homelands that form their personal root system' (Ostrander, 2015, p. 25). This statement attests at once to the possibilities of inclusivity and the gaps in theoretical discourse. It also suggests the risk that museology could still be a tool for colonial retentions in the Anglophone Caribbean, when considering which knowledge system is being drawn on and who created those epistemologies. All of these are key elements in understanding a rhizomatic approach to museum work, both in their application and in the gaps they highlight in debates around the Anglophone Caribbean social context, to which they claim to contribute. In my research, I take this one step further by investigating the notion of rhizomes as an accessible tool for community inclusion in museum practice.

When considering a practice-based approach to research or the creation of a research framework, we often encounter Glissant's rhizomes in pedagogical theory. For instance, in identifying a 'heterogeneous (and open)' assemblage system for reading research data, Yu and Lee (2008, p. 255) 'call it a "rhizome" or rhizomatic system'. Clarke and Parsons (2013, p. 36) suggest a 'research as agency' model by becoming 'rhizome researchers' to dismantle binaries in educational research. Masny (2014, p. 351) has proposed a disruption of ethnography in educational research by 'rhizoanalysis', which is described as 'not a method; in other words, there is no one way to do rhizoanalysis'. These previously structured research frameworks that have utilised a rhizomatic lens provide insight into how the rhizome could be translated to the conducting of museological research.

With these frameworks in mind, I argue that a rhizomatic approach to museological research best aligns with post-critical museological theory and can be outlined through the following principles:

1 *Multi-vocality in museum research and practice through assemblage.* Within a rhizomatic approach, multi-vocality functions to distribute authority within meaning-making. Knowledge exchange among museologist, researcher, curator, publics and communities occurs in a manner that is lateral and relational, rather than hierarchical and fixed. Contributions to museum projects from publics and communities are visibly integrated as assemblages, with little or no interpretations by curators or researchers.
2 *Research methodology as a relational ecology rather than a linear framework.* A rhizomatic approach suggests that museological research be relational and transformative, rather than aiming to carve out linear frameworks for museum research and practice. It facilitates micro-level case study research over large-scale survey projects, interrogates definitions and demonstrates sustainable relationship-building within museum theory rather than data collection.
3 *Accessible research outputs.* A rhizomatic approach articulates academic writing within museology through accessible, ongoing projects, contributing to ecologies of museum theory intersecting practice that are continually relational and participatory. Ideally, these are community-led and reflexive. Such practices include the creation and continual revision of toolkits and aids for museum practitioners and communities as active products of research projects, which are shared across open access platforms.

Glissant (1997) speaks of the Caribbean experience as a continual process of transformative creolisation, a multitude of relationships extending through intersecting identities. A consideration of the notion of *Caribbeanness* suggests a case for flexible tools in qualitative research analysis; a rhizomatic research approach in museology is just such a tool and could facilitate further understanding of the Caribbean within this field.

The research project: methodology

The principles of a rhizomatic approach in museology were implemented in my ongoing doctoral project in the PhD Cultural Studies Programme at the University of the West Indies Cave Hill Campus (2016–). The central focus of this research is on community inclusiveness within museological practice in Barbados and Trinidad and Tobago, and the applicability of rhizomatic research principles.

Sites of research included postcolonial institutions as well as artist-led initiatives and community-run cultural spaces. This range of subjects was chosen to examine the notion of a relational cultural ecology not only within each country but also in the networks that exist across them. Moreover, this research investigates these connections beyond the content of public-facing exhibitions or surveying activities, asking further how the museum space creates connections in its relationship with the communities who encounter these representations.

Although there are limitations to examining only two Caribbean countries, the multifaceted nature of Caribbean identities enables this process to align with the rhizomatic research principle of preferring micro case study research to large surveys. Given the methodologies used in this study to gather in-depth, individual vignettes of data, any attempt to span the entire Caribbean would risk subjecting the territories to generalisations. As outlined in the rhizomatic approach, this research did not aim to generate definitive conclusions in any way, even within the Anglophone Caribbean context. Rather, it hoped to act as a springboard for discussions about agency in these spaces, both within other Anglophone Caribbean countries and in future research in other linguistic territories.

In the following subsections, I address each principle used in this rhizomatic approach through the knowledge-sharing sessions, community-based projects and toolkits.

Participation: knowledge-sharing sessions

In a rhizomatic approach, the notion of an interview was shifted to a knowledge-sharing session, placing the researcher on the same level as the participant. Participants in the knowledge-sharing sessions generally comprised those that use the museum spaces detailed in the case studies, both staff and visitors. The research project would be discussed with them usually in an informal setting, with a follow up interview being organised at the expressed interest of participants to take part. The interviews were all video recorded and had three base questions posed to participants:

1 Please introduce yourself and your practice/interests.
2 Is there a cultural space in Barbados/Trinidad that you feel connected to, and if so can you expand on the nature of that connection?
3 What does the term community mean to you?

These questions were selected as examples of entry points into discourse regarding community voice in understanding museology in the Caribbean. An aim of this research is to explore the use of the 'discursive', which describes approaches to meaning-making in representation and culture (Hall 2003). Or, put differently, how are we formulating meanings about places and objects with the knowledge sets we bring, and what is being enforced or excluded by museum spaces? This study also sought to engender reciprocal participation, with benefits to contributors as outlined in Yu and Lee's (2008, p. 255) 'cascade' approach, which allows for the creation of 'virtual multiplicity, defined by the "many" constituting an "assemblage"' of data.

FIGURE 9.1 QR code. Video sample of responses to the term 'community' from knowledge-sharing sessions in Trinidad and Tobago, March 2018.

The sessions allowed for each participant to direct responses to aspects relevant to their experiences and interests. Under the principles of a rhizomatic research approach, the aim of the knowledge-sharing sessions was not to construct any definition of community relations in the museum space. Rather, it was both to show that the researcher valued their interpretation of terminology and to indicate that, with such a vast diversity of terminology, it is impossible for museum practitioners to create viable programming that uses these terms without first engaging with their communities and publics about them. Here, following the principle of multi-vocality, community voices were prioritised in generating understandings of the term 'community'—a word that is heavily saturated with meanings related to inclusion and agency in museums within scholarly literature. Participant responses were incorporated as assemblages into the research project through direct quotes, rather than interpretations or summaries.

There was also an attempt to reduce any authority ascribed to the researcher to unpack participants' responses to this term; instead, they were given the agency to interpret the term themselves. The invitation to participants, especially those from traditionally marginalised communities, to evaluate the meaning of terminology has a precedent in focus groups by museums such as the Immigration Museum in Melbourne, Australia (The Immigration Museum Consultancy Group 1994, pp. 2–6), which asked participants 'what does "museum" mean to you[?]' However, these archives of community voices are often filed away or shared only in internal reports. What discourses could arise by sharing these interpretations more publicly and within the museum spaces themselves? Taking into consideration the post-critical museology notion of collaborative research as being democratised, with participant and researcher as equal stakeholders, all footage from videos was shared with those interviewed, as was the copyright.

Participation: community-based projects

As part of the study, I implemented a rhizomatic approach to museology through three projects that were co-created with communities as accessible research outputs. These comprise the community-informed international video art programme

Transoceanic Visual Exchange (*TVE*), installed in 2017 and 2019; *Research Yard*, a pop-up community reading room held in Alice Yard, Trinidad and Tobago, in 2018; and *Artistic Interventions*, held at the Barbados Museum & Historical Society (BMHS), Barbados, 2018.

Transoceanic Visual Exchange

TVE aimed to negotiate the in-between space of cultural communities that exist outside of the traditional geopolitical zones of encounter and trade. The exhibition was centred on developing a survey of recent film and video works by contemporary artists (screenings, installations and expanded cinema) that were shown in multiple participating regions in every edition of the programme, with an accompanying digital exhibition space. A key aspect of *TVE* was the introduction of a community of curatorial practice to integrate community voice into its curatorial framework (McGuire, 2015). Public round tables made up of artists, curators, researchers and interested members of the local communities were held in each participating country to discuss themes and practices in current new media, film and video art works. The outcome of these discussions then formed the curatorial framework for each iteration

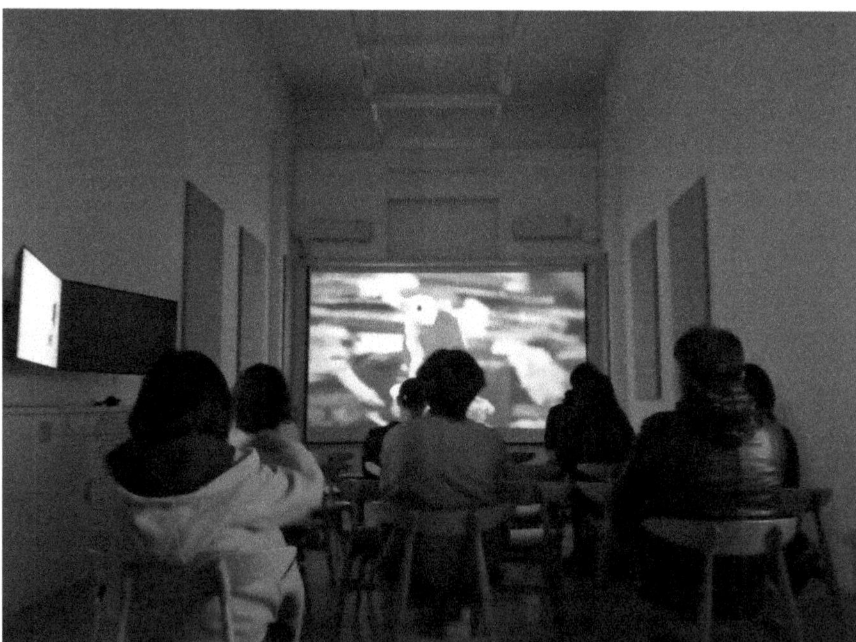

FIGURE 9.2 'For Peace' (2019) by Barbadian artist Versia Harris on display in Cache Space, Beijing, November 2019.

Image courtesy of the TVE project, China Residencies and the artist.

and the selection of works from the open call. This procedure was used in order to explore the effectiveness of a lateral approach to curatorial practice, as opposed to the traditional hierarchical approach where the curators alone decide the theme that informs the curatorial framework and invite relevant artists.[2] Within a rhizomatic research approach, each iteration of *TVE* shifted in its curatorial methodology, based on the interests of each participating space and community, and ownership of the project by these spaces was encouraged. As a result, this process saw shifts in the logo, installation approaches and media associated with the project.

Research Yard

Research Yard was a pop-up community reading room that I facilitated during my time as a doctoral researcher in residence at Alice Yard, Trinidad and Tobago (March 2018). The project was born out of informal conversations in the yard around the (lack of) access to critical texts within arts and culture in the Caribbean and internationally. This prompted a collaborative idea for an open day at Alice Yard. Cultural practitioners, artists, cultural collectives and academics were invited to share texts from their personal libraries, selected by them based on cultural topics in which they were currently interested.[3] A similar community reading room was installed at the BMHS in May 2019.

Artistic Interventions

In *Artistic Interventions*, the intersection between theory and practice is prominent. This project was developed within my role as both PhD candidate and curator at the BMHS. Members of the creative community were invited to critically engage with the BMHS's collections through a series of interventions. Six artists responded, and their works were interwoven throughout the museum's galleries, interrogating and recontextualising the historical narratives on display. These interventions were accessible in the BMHS from May to June 2018. This project avoided entrenching dichotomies between the grassroots and government, and art and history; instead, it engaged with the experience of living the practice in these entities, encouraging a museological ecosystem (see Davis, 2017, pp. 152–3).

Artistic Interventions also aligned with what Golding (2013, p. 97) describes as an 'affective museum', which 'works with poetics to assist visitors to look through that which was hidden and rendered opaque in traditional linear displays, such as the colonial histories that have disadvantaged Black people and women, but which creolised voices can bring to the fore'. For example, in the knowledge-sharing sessions at the BMHS, one of the key issues raised was the enduring perceived lack of equitable representation. Llanor Alleyne, one of the artists, stated:

> I was really affected by the Cunard gallery, [4] because it represented plantation life across the region. And that stuck with me [...] I didn't feel like the prints

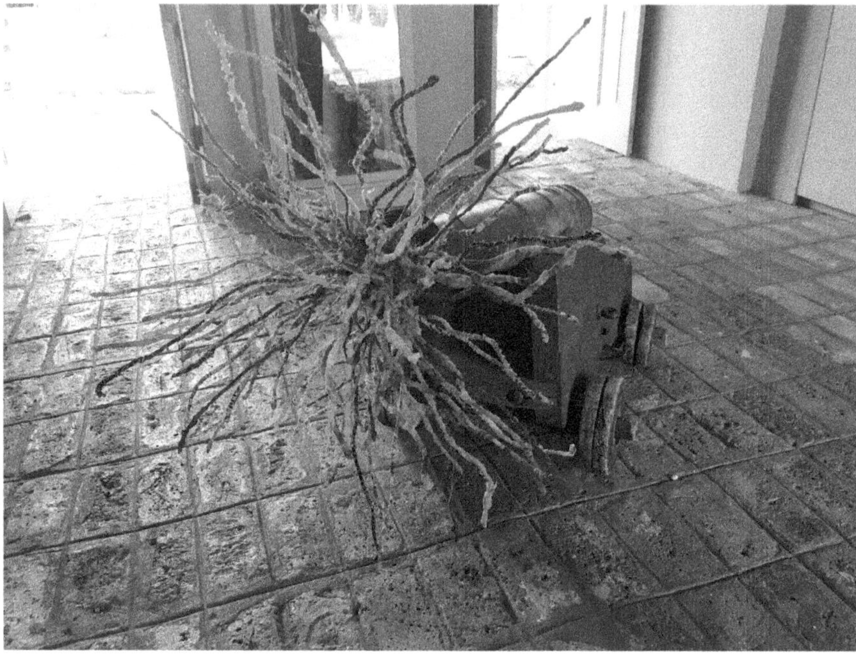

FIGURE 9.3 ‘Burst’ by Katherine Kennedy installed in the Military Gallery of the Barbados Museum & Historical Society (BMHS), May 2018.

Image courtesy of the BMHS and the artist.

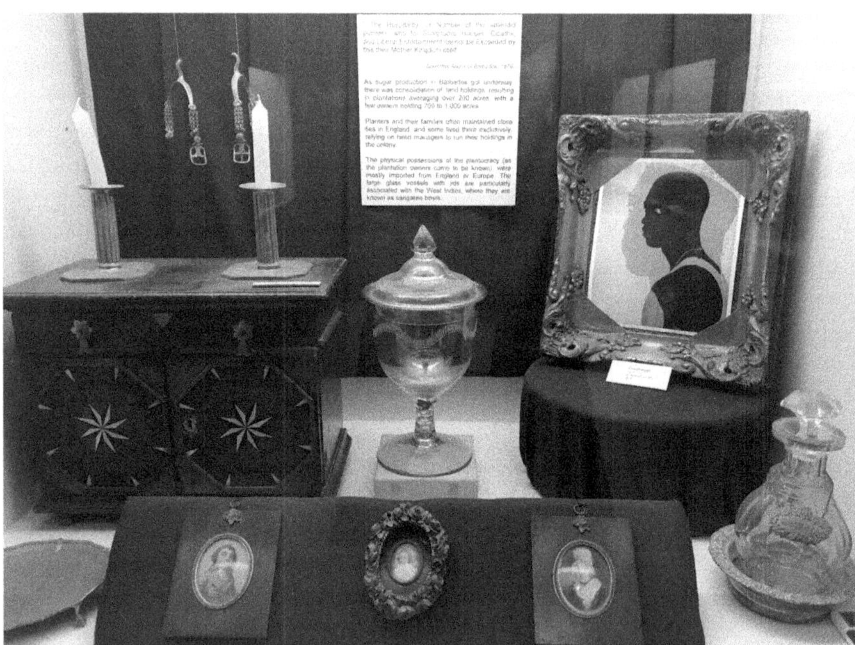

FIGURE 9.4 ‘King’ by Adrian Richards installed in the Jubilee Gallery of the Barbados Museum & Historical Society (BMHS), May 2018.

Image courtesy of the BMHS and the artist.

and lithographs included in that collection really reflected the life of the slaves, or the enslaved or even the free mulattoes that were in that society. So I wanted to just give another side to that, where these enslaved people in these lithographs did have a life, did have a sexual life as well, and a hidden life.[5]

Participation: toolkits

The community projects were generative in terms of implementing a rhizomatic research approach. However, how can these projects and the data collected be sustainable and inclusive? Under a rhizomatic research approach, one avenue might be through toolkits. One of the outputs from the interviews and community-based projects was the creation of toolkits that combine practitioner and community voices, which can be shared and adapted by users to fit their particular context and content. These are available in centralised locations online, such as the website Caribbean Museum Toolkits (n.d.), which was created in 2018. Post-critical museology includes the use of new media, or digital tools, as a viable avenue for fostering a more democratic relationship between researcher and researched. Dewdney, Dibosa and Walsh (2013, p. 195) state that digital tools and content 'places the producer and the consumer in the same position in relationship to all that metadata has so far remediated'. Barrow (2017) speaks of the significance of digital networking as a way of connecting Caribbean cultural spaces across new cartographies. However, even with heavy engagement in the digital sphere, consideration still needs to be given to *who* is participating and at what level. How many of the community participants also created online content from their encounters in this research? How are the products of this research being interpreted in the digital realm? These are ongoing investigations as the project develops.

The year 2020 and the global COVID-19 pandemic accelerated a rethinking of what a museum is. Almost overnight, thousands of museums worldwide had to shut their doors in national lockdowns and shift their programming and collections online. In April 2020, the Museums Association of the Caribbean conducted a survey into how COVID-19 was affecting Caribbean museums. One striking finding of this survey was the high interest among respondents in webinars for digital strategies (89.8 per cent) and in internet access as a human right (53.06 per cent) (Museums Association of the Caribbean, 2020). These results seem to reflect accelerated museum engagement in the digital realm and may have promoted inclusivity and intersectionality in museum-related programming. A wealth of museum organisations began to offer free webinars on current museum issues and opportunities, which, although they existed prior to the pandemic, now increased. This, arguably, has charted a path for more democratised access to international museological discussions, particularly for practitioners in the Global South. Caribbean Museum Toolkits, in turn, also adapted its platform to further serve the needs of communities by providing a calendar of webinars and possible funding opportunities available to Caribbean organisations (Caribbean Museum Toolkits, n.d.).

In the eventual development of policy around crisis recovery, it is important that a full understanding is obtained of how museums are shifting their perspectives due to the COVID-19 pandemic. This means that the voices, open access initiatives and qualitative data must be as prominent as surveys and statistics. Indeed, what has been clear is the accelerated interest by museums in inclusivity and engagement with communities in multi-vocal ways, which has countered the near silencing of narratives that might have been the outcome of physically shutting museums. Here, the possibilities for alternative ways of conducting research within museology become even more apparent, including the possibility for a rhizomatic research method.

Reflections

> The most important thing is that we take responsibility for our research and explain where we're coming from, because not everyone sets out on museum research from the same place or arrives at the museum using the same terms of reference.
>
> *(Grewcock, 2014, p. 166)*

This research has developed out of an aspiration to understand museological approaches in the Anglophone Caribbean in relation to community agency. The focus on case studies in Barbados and Trinidad and Tobago was aimed at exploring these approaches at a micro-level as well as offering a tangible outlet to apply theories of institutional responsibility in terms of cultural inclusion and sustainability.

Mason (2002, p. 177) states that, within qualitative research practice, employing multiple voices 'shows a sensitivity to a range of interpretations and voices, and a willingness to critique and question [our] own as well as others'. With that in mind, I am aware that I am both a researcher and a museum practitioner, and that my own position and voice may persist in the research and have influenced my perspective and use of case studies. Under a rhizomatic research approach, it is important to ensure the accountability of the researcher and document their perspective in the work.

It is difficult to try to draw concrete general conclusions about the relationships of museums with communities (especially postcolonial communities) from a case-by-case analytical perspective, as each process is a product of a particular time and institution. As the discipline of museology continues to develop, institutional understandings of community needs and the role museums play in shaping national identity through representation will no doubt expand. What is commonly evident from the cases presented in this research is that Caribbean communities find strength in being part of the exhibition development process. Practitioners and communities therefore have an opportunity to think critically about cultural identity and forge collaborative relationships with museum spaces or strengthen those that had previously been fragile.

Yet, I do not intend to suggest that a rhizomatic approach to museology should be held up as a singular or ideal framework. Rather, it provides another possibility for broadening the scope of our understanding of Caribbean museology. The sustainability of rhizomatic research practices, and their efficacy in developing a post-critical museology in the Caribbean as well as encouraging greater inclusivity in museums, will be charted as the project develops. There is no all-encompassing model for better inclusion that can guarantee an idyllic museum–community relationship. Instead, curators and community representatives learn through interpreting individual case studies and listening to, rather than analysing, data. It may be that the future of understanding museology lies not in the boundaries of museum practice, but in a shifting ecology of multiple entry and exit points of encounter between museums and communities.

Notes

1 Original author's emphasis.
2 Please see www.transoceanicvisualexchange.com for more information on the 2017 and 2019 editions of the programme, as well as the 2021 edition.
3 For more on *Research Yard*, please see http://aliceyard.blogspot.com/2018/04/an-eleventh-country.html.
4 The Cunard Gallery is one of the permanent exhibitions at the BMHS, showcasing historical prints, paintings and maps.
5 Knowledge-sharing session with the author, June 2018.

References

Barrow, M. (2017). 'Policy entrepreneurship: expanding multimodality and cultural development through Caribbean Intransit', in Wainwright, L. and Zijlmans, K. (eds.) *Sustainable art communities: contemporary creativity and policy in the transnational Caribbean.* Manchester: Manchester University Press, pp. 135–50.

Beans, S. (1994). 'Museums and the reformulation of ethnographic practice', *American Ethnologist*, 21(4), pp. 886–91.

Bennett, T. (1995). *The birth of the museum: history, theory, politics*. London: Routledge.

Boast, R.B. (2011). 'Neocolonial collaboration: museum as contact zone revisited', *Museum Anthropology*, 34(1), pp. 56–70.

Britton, C. (1999). *Edouard Glissant and postcolonial theory: strategies of language and resistance*. Charlottesville: University Press of Virginia.

Brown, K. (2018). 'Revisiting the ICOM definition of a museum through the lens of EU-LAC museum relations', in Soares, B.B., Brown, K. and Nazor, O. (eds.) *Defining museums of the 21st century: plural experiences*. Paris: ICOFOM.

Busse, M. (2009). 'Epilogue: anxieties about culture and tradition – property as reification', *International Journal of Cultural Property*, 16, pp. 357–70.

Caribbean Museum Toolkits (n.d.). *Caribbean museum toolkits*. [Online]. Available at: https://caribbeanmuseumtoolkits.wordpress.com/ (Accessed: 23 June 2021).

Cassidy, C.A., Fabola, A., Miller, A., Weil, K., Urbina, S., Antas, M. and Cummins, A. (2018). 'Digital pathways in community museums', *Museum International*, 70(1–2), pp. 126–39.

Clarke, B. and Parsons, J. (2013). 'Becoming rhizome researchers', *Reconceptualizing Educational Research Methodology*, 1(4), pp. 35–43.

Cummins, A. (2004). 'Caribbean museums and national identity', *History Workshop Journal*, 58, pp. 224–45.
Cummins, A. (2013). 'Natural history = national history: early origins and organizing principles of museums in the English-speaking Caribbean', in Cummins, A., Farmer, K. and Russell, R. (eds.) *Plantation to Nation: Caribbean museums and national identity*. Champaign, IL: Common Ground Research Networks, pp. 11–46.
Davis, A. (2017). 'Champagne tastes and mauby pockets: towards healthy cultural ecosystems in Barbados', in Wainwright, L. and Zijlmans, K. (eds.) *Sustainable art communities: contemporary creativity and policy in the transnational Caribbean*. Manchester: Manchester University Press, pp. 151–68.
Davis, P. (1999). *Ecomuseums: a sense of place*. London: Leicester University Press.
Deleuze, G. and Guattari, F. (1987). *A thousand plateaus: capitalism and schizophrenia*. Minneapolis: University of Minnesota Press.
Dewdney, A., Dibosa, D. and Walsh, V. (2013). *Post-critical museology: theory and practice in the art museum*. London: Routledge.
Farmer, K. (2013). 'New museums on the block: creation of identity in the post- independence Caribbean', in Cummins, A., Farmer, K. and Russell, R. (eds.) *Plantation to nation: Caribbean museums and national identity*. Champaign, IL: Common Ground Research Networks, pp. 169–77.
Freeman, K. (2005). 'Black populations globally: the costs of the underutilization of blacks in education', in King, J.E. (ed.) *Black education: a transformative research and action agenda for the new century*. Mahwah, NJ: Lawrence Erlbaum Associates, pp. 135–58.
Freire, P. (1973). *Pedagogy of the oppressed*. New York: Seabury Press.
Freire, P. (1974). *Education for critical consciousness*. London: Continuum.
Glissant, E. (1989). *Caribbean discourse: selected essays*. Translated by J.M. Dash. Charlottesville: University Press of Virginia.
Glissant, E. (1997). *Poetics of relation*. Translated by B. Wing. Ann Arbor: University of Michigan Press.
Golding, V. (2009). *Learning at the museum frontiers: identity, race and power*. Farnham: Ashgate.
Golding, V. 'Introduction', in Golding V., and Modest, W. (eds.) (2013). *Museums and communities*. London: Bloomsbury.
Grewcock, D. (2014). *Doing museology differently*. London: Routledge.
Hall, S. (2003). *Representation: cultural representations and signifying practices*. Sage Publications.
Hickling-Hudson, A. (2004). 'Towards Caribbean "knowledge societies": dismantling neo-colonial barriers in the age of globalization', *Compare*, 34(3), pp. 293–300
Hume, Y. and Kamugisha, A. (2013). 'Introduction: Caribbean cultural thought in pursuit of freedom', in Hume, Y. and Kamugisha, A. (eds.) *Caribbean cultural thought: from plantation to diaspora*. Kingston: Ian Randle Publishers, pp. ix–xxx.
ICOM (2019). *Plenary session: the museum definition – the backbone of ICOM*. [Online video]. Available at: https://youtu.be/fSDP8DXdwrA (Accessed: 23 June 2021).
The Immigration Museum Consultancy Group (1994). *Museum of Immigration at Station Pier, October, strategic management – strategic partnerships – Immigration Museum at Station Pier proposal – 1994–1995* (SM/094/5). Museum Victoria Archives.
Kamugisha, A. (2019). *Beyond coloniality: citizenship and freedom in the Caribbean intellectual tradition*. Bloomington: Indiana University Press.
Maréchal, J-P. (1998). 'For an island museology in the Caribbean', *Museum International*, 50(3), pp. 44–50.

Masny, D. (2014). 'Disrupting ethnography through rhizoanalysis', *Qualitative Research in Education*, 3(3), pp. 345–63.
Mason, J. (2002). *Qualitative researching*. London: SAGE.
McFarlane, D. (2012). *Representing blackness: Marcus Garvey and the politics of museology in post-colonial Jamaica*. PhD Thesis, University of Leicester.
McGuire, N. (2015). 'A community of curatorial practice', *Transoceanic Visual Exchange*. [Online]. Available at: https://transoceanicvisualexchange.com/about-tve-2/a-community-of-curatorial-practice/ (Accessed: 23 June 2021).
Modest, W. (2012). 'We have always been modern: museums, collections, and modernity in the Caribbean', *Museum Anthropology*, 35(1), pp. 85–96.
Morales Lersch, T. and Camarena Ocampo, C. (2018). 'Why is the definition of a community museum important?' in Soares, B.B., Brown, K. and Nazor, O. (eds.) *Defining museums of the 21st century: plural experiences*. Paris: ICOFOM.
Museums Association of the Caribbean (2020). Available at: https://caribbeanmuseums.com/ (Accessed: 23 June 2021).
Ostrander, T. (2015). 'Placing poetics', in *Poetics of relation*. Miami: Perez Art Musuem, Exhibition Catalogue, pp. 22–32.
Rússio, W. (1981). 'Interdisciplinarity in museology', *Museological Working Papers*, 2, p. 56.
Rússio, W. (2010). 'Quem são e o que são os museólogos?', in Bruno, M.C.O. (Coord.) *Waldisa Rússio Camargo Guarnieri: textos e contextos de uma trajetória profissional*. V. 1. São Paulo: Pinacoteca do Estado de São Paulo / ICOM Brasil: Brazilian National Committee of the International Council of Museums, p. 241.
Sandell, R. (2002). *Museums, society, inequality.* London: Routledge.
Soares, B.B., Valentino, A.C. and Limoeiro, D. (2019). 'Waldisa Rússio', in Soares, B.B. (ed.) *A history of museology: key authors of museological theory*. Paris: ICOFOM, pp. 100–10.
Soares, B.B. (2019). 'Introduction', in Soares, B.B. (ed.) *A history of museology: key authors of museological theory*, ICOFOM. Paris.
Szekeres, V. (2002). 'Representing diversity and challenging racism: the Migration Museum', in Sandell, R. (ed.) *Museums, society, inequality.* London: Routledge, pp. 142–52.
Tancons, C. (2012). 'Curating carnival? performance in contemporary Caribbean art', in Bailey, D., Cummins, A., Lapp, A. and Thompson, A. (eds.) *Curating in the Caribbean*. Berlin: The Green Box, pp. 37–62.
Trouillot, M.R. (1995). *Silencing the past: power and the production of history*. Boston: Beacon Press.
Vergo, P. (ed.) (1989). *The new museology*. London: Reaktion.
Watson, S. (ed.) (2007). *Museums and their communities*. London: Routledge.
Weil, S. (1999). 'From being about something to being for somebody: the ongoing transition of the American museum', *Daedelus: Journal of the American Academy of Arts and Sciences*, 128(3), pp. 28–52.
Witcomb, A. (2003). *Re-imagining the museum: beyond the mausoleum*. London: Routledge.
Yu, J.E. and Lee, J.W. (2008). 'Creating rhizomatic networks and ethics for the marginalized group', *Systemic Practice and Action Research*, 21(4), pp. 253–66.

10

CO-CURATING MEMORY

Deconstructing the silences around Caribbean migration to Britain

Kaye Hall and Natalie McGuire

In 2018, the University of the West Indies (UWI) and the Barbados Museum & Historical Society (BMHS), with funding support from EU-LAC Museums, embarked on a project coordinated by the University of St Andrews to facilitate a community-led composite history of the post-World War II Caribbean migratory experience to Britain and its role in multi-regional exchanges. Its importance to world history and politics at the time could not be overstated, given the ongoing Windrush migration scandal within Britain and the worldwide outcry of the Caribbean diaspora communities beyond. The project therefore set out to move beyond the political and the sensational to get to the real histories—the real experiences and the implications of those experiences for the diverse communities affected in both Britain and the Caribbean—and the cultural impact of those experiences on these affected communities.

Within this framework, it becomes critical to reconsider models of museum practice, especially exhibition development, to explore what could empower more Caribbean ways of knowing by making visible historical narratives that have previously been submerged. The two projects considered in this chapter, the Virtual Museum of Caribbean Migration and Memory (VMCMM) and the exhibition *The Enigma of Arrival: The Politics and Poetics of Caribbean Migration to Britain*, are both outputs from the 'EU-LAC Museums: Museums and community: concepts, experiences and sustainability in Europe, Latin America and the Caribbean' group—a consortium of eight institutions investigating the social, technological and cultural relations between Europe, Latin America and the Caribbean through community museology (Cassidy et al., 2018). In this chapter, we argue that the participatory methods used to implement these projects encourage ethical museum practice through multi-vocality and community inclusion, contributing to the development of a best practice in the decolonisation of the Caribbean museum as an institution.

DOI: 10.4324/9781003288138-13

Trouillot (1995, pp. 26–7) notes:

> Silences enter the process of historical production at four crucial moments: the moment of fact creation (the making of sources); the moment of fact assembly (the making of archives); the moment of fact retrieval (the making of narratives); and the moment of retrospective significance (the making of history in the final instance) [...] To put it differently, any historical narrative is a particular bundle of silences, the result of a unique process, and the operation required to deconstruct these silences will vary accordingly.

Our approach combined multi-vocal, co-curatorial methods with 'poetics'—language and cultural expressions through literature, music, theatre and visual art—in the development of these exhibitions.[1] This arguably allowed for the further development of distinctly *Caribbean* ways of knowing in exhibition-making and agency in the sharing of the Windrush generation narratives, thus mitigating these silences or gaps in the telling of these histories (see Cummins, 1992, 1994, 2004, 2013; Cummins, Farmer and Russell, 2013; Farmer, 2013).

The Windrush generation and importance of memory

The Windrush migration phenomenon is named after the *Empire Windrush*, the famous decommissioned troop ship that featured in the first widely publicised landing of West Indian migrants to Britain in 1948. It refers to a period of history between 1948 and 1980 that saw a mass migration of communities from the Anglophone Caribbean to Britain. This period represented major shifts in the economic, cultural and social landscapes of both Britain and countries within the Caribbean, culminating in the 1981 British Nationality Act, whose implementation and interpretation have been responsible for the unethical deportation of Caribbean people in Britain as recently as 2018.[2] Overviews of the Windrush generation as part of Britain's history have been shared through media and scholarship. However, they have not been as prominent in the field of memory for local Caribbean narratives. When approaching the sharing of stories from this era, the question arises of how a museum-based reading of this history for an international project can demonstrate empathy and inclusiveness for the communities involved as well as ethically collaborating with the living communities who continue to experience that history and memory.

Trouillot's (1995) perspective on the entanglement of silence and history holds particular significance for Caribbean museum practitioners because it speaks eloquently of the Caribbean's particular historical context, which has been shaped by the experiences of enslavement and colonisation. In fact, it could be argued that any authentic representation of a historic phenomenon that forms part of the Caribbean experience must always bear the legacy, or the taint, of that experience. Migration, by its nomadic nature, is poignantly a part of this experience of colonisation. The deliberate separation of the colonised from their cultural roots; the overlay of Eurocentric systems of law, governance and process; and the adherence to a binary perception of being and power, with the empire at the centre and the colonies and colonial subjects on the margins, created circumstances so unique that they must be a primary consideration in any interpretation of history originating from these regions. This is so both in terms of inherent bias within these systems and any attempt to redress the imbalance of power and presence historically. We cannot tell our history unless we examine how it has been shaped by past 'outside' interpretations.

In the light of Trouillot's silences, a crucial component of mitigating these imbalances in historic power is the telling of untold stories from our pasts that render visible, new historic information that has been previously overlooked in the historical narratives originating in the colonial centre. Indeed, Trouillot is not alone in his analysis of the relationship between the centre and margin, which places the imbalance of power between the two at the crux of our modern understanding of our historic circumstance. The margins, and their access to history, continue today at the pleasure of the centre and it is only an equalisation—as it were—of this imbalance that will see a more equitable recounting of history at the margins, devoid of the psychic and emotional burden of lesser representation in histories emanating from the Global North. Earlier texts by Frantz Fanon (1952, 1961) and C. L. R. James (1938), for example, identify this paradox of being—expressed daily in our culture, our literature and our very values created within this state—as problematic at best and critically destructive at worst.

The telling and retelling of Caribbean stories of migration experiences in colonial centres, and the effects on these distinct but parallel cultures at the centre and margin, have previously remained largely unexplored by museums throughout the Caribbean region. They must therefore form a key part of any new explorations and alternate perspectives or narratives. As we shall see, the two outcomes of the EU-LAC Museums project discussed here—*The Enigma of Arrival: The Politics and Poetics of Caribbean Migration to Britain* and the VMCMM (VMCMM, n.d.)—offered an opportunity to examine in detail the influence of a specific instance of the Caribbean past, beyond peripheral and superficial British interpretations most recently coloured by the Windrush scandal, to the lives of the nations, families and individuals who were the key actors in the story: the Caribbean migrants.

Social responsibility of museums

The question of the social responsibility of museums is imperative when considering exhibition development in the Anglophone Caribbean. The new museology has brought about a series of philosophical shifts regarding the role of the museum, specifically pertaining to policies around the display, to inform and expand museum practice (Hooper-Greenhill, 1992). Cummins, Farmer and Russell (2013) and Vawda (2019) have explicitly addressed the implications of these shifts in a postcolonial context and, in particular, the necessity of addressing aspects of power and social responsibility within Caribbean-based practices. They emphasise the need to move away from colonialist motives governing approaches to collections.

Within the EU-LAC Museums project (2021), there have been many layers of community inclusion and engagement, including a database of community museums, digital technologies for collection preservation and democratisation, youth exchanges and a significant contribution to museological scholarship and policy on frameworks for community museums. In reporting on the recommendations arising from a policy round table, Barriteau, Cummins and Keohane (2019, p. 16) outline the significance of considering migration and memory for the Anglophone Caribbean EU-LAC Museums partners:

> The conditions under which post-Columbian Caribbean societies were forced into being constructed a vastly different psychic profile for the women and men from Africa and Asia who peopled the region. The confluence of the forced, unfree or conscripted conditions of Caribbean arrivals, the hybridity of cultures and the brutality of their existence over four hundred years of slavery and indentureship left Caribbean people with a deep sense of uprootedness. They existed but never felt settled. Many Caribbean people therefore operate with a discernible ideology of return, coexisting with an easy willingness to roam. While we agree with Brian Hudson that migration fosters ambiguity, ambivalence and destabilises conventional notions of identity, it can also foster conditions of solidarity and the constructions of more radical notions of identity and belonging.

The BMHS has worked to enable the UWI, and the project at large, to develop capacity around project activities and outcomes, especially in the digital realm. The BMHS was established in 1933. However, Cummins (2004) situates the institution as central to debates around nationalism and cultural identity in the post-independence era, detailing ministerial involvement in the shift in the BMHS's role in society and advocacy for more locally oriented content within its permanent galleries from 1980. This shift resulted in an expansion of the historical and cultural narratives within the Museum to be more reflective of the wider Barbadian population's experiences and interests. Given the context described above, it was essential to actively pursue integration of migration stories through:

1 an exhibition that could travel and contribute to wider narratives within the permanent galleries of museums in the Anglophone Caribbean; and
2 that this integration also occurred in a 'living' form through a virtual platform, allowing for independent, ongoing contributions to these narratives.

Sandell (2002, p. 6) provides a positive outlook on the role of museums within communities, arguing that they 'have provided an enabling, creative, perhaps less threatening forum through which community members can gain the skills and confidence required to take control and play an active, self-determining role in their community's future'. Morales Lersch (2019, p. 40) also states the importance of this form of participation for communities within the EU-LAC Museums project, stating that:

> The act of collective self-interpretation is a creative act, which is both affirming and transformational. It affirms the existence of a way of life, with its unique development and significance, and it affirms community members' connection to this way of life.

This is an important guiding ethos in the transition from passive to active museum practice. One of the methods of facilitating this shift is to employ a participatory methodology, which we utilised in different ways in the development of the VMCMM and the *Enigma* exhibition.

Participatory methodology in exhibition-making

One of the most significant developments regarding the role of museums has been a move towards more reflexive and people-centred spaces, like the 'living' museums introduced by Davis (1999), Karp (1992), Sandell (2007), Simpson (2001) and Watson (2007), which often emphasise the importance of community inclusion. Recently, the American Alliance of Museums has acknowledged a need for what it describes as 'community engagement curators' who can respond 'to changing demographics, climate change, and the need to be relevant to their full communities' (Berlucchi et al., 2018). It is thought that this can be achieved by personnel who can 'engage local communities in an open dialogue about their needs, elevate the museum's connection to local communities, and effect meaningful change through educational programming'.

Nina Simon (2010) has proposed a similar process that uses a participatory methodology, stating that, within museums, all participatory projects are based on three institutional values:

- Desire for the input and involvement of outside participants
- Trust in participants' abilities
- Responsiveness to participants' actions and contributions

She suggests four models for best practice participation by communities in museums. These are contributory, collaborative, co-creative and hosted (Simon, 2010). Contributory participation consists of participants from the public submitting limited material towards a larger project, which is controlled and interpreted by the curator or researcher. Collaborative projects often include community groups who inform the direction of an exhibition of programming; they commonly involve focus groups and other behind-the-scenes data collection. The third model in participatory practice is co-creation. This goes a step further than collaboration: participating individuals and groups from communities work with a museum from the inception of a project and assist in its design, interpretation and development, as well as being visible in the project itself. In the fourth hosted model, the museum hands over a space and resources to a group or individual for an exhibition of their choosing. While acknowledging that focus groups constitute an attempt at participation, Simon emphasises that more can be done for visible exchange and input both through the tools of technology and by creating tangible, safe spaces for visitor participation. According to Simon (2010), these participatory techniques of inclusion are able to 'address particular institutional aspirations to be relevant, multi-vocal, dynamic, responsive, community spaces'. As seen in the discussion regarding the *Enigma* exhibition, collaborative and co-creative techniques from Simon informed participatory aspects of the community of curatorial practice.

With regard to implementing a participatory methodology in museum practice, Diamantopoulou, Insulander and Lindstrand (2012) suggest that the best tool for community inclusion is a social semiotic approach to visitor experiences in museums. In their case, they asked visitors to draw maps of what they encountered and then analysed the content to gauge visitor agency in the interpretation of collections. Although the researchers collected data across a range of exhibition media and visitor demographics, this methodology arguably still does not allow the formation of conclusions regarding co-creation (using Simon's model). First, their theory continues to position the institution as the centre of power for driving diverse social conditions (Diamantopoulou, Insulander and Lindstrand, 2012, pp. 12–13), which arguably removes a fundamental aspect of agency from the community of visitors. Second, although visitors are given only loose instructions for constructing their maps, the *interpretation* of the content still lies with the curators or museum professionals (Diamantopoulou, Insulander and Lindstrand, 2012, p. 14). They draw connections and postulate meanings, thereby running the risk of not correlating with the narratives of those who created the maps. With these two points, then, the 'inclusion' of community voices remains superficial and only at the direction and control of museum staff or researchers.

How then can assessing community voice through co-creation challenge authority in the collection and analysis of data? One example of this can be seen in critical pedagogy, for example, the learning maps model developed by Annan et al. (2014), which employs a strong community-of-practice ethos. Learning maps involve a series of ongoing documentation by participants on their current learning environment. What is different, however, is the emphasis on a community of practice through the map. Drawing on Lave and Wenger's (1991) concept of community of practice, Annan et al. (2014, p. 18) extend its application to learning environments, giving equal value to the participation of the learner, teacher, leader and community in the process of learning. This idea departs from a hierarchical approach to learning, where an individual (teacher, leader) holds the knowledge and passes it on to a receiver (learner, community).

The Enigma of Arrival: The Politics and Poetics of Caribbean Migration to Britain

In 2018, UWI Cave Hill Campus in partnership with EU-LAC Museums developed a travelling exhibition on Caribbean migration, facilitated by the BMHS. *The Enigma of Arrival: The Politics and Poetics of Caribbean Migration to Britain* aimed to create awareness about the active migration crisis in Britain by focusing on Caribbean migration from the 1940s to the 1970s. Designed in the year of the seventieth anniversary of the *Empire Windrush*'s landing, it sought to inform the general public across both the Caribbean and Caribbean communities in the United Kingdom about the current state of affairs concerning Caribbean migrants in Britain, as well as the origins and circumstances under which these conditions had developed. The title, *The Enigma of Arrival*, was loosely linked to V. S. Naipaul's novel of the same title in order to align with his exploration of West Indians' disillusionment with 'Britishness' and their survival as a migrant community.

Using interpretive panels and interactive audio-visual materials, the audience was introduced to the historical parameters and scope of post-war Caribbean migration to Britain and its legacy among later generations, both at home and abroad. The exhibition themes were developed by the researchers at UWI Cave Hill Campus as follows:

- Caribbean Conditions Pre-migration
- Arrival in the United Kingdom
- Accommodation
- Settling-in Conditions
- Cultural Integration
- Caribbean Resettlement
- Generational Experiences

In developing the *Enigma* exhibition, the project team adopted a curatorial framework that incorporated aspects of both a co-creation approach and a community of practice.[3] Here, they were informed by the paradigm of museums as responsible social actors that necessarily evolve as they seek to improve their engagement with community concerns and global challenges. They therefore sought to provide new opportunities for the region's museums and communities to collaborate in the co-curation and co-creation of previously unarticulated national and regional narratives by inviting these groups to add relevant local content. These experiences of working together to establish curatorial content and text, and to share images and (digital) artefacts, allowed for a design process that yielded a richer experience for local and diasporic communities as well as host museums, as together they explored new ways to acknowledge their routes to their roots.

The project team for this exhibit was jointly headed by the BMHS's key personnel who are engaged in community activities on a day-to-day basis: the Education and Community Outreach Officer and the Curator—Social History and

Engagement. These two officers are fully committed to community co-curation as a means of actively engaging the BMHS's audiences in the dissemination of these new historical perspectives, which promote greater inclusivity in museum and heritage studies. Under their direction, the exhibition offered opportunities to take inclusivity beyond the mere (re)telling of histories via a single curatorial or institutional voice to a multi-vocal approach via a community of practice. This was aimed at allowing those who actually lived through the experience of migration to contribute their stories and therefore their voices and direction to the evolution of the research and exhibition development. Several approaches were implemented to encourage this, which we will now explore in turn.

Collaborative research

A preliminary research team from UWI Cave Hill Campus established the key areas of interest for the project. They, alongside the BMHS, worked on developing a framework for themes that might be key in sharing the histories of the Windrush migrants. This process drew on diverse resources within the Anglophone Caribbean region that might allow the voices of those involved in the migratory movement to emerge.

One of the significant stakeholders in this process, the West Indies Federal Archives at UWI Cave Hill Campus shared a collection of correspondence between the central government in London and the various colonial offices across the region. The collection addressed many of the themes that later appeared in the exhibit and its corresponding education programming, including policy-making, accommodation conditions and the effects of the Windrush migration on local Caribbean communities. Through this content, researchers were able, for instance, to trace the migration movement from post-war recruitment through to reports of racial tensions. The documentary story also highlighted the resulting enactment of restrictive legislation to stem the flow of migrants, followed by the consequences of these developments in both the metropolitan centre and the colonies.

Another significant collaborator for research, as well as media content, was Claude Graham and the Caribbean Broadcasting Corporation. While working as a freelance journalist in the period 1990–2000, Graham had developed a series of television programmes that investigated the Windrush generation and particularly the experiences of the 'second generation' (the migrants' children).[4] These programmes tackled topics of cultural identity, racism, economic mobility, repatriation and social integration. Although they focused primarily on Barbadian experiences, the breadth of programming also included diasporic stories of Antiguan and Jamaican communities. This content contributed greatly to humanising the historical and political aspects of the Windrush migration.

The collaborative research for the *Enigma* exhibition also extended into the documentation and examination of the roles of notable migrants and people of Caribbean descent who have contributed significantly to the development of modern Britain and the Caribbean beyond their migratory experiences in Britain. This aspect of the exhibition was proposed by a community participant as a necessary area to cover. It was developed as an open-ended and constantly growing resource, whereby anyone working on the project, regardless of their place within the diaspora, could identify a person who met the relevant criteria, and their biography would be added to the database. The primary entries included more comprehensive biographies of people already featured prominently in historical narratives, which were documented for the development of the exhibit. This preliminary group was supplemented and expanded to include those who had been formally recognised by either the British or Caribbean governments for their contributions, as well as those similarly highlighted through social media by professional or other meritorious institutions or entities from the diaspora.

Co-creation and co-curation of content

The co-creation and co-curation of exhibition content were implemented in four layers. First, a public call was made to the Caribbean region and diaspora asking people to contribute their memories, stories and unique perspectives on the Windrush migration. The community was invited to submit content digitally, loan or donate artefacts, contribute oral histories in person and, of course, comment on all content shared publicly prior to the completion of the exhibit.

Second, as the VMCMM developed, this was shared as an accessible tool for communities to contribute and engage with migration stories. Using a simple form on the VMCMM website, individuals in communities could write personal histories, share images or video recordings and then pin the location in the Caribbean from which they had migrated and where in Britain they had migrated to. The aim of this was to encourage organic contributions beyond the curatorial vision of the project team alone.

Third, the project team employed a community-of-practice approach shaped by lateral learning within the shared domain of the Windrush period and associated scholarship. Drafts of the interpretive panels were shared with local, regional and international colleagues with a range of relevant expertise in the historical, cultural or poetic context as well as museum expertise. The process was intended not just for feedback but also for the co-creation of the content as it developed. This resulted in the final interpretive panels being a product co-created by a community that extended well beyond the initial research team.

Finally, co-creation existed within the exhibition itself, as installed at the BMHS. Visitors to the exhibition could contribute to its content in various ways: there was a feedback board where comments, reactions and ideas could be shared, and the VMCMM was installed in the exhibition on a touchscreen for visitors to view uploaded stories, upload their own or contact the project team for a more formal interview. A children's activity booklet was also created for younger visitors to draw or write about their experiences through guided activities in the exhibition.

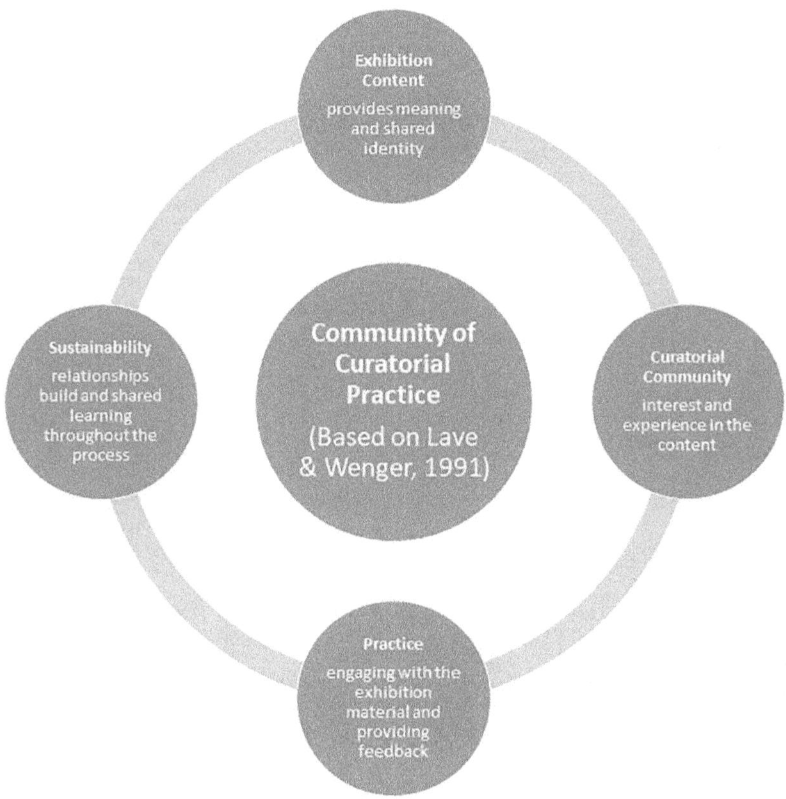

FIGURE 10.1 Community-of-practice cycle based on Lave and Wenger (1991).

Source: EU-LAC Museums and BMHS.

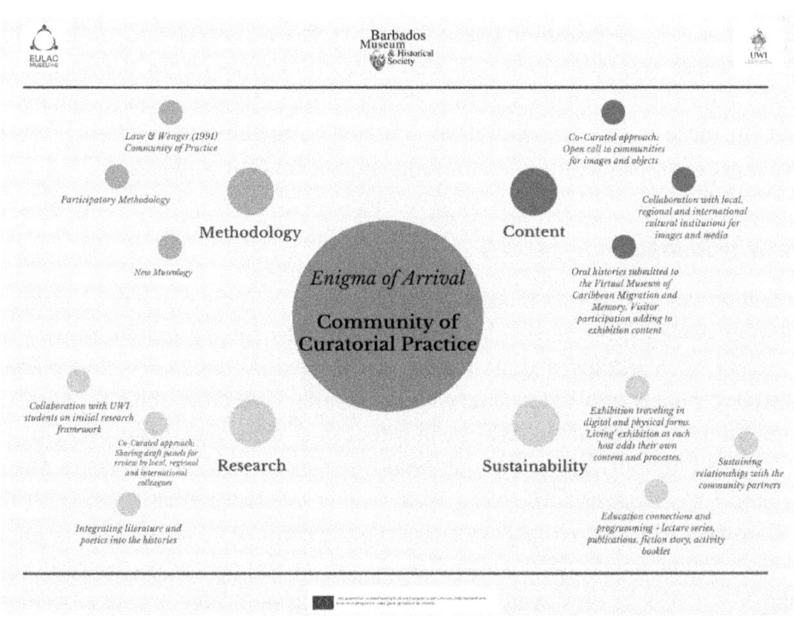

FIGURE 10.2 Community of curatorial practice for *The Enigma of Arrival* exhibition.

Source: EU-LAC Museums and BMHS.

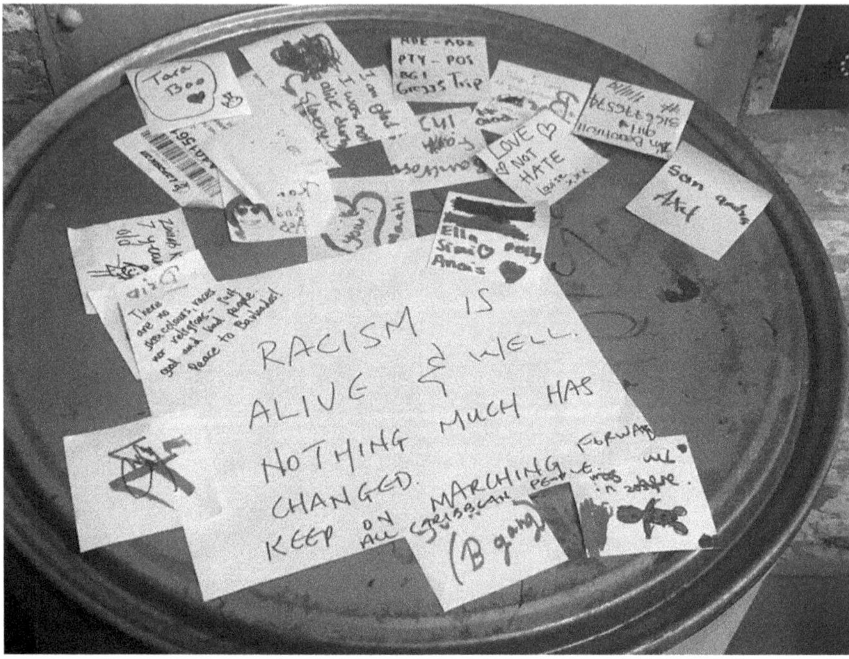

FIGURES 10.3 Visitor engagement in the *Enigma* exhibition, January 2020.

Source: EU-LAC Museums and BMHS.

This booklet could be taken with them or left in the exhibition as part of the content. Additionally, a slideshow based on the database of notable Caribbean immigrants was also created, which provided contact information for the project team so that visitors could make suggestions for additional figures to be researched and added. In these ways, the exhibition emerged as an interactive, living space in which a growing collection of community stories was encouraged.

Reflexivity in the design process

Co-creation and co-curation also extended as a reflexive practice during the exhibition design process. As the ultimate aim was for the exhibition to travel, the project team piloted a draft format for the exhibition that included mainly interpretive panels and video interviews with the VMCMM—that is, emphasising the human stories of the Windrush rather than the object histories. This pilot was installed digitally at the International Museums *Itinerant Identities: Museum Communities/Community Museums* conference (7–9 November 2018, UWI Cave Hill Campus). Museum colleagues attending the conference as well as community audiences volunteered anecdotes and feedback. The panels also remained visible on the VMCMM as they were developed. This allowed for an arguably deeper exploration of a co-curation process, as public feedback was given during the development period, rather than after the exhibition had launched.

This feedback informed the use of image content within the exhibition design and gave the project team a better understanding of the public's engagement with the interpretive panels, which led to a more integrated approach to the poetic aspects of the written content. Quotations from literary arts of the period—songs, poems and novels—were combined with journals, speeches, letters and memoranda, top-secret telegrams and migration manuals to more effectively structure the story according to the loose sequence we had initially devised. These elements were brought together to tell a comprehensive narrative that acknowledged not just facts and figures but also the lives and lived experiences of our people, their thoughts and feelings expressed in literature, which helped to set the scene and tell our stories.

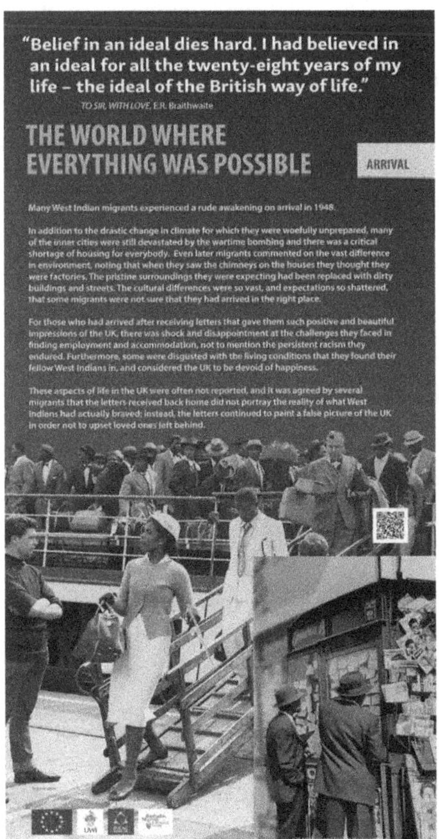

FIGURE 10.4 Exhibition panel for *Enigma*, displaying use of literature, January 2020.

Source: EU-LAC Museums and BMHS.

The resulting panels were richly interpretive. Each began with a quotation from a literary work or experiential oral history to set the thematic and emotive tone for the panel, which then provided a concise and evocative snippet of an aspect of the whole that was being described and examined. Pictures, charts and diagrams were interspersed with the text where appropriate and most impactful. In addition, videos and other interactive elements, along with the few objects we obtained, were placed closest to those areas where they could best complement panel text. The overall impact was of a dense and vast story that encouraged multiple visits to absorb all of its elements but which proved fascinating to visitors from all backgrounds, given the pervasive contemporary influence of the Windrush migration.

The arguably unusual incorporation of literature (that is, deliberately creative and fictional content) into the exhibit made for an interesting twist on the community co-curation approach. We had noted on many occasions that the stories shared by migrants often paralleled fictional accounts so closely that they were indistinguishable in terms of the history they documented. However, one key difference and advantage of the fictional and creative accounts was the emotional and emotive, and therefore humanising, nature of the language utilised in them. It was felt that this 'human-feeling' language would, if used effectively, form a gateway for connecting audiences with the more formal aspects of the exhibit content. This strategy proved more successful than anticipated and was perhaps the most

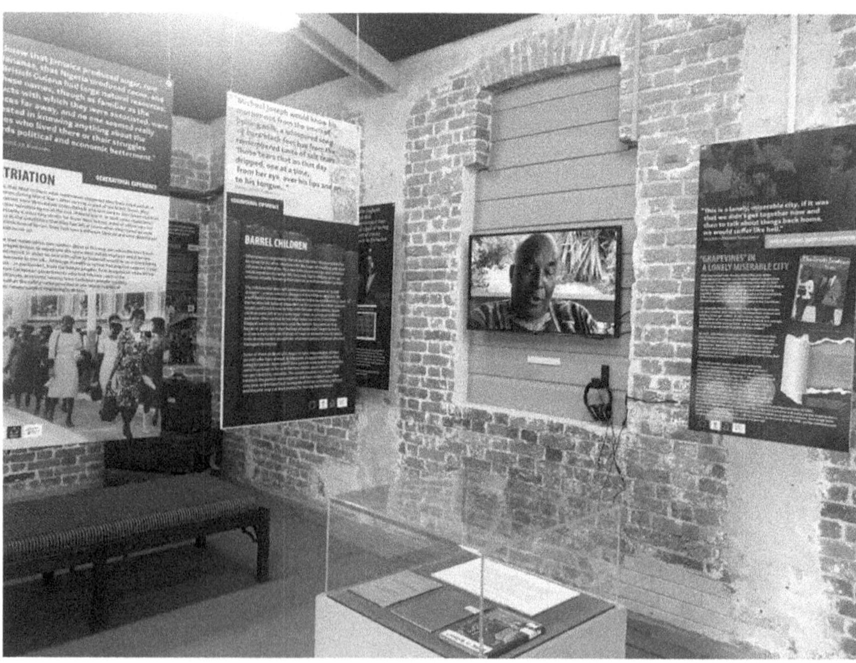

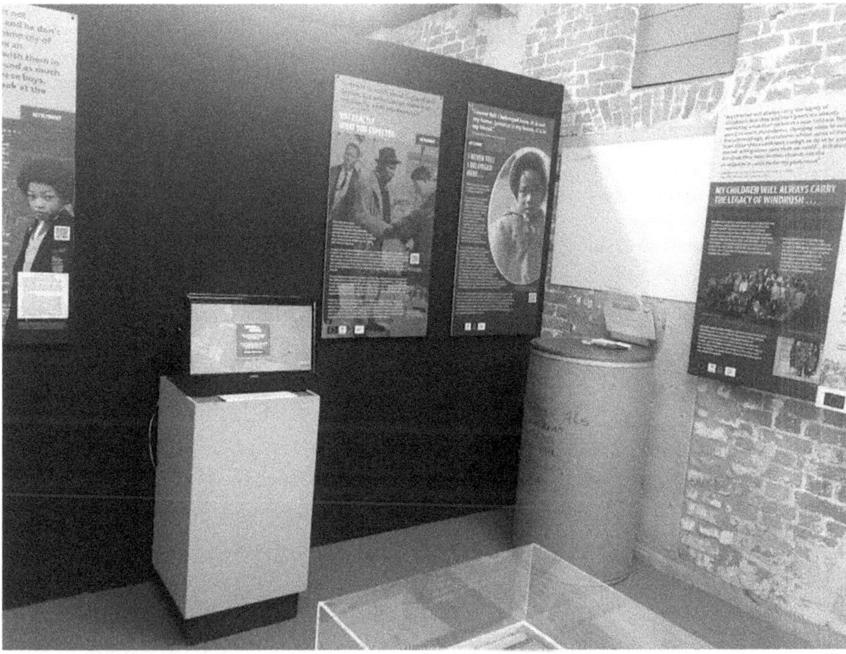

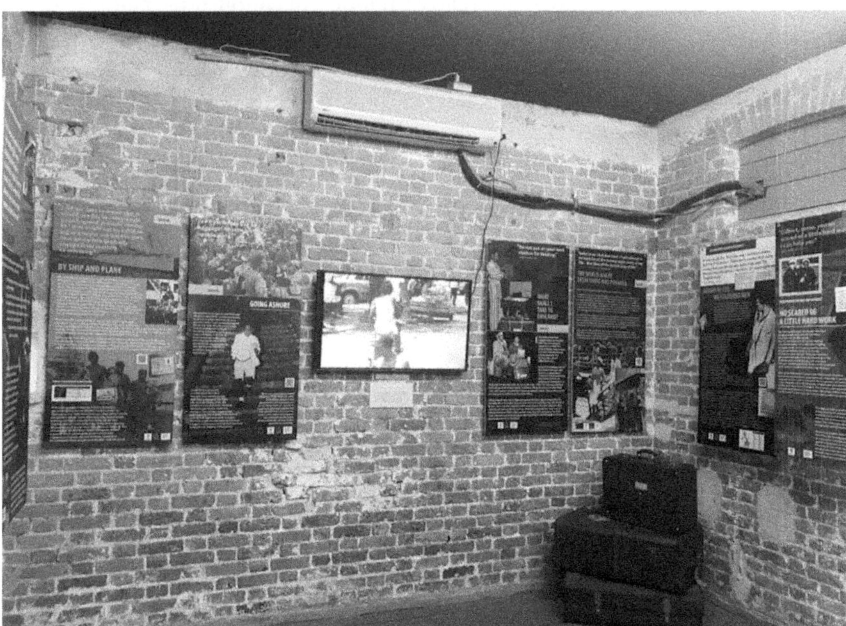

FIGURES 10.5 TO 10.7 *Enigma* exhibition installation at the BMHS, June 2019.

Source: EU-LAC Museums and BMHS.

commented-on feature of the exhibit, both by the audience at the launch and in successive feedback received.

The physical exhibit was formally launched at the BMHS on 21 June in time for the public to view it on Windrush Day on 22 June 2019.

Educational programming for the exhibition

Utilising museums as an educational resource originates in constructivist pedagogy, which is based on a concept of actively engaging learners in the process of creating meaning and knowledge. While constructivism is itself not a new concept in education, having roots in the work of Dewey (1938), Piaget (1936) and Vygotsky (1978), it has experienced some resurgence in more recent times due to support from the field of cognitive psychology (Hein, 1991, p. 1). Hein (1991, p. 7) notes:

> The principles of constructivism, increasingly influential in the organization of classrooms and curricula in schools, can be applied to learning in museums. The principles appeal to our modern views of learning and knowledge but conflict with traditional museum practices.

This framework accorded well with the community co-curation model that we followed in developing a number of creative public education programmes for the *Enigma* exhibit. Our process of creating narratives relevant to the community, which would also engage and mitigate the silences in the existing histories, was also compatible with this constructivist education outlook and worked well within the key aims of the project. These were:

1 to provide another avenue for public engagement with, and co-curation of, the exhibit; and
2 to disseminate the emerging research content and historical narratives in ways that actively engaged informal audiences in the process of managing and building their own knowledge and that of others.

A constructivist pedagogy encouraged a great deal of flexibility in devising innovative and customisable educational solutions that complemented the project, and its audiences, as it developed. Public education therefore began well prior to the launch in order to facilitate the first aim. It then continued throughout the process of exhibition development and display, with new ideas for engagement and dissemination being developed right up to the writing of this chapter.

Theatrical representation

The BMHS is actively involved in an applied theatre programme collaboration with the Barbados Community College (BCC) that is aimed at preserving aspects of heritage and culture, particularly intangible cultural heritage, through theatrical exploration. Prendergast and Saxton (2009, pp. 153–4) describe the use of museum theatre as an effective medium for 'mediating knowledge and understanding in a museum setting', including a number of opportunities for development:

- first-person interpretation or role-playing by exhibitors or guides
- theatre performance aimed at understanding a particular time period or issue
- historical re-enactments allowing participation by large audiences of persons
- role-playing or second-person interpretation

The community theatre programme actively utilises all the above techniques except historical re-enactments in order to connect with multiple historic voices and audiences. Much of the content developed in these activities feeds directly into exhibition development and community practices. The programme also operates as a way to increase youth engagement and access audiences who would not ordinarily be drawn to a museum except via its entertainment value. In this instance, it was thought that allowing the students to interact with the content being developed would provide them with the opportunity to connect with that particular aspect of our history with which they were largely unfamiliar, as well as provide young people's feedback as part of our community co-curation process.

The group was provided with not just raw data content but also a list of the literary content that we had identified as relevant to the period. In addition, the students were encouraged to do their own research online as well as in the local community with those of their elders who had been a part of the Windrush migration experience. They then worked to develop a theatrical performance based on their own interpretations of these composite inputs. The resulting theatrical work consisted of a series of loosely linked vignettes that explored the themes of the exhibition through the associated existing literature as well as some original work. It was performed at the Association of Caribbean Historians conference held in Barbados in June 2018 and then again at the *Itinerant Identities: Museum Communities, Community Museums* conference in November of the same year, where it was also recorded. The work was well received by both audiences, who took the opportunity to interact with the students and garner their unique perspectives on this history. Feedback from both audiences and students was incorporated into the exhibition development.

The students shared their experiences in conversation with the exhibition development team. These involved both the discovery of family history through the informal oral histories that they gathered from within the community and their own journeys of self-discovery as descendants of those who would have lived through the Windrush experience, both at home in the Caribbean and abroad. Many of them commented that they appreciated having been exposed to this previously unknown or hidden aspect of the historical narrative that is not linked to the stories of enslavement, rebellion or independence that characterise the general Caribbean history curriculum in schools. Students' comments also highlighted the opportunity to connect with history in a more direct, personal and meaningful way, outside the typical classroom or guided tour session. They further indicated that 'living history'—interacting with those who have lived history—was far more engaging than history learned from books. They also commented that their theatrical practice was enhanced by being able to engage directly with the subject matter through the community. In fact, they identified this as a process that would inform their practice and craft going forward, which would add additional dimensions and complexity to their portrayals.

Public lecture series

A public lecture series, entitled 'From Importation to Deportation: 70 Years of the Windrush Generation' and produced by the BMHS and UWI Cave Hill Campus, was held from March to May of 2019. The series consisted of a programme of expositions by academics and professionals, such as journalists and diplomats, as well as other persons from within the community. The topics of their presentations included various aspects of the migrants' experiences that were considered of interest to the community in light of the burgeoning public interest engendered by the Windrush scandal. The series was well attended with over 100 audience members on its best night. The Question-and-Answer sessions in the series sparked community debate and encouraged many Windrush returnees in the audience to share their stories as part of the activity.

The lectures were also livestreamed to increase public dissemination and feedback. Although the series ended in June 2019, these livestreams continue to be watched online and receive feedback. In addition, one of the presentations from the lecture series was adapted in volume LXV of *The Journal of the Barbados Museum & Historical Society* (December 2019) under the title 'The Windrush Scandal: An Insider's Reflection' by Guy Hewitt. In this chapter, Hewitt (2019, p. 151) discusses the premise of the Windrush scandal and the role of Caribbean governments in highlighting what he regards as 'a blot on Britain's socio-political landscape'.

Children's exhibition workbook

An interactive workbook developed around a play-and-learn model was designed to facilitate young visitors' engagement with the exhibition's content. It encouraged them to think about the specific activities in which their ancestors would have taken part when preparing to migrate. The book was also designed along constructivist lines, allowing children to learn and engage with the exhibit on their own terms and at their own pace, utilising the activities within the booklet as a guide.

Museum internships

The BMHS also takes a constructivist approach to learning within its internship programmes, allowing interns to freely engage with the development of content and programming activities for its exhibitions. Over time, this tactic has resulted in the development of some engaging interactive programming content created by interns, who bring varying experiences and skill sets to each development process. In the case of *Enigma* specifically, the Museum benefitted from the creativity of a UWI History major with an interest in creative writing, which resulted in the production of a serialised story entitled *Barrel Child*, which was illustrated by yet another intern, a Sociology and Psychology major. The story was targeted at young audiences visiting the exhibit who may not have been able to engage with the weightier content and presentation of the main exhibit. It chronicles and explores the experiences of Caribbean families, parental migration and family separation, which led to the existence of so-called barrel children who were left behind in the Caribbean to be raised by grandparents and other relatives when their parents migrated for better opportunities.

The book was published internally and added to the exhibit as a free takeaway for interested parties. The BMHS received mainly positive feedback from visitors, including requests for further instalments of the story. It was therefore considered a necessary tool for engaging young audiences with the material and a good alternative to Western-centred storytelling narratives in general.

The two interns subsequently worked on a series of poems, a colouring book and an interactive workbook for older children. Both indicated that they valued the opportunity to make concrete contributions to the exhibit content, as well as to engage directly with history in a creative fashion. They, like the theatre group, were encouraged to interact not only with the research material but also with the members of the community who had experienced the Windrush phenomenon directly. The pair reported that it allowed them to forge stronger familial and community bonds.

Social community-based activities

The final major activity accompanying the exhibit was aimed at encouraging interactive engagement with the exhibition content in a social setting. The Museum therefore hosted an event entitled 'Tea and Conversations', which took the form of an afternoon tea party. The goal of this event was to encourage members of the Windrush generation and their families to come and share their stories with one another, and also hopefully with the VMCMM, further expanding the scope of the exhibit content. The proceedings, which were intended to encourage conversation and discourse, included snippets of Windrush story narratives, as well as readings of poetry by one of the Museum's interns, which were based on her interactions with the exhibition content as it developed. Those who attended the event came ready and prepared to share stories and connect with one another, rendering the preparation of the narrative story somewhat unnecessary, although they did listen and make comments and ask questions. The poetry was very well received by the audience and they passed several hours in companiable discourse (though they still proved reluctant to have their exchanges recorded), which encouraged the Museum to extend the event past the originally intended closing time. Some exhibition visitors indicated that they would share their stories directly with the VMCMM at a later date.

Challenges in the participatory method and reflections on *Enigma* as a travelling exhibition

One early challenge with implementing a participatory methodology in this exhibition process was doing justice to the many voices and stories within Windrush history, and how they could be effectively developed into a coherent yet fluid whole that would be accessible to the regional and diasporic audiences for which it was being prepared. Inundated with facts, figures, photographs, original documents, oral histories and the various paraphernalia of one of the greatest movements of people across the globe, the team struggled to find a way to connect with intended communities and audiences. It was then that we realised that the very cultural traditions through which our peoples have always carried their stories could be utilised to link their experiences and that they would enable us to provide a pathway to understanding the layers of this history. The development of the VMCMM then provided opportunities for the exhibition to be 'living', with stories and subthemes continually added as the exhibition travelled.

A second challenge was that although in the exhibition the visitor feedback board was well utilised, there seemed to be a particular reluctance among members of the public to share their stories in an international digital format. We attempted to counter this reticence through educational programming. Nonetheless, while those who had experienced the Windrush migration were willing to talk—oftentimes passionately—about their experiences at in-person events, they were challenged when faced with online sharing facilities. This public reticence is something of a cultural norm in Barbados, where the bulk of the interviews were conducted. However, it is hoped that, over time and with further community interaction, the observed culturally learned 'shyness' and technology bashfulness could be overcome in order to increase the recorded archive of these stories.

The exhibition is fulfilling its aim to travel both within the Caribbean and in the United Kingdom. There has been a digital iteration at Goldsmiths University[5] (October 2019), and an installation at Birmingham by the 2nd Generation Barbadian & Friends[6] Association (December 2019–January 2020). In February 2020, the exhibition opened at the UWI Museum,[7] Mona Campus, Jamaica, and in June 2020, it was launched on Windrush Day at Reading Museum in England.[8] The Reading Museum iteration in particular has yielded significant new content in terms of interviews with Windrush migrants still living in Reading and their descendants. This will be added to the VMCMM and will also become part of additional interactive content and online materials for both young and old audiences.

The UWI Museum has also sourced additional content which, while too vast for inclusion in the exhibit in the limited space in the museum, may be suitable for inclusion in the VMCMM. This continued scholarship is encouraging and bodes well for the sustainability of the project and the history it represents.

Conclusion

As the exhibition travels, communities will be encouraged to share their stories and contribute to the content, furthering the ethos that *Enigma* is a living and co-curated platform, as well as developing and expanding the VMCMM as a growing and self-sustaining project. The project's realisation of a community of curatorial practice has yielded significant results. Most significantly, it has opened possibilities for museums in the Caribbean region, and those with diasporic communities, to effectively devise exhibition frameworks that draw on Caribbean ways of knowing. The model of co-curation has been successful enough that almost every exhibit displayed thus far has resulted in an increase in the knowledge collected, developed and archived by the project as well as unplanned extensions in the public's interaction with the exhibit.

This particular method shows potential for such frameworks to be established and formalised by Caribbean museum professionals seeking to develop exhibitions that put community voices at the forefront of the design and that revisit historical narratives from the perspectives of those living in Caribbean countries. The BMHS has made good use of the methodologies developed in this project. Co-curatorial practices have informed several of its projects going forward, with visible gains in public interaction and interest, as well as community engagement with the Museum and its interpretive projects. For the team, this is a heartening indication of the validity, future success and further development of this model. It also suggests that Caribbean museology is on the right path with regard to relevance, significance, representation and engagement within its communities as we fill Trouillot's silences in our (hi)stories. Research and development of previously undocumented or poorly documented histories from within the Global South provide many opportunities for the (re)writing of heritage discourses to include more holistic and inclusive visions and versions of history. In addition, techniques such as oral history, and interpretation based on artistic expression of the period under study, expand the scope of the telling of history beyond traditional methodologies for a more inclusive, multi-focal and multi-vocal recounting. This is especially important for Caribbean peoples, who, as descendants of forced migration and enslavement, are largely dependent on stories inherited through oral traditions. In order to continue to make space for such methodologies in exhibition-making, research and interpretation must be open to these new inputs into historical discourse. In turn, these can function as new means of engaging with the community to establish relevance and provide new ways of seeing and understanding the politics and poetics of our past.

Notes

1 This project involved the primary curatorial team at the BMHS (Cummins, McGuire- and Hall), with collaborations from the UWI departments of Cultural Studies and History

and Philosophy. It utilised a number of research resources including, but not limited to, the West Indies Federal Archives, the Sidney Martin Library, the UWI (Mona) Museum and the University's Office of Research as well as the performing arts ensemble from the BCC. Also involved were a number of independent researchers with an interest in migration history (Emerita Mary Chamberlain, Rosalie Mayers, Marcia Burrowes and Kenneth Walters), as well as journalist Claude Graham. The project also benefitted from the influence of the project team at St Andrew's University in Scotland, whilst the exhibition has also expanded to include content from the venues to which it has travelled, particularly Reading Museum and the UWI Museum, as well as contributors to the lecture series, Henderson Carter, Alan Cobley and others.

2 Further reading on this can be found in Hewitt (2019).

3 Although adapted and formalised for the EU-LAC Museums project, a 'community of curatorial practice' was first designed and implemented in a Caribbean exhibition context by McGuire (2015) for Transoceanic Visual Exchange. More on this can be found at *TVE 1: A Community of Curatorial Practice*. https://www.transoceanicvisualexchange.com/tve-1, Accessed 20 January 2022.

4 Such as *Second Generation* (1997–2000), *London Link* (1982) and *Unshackled?* (1997–2000).

5 Advertised at this link https://www.gold.ac.uk/calendar/?id=12821

6 This exhibition page can be located at https://2ndgenbdosbham.org.uk/enigma-arrival-exhibition

7 The virtual exhibition links can be seen here, although the physical exhibition has since been decommissioned https://uwi.edu/museum/exhibitions/enigma-arrival-politics-and-poetics-caribbean-migration-britain

8 This exhibition page can be located at https://www.readingmuseum.org.uk/explore/online-exhibitions/windrush-day-2020

References

Annan, B., Annan, J., Wotton, M. and Burton, R. (2014). 'Facilitated networks of learning', *Seminar Series*, 237, pp. 3–19.

Barriteau, E., Cummins, A. and Keohane, K. (2019). 'Heritage, gender and migration', in Brown, K., Cummins, A. and Weil, K. (eds.) *EULAC Museums. Museums and community: concepts, experiences and sustainability in Europe, Latin America and the Caribbean. Report on a policy round table held at the European Commission offices, Brussels, 29 April 2019*. EU-LAC Museums, pp. 15–21.

Berlucchi, M., Bonnici, S., Burns, W.J. and Monenerkit, M. (2018). 'Community curating: a macro to micro view', *American Alliance of Museums*, 13 March. [Online]. Available at: https://www.aam-us.org/2018/03/13/community-curating-a-macro-to-micro-view/ (Accessed: 22 June 2021).

Cassidy, C.A., Fabola, A., Miller, A., Weil, K., Urbina, S., Antas, M. and Cummins, A. (2018). 'Digital pathways in community museums', *Museum International*, 70(1–2), pp. 126–39.

Cummins, A. (1992). 'Exhibiting culture: museums and national identity in the Caribbean', *Caribbean Quarterly*, 38(2), pp. 33–5.

Cummins, A. (1994). 'The "Caribbeanization" of the West Indies: the museum's role in the development of national identity', in Kaplan, F. (ed.) *Museums and the making of ourselves: the role of objects in national identity*. Leicester University Press, pp. 192–221.

Cummins, A. (2004). 'Caribbean museums and national identity', *History Workshop Journal*, 58, pp. 224–45.

Cummins, A. (2013). 'Memory, museums and the making of meaning: a Caribbean perspective', in Stefano, Michelle L., Davis, Peter and Corsane, Gerard (eds.) *Safeguarding intangible cultural heritage*. The Boydell Press, pp. 2–32.
Cummins, A., Farmer, K. and Russell, R. (eds.) (2013). *Plantation to nation: Caribbean museums and national identity*. Champaign, IL: Common Ground Research Networks.
Davis, P. (1999). *Ecomuseums: a sense of place*. London: Leicester University Press.
Dewey, J. (1938). *Experience and education*. New York: Simon & Schuster.
Diamantopoulou, S., Insulander, E. and Lindstrand, F. (2012). 'Making meaning in museum exhibitions: design, agency and (re-)representation', *Designs for Learning*, 5(1–2), pp. 11–29.
EU-LAC Museums (2021). *EU-LAC Museums*. [Online]. Available at: https://eulacmuseums.net/index.php (Accessed: 22 June 2021).
Fanon, F. (1952). *Black skin, white masks*. New York: Grove Press, 1991.
Fanon, F. (1961). *The wretched of the earth*. New York: Grove Press, 1963.
Farmer, K. (2013). 'New Museums on the Block: Creation of Identity in the Post-Independence Caribbean', in Alissandra Cummins, Kevin Farmer & Roslyn Russell (eds.) *Plantation to Nation: Caribbean Museums and National Identity*, Common Ground Publishers, pp. 169–77.
Hein, G. (1991). 'Constructivist learning theory: the museum and the needs of people', *CECA (International Committee of Museum Educators) conference*, Jerusalem, 15–22 October. [Online]. Available at: http://beta.edtechpolicy.org/AAASGW/Session2/const_inquiry_paper.pdf (Accessed: 14 January 2022).
Hewitt, G. (2019). 'The Windrush scandal: an insider's reflection', *The Journal of the Barbados Museum & Historical Society*, LXV, pp. 149–82.
Hooper-Greenhill, E. (1992). *Museums and the shaping of knowledge*. London and New York: Routledge.
James, C.L.R. (1938). *The black Jacobins: Toussaint L'Ouverture and the San Domingo Revolution*. New York: Vintage Books, 1963.
Karp, I. (1992). 'Introduction: museums and communities: the politics of public culture', in Karp, I., Kreamer, C.M. and Lavine, S.D. (eds.) *Museums and communities: the politics of public culture*. Washington, DC: Smithsonian Institution, pp. 1–17.
Lave, J. and Wenger, E. (1991). *Situated learning: legitimate peripheral participation*. Cambridge: Cambridge University Press.
McGuire, N. (2015). 'TVE 1: a community of curatorial practice', *Transoceanic Visual Exchange*. Transoceanic Visual Exchange. Available at https://www.transoceanicvisualexchange.com/tve-1 (Accessed: 20 January 2022).
Morales Lersch, T. (2019). 'Community museums: telling a story, building a future', in Brown, K., Davis, P. and Raposo, L. (eds.) *On community and sustainable museums*. EU-LAC Museums, pp. 38–53. [Online]. Available at https://doi.org/10.5281/zenodo.2646479
Piaget, J. (1936). *Origins of intelligence in the child*. London: Routledge & Kegan Paul.
Prendergast, M. and Saxton J. (eds.) (2009). *Applied theatre: international case studies and challenges for practice*. Bristol: Intellect.
Sandell, R. (2002). *Museums, society, inequality*. London: Routledge.
Sandell, R. (2007). *Museums, prejudice and the reframing of difference*. London: Routledge.
Simon, N. (2010). *The participatory museum*. [Online]. Santa Cruz: Museum 2.0. Available at: http://www.participatorymuseum.org/read/ (Accessed: 22 June 2021).

Simpson, M. (2001). *Making representations: museums in the post-colonial era*. London: Routledge.
Trouillot, M.R. (1995). *Silencing the past: power and the production of history*. Boston: Beacon Press.
Vawda, S. (2019). 'Museums and the epistemology of injustice: from colonialism to decoloniality', *Museum International*, 71(1–2), pp. 72–9.
VMCMM (n.d.). *Virtual museum of Caribbean migration and memory*. [Online]. Available at: https://eu-lac.org/vmcarib/ (Accessed: 22 June 2021).
Vygotsky, L.S. (1978). *Mind in society: the development of higher psychological process*. London: Harvard University Press.
Watson, S. (2007). *Museums and their communities*. London: Routledge.

11

A CASE STUDY OF COMMUNITY VIRTUAL MUSEUMS IN THE AGE OF CRISIS

Designing a Virtual Museum of Caribbean Migration and Memory

Catherine Anne Cassidy, Alan Miller and Alissandra Cummins

The Virtual Museum of Caribbean Migration and Memory (VMCMM) tells stories of migration from a Caribbean perspective and is focused on the experience and impact of the Windrush generation. The VMCMM was created in part to commemorate the 70th anniversary of the HMT *Empire Windrush*'s landing in the United Kingdom, carrying approximately 600 migrants from the Caribbean.

This chapter places the VMCMM within a wider context of future capacity and challenges for virtual museums in the Caribbean by reporting and reflecting on the round table held at the MAC Annual Conference in 2018. The panel included the University of West Indies (UWI) Museum, the Barbados Museum & Historical Society (BMHS), the Museums Association of the Caribbean, the Universidad Austral de Chile and the University of St Andrews. We discuss our experiences in developing digital architecture to support workshops held in communities and museums in Jamaica, Trinidad and Tobago and Barbados. These were organised through the EU-LAC Museums project and involved hundreds of participants, including school students, museum volunteers and professionals. The workshops streamlined and simplified photogrammetry for community museums to create, archive and curate representations of heritage using 3D and 360-degree media.

In the discussion that follows, we detail the design and architecture of the VMCMM, a virtual museum, based on experience of working with community museums. At its core is the open-source Omeka archive and exhibition-building content management system framework for museums and collections. The virtual museum framework supports interactive mapping, digital galleries, timelines and digital panels that can be integrated into web narratives. The framework connects with Social Archive sites, the World Wide Web and mobile applications. It makes use of open standards and provides powerful but simple metadata and archives. Our design philosophy sees virtual museums as an active resource, supporting

DOI: 10.4324/9781003288138-14

content generation as well as enabling content to be deployed in multiple exhibition contexts.

The chapter discusses how the virtual museum architecture was developed and combined with content to create the VMCMM. The mapping capabilities of the VMCMM enabled spatial visualisation of migrations, showing the scale of departure, movement and final terminus. Spatial context visualisations help viewers to understand the significance of a people in transition and the effects that migration has on society and culture. The combination of virtual maps, historical narratives, first-person accounts and digital heritage generates an understanding and appreciation of humanity during major events. In addition, its integration of emergent media, such as 3D models, 360-degree images and virtual reality (VR), adds depth and authenticity. The VMCMM infrastructure connects interactive mapping with media through a content management system to achieve synthesis between the two.

The VMCMM was unveiled as part of *The Enigma of Arrival* exhibition at the Museums Association of the Caribbean's 2018 conference and continued as a touring exhibition across the Caribbean. *The Enigma of Arrival: The Politics and Poetics of Caribbean Migration to Britain* mixed physical exhibits with projections of historical audio-visual recordings, digitised 3D objects and interactive panels with embedded associated media. It created an immersive space that allowed visitors to experience Caribbean perspectives on migration. The virtual museum, available online as well as within the exhibition site, further became an active platform where users can upload personal migration stories, adding to the breadth of co-curated content and known cultural narratives. The online element of the exhibition supported the transition to strictly virtual content during the COVID-19 pandemic and has bolstered the digital content and resources on this topic that are available to a global audience.

The Virtual Museum of Caribbean Migration and Memory

When the HMT *Empire Windrush* docked in the Port of Tilbury outside London in June 1948, it carried one of the largest groups of West Indian migrants to the United Kingdom. Its journey has become an iconic symbol of Caribbean post-war migration to Britain. The impact of West Indian migration was transformational for British society, touching all aspects of life, including work, culture, politics and sport. The following case study examines the practice of designing a virtual museum to capture and communicate the Windrush migration experience from a Caribbean perspective.

The design for the VMCMM began in 2017 as part of the EU-LAC Museums Horizon 2020 project. Three motivations were key in its development:

- The topic of migration goes to the heart of cultural and political issues around the globe. The Caribbean experience, from the Panama Canal, through the Windrush generation to today, provides a complete insight into both the long-term benefits to the receiving localities and the challenges posed to communities left behind.

- There are abundant museums, archives and resources relevant to migration and gender but are often difficult to locate, presenting a challenge to gaining an overall sense of the resources available. A virtual museum on the topic offers a single destination and framework for those interested in the subject.
- New methods of presentation and associating data are regularly emerging, which herald the potential of a qualitative enhancement of virtual museum experiences. Through 3D digitisation processes, digital artefacts produced preserve and enhance the original physical media. Digital objects can fill a digital gallery, replicating a physical gallery, with associated interpretation and metadata to provide dissemination. They can be placed into scenes that situate the artefacts within real or virtual representations of their original context.

The virtual museum was paired with a complementary physical exhibition designed by the BMHS and the UWI. The exhibition travelled to key locations on main Caribbean migration routes, as well as prominent final ports of call, in order to encompass the totality of the migration experience, obtain stories from those affected and show connections with modern immigration topics.

The virtual museum combines physical and digital exhibits, offering greater audience accessibility and participation in order to develop a collection of previously undocumented individual stories. It uses an archiving system designed specifically for the project, which aligns with open heritage aggregators and schema. It also supports the creation of new media, such as video entries, photogrammetry for the preservation of tangible heritage and virtual tours of locations associated with stories. A focus on telling individual stories of the Windrush generation using images, text and video allows previously unheard voices to contribute to a significant historical moment. Collected stories, associated media and resources and featured individuals are accessed by search, galleries and an interactive mapping interface, displaying the virtual museum's collection in a comprehensive design.

As access to digital technologies and growth of digital literacies progress, experimentation to discover novel ways to preserve and record history is paramount for heritage organisations, and within reach (Cassidy et al., 2018). The VMCMM models an innovative approach to collecting memory while disseminating academic research and narratives and offers support for further media creation to enhance the overall understanding of Caribbean migration. The VMCMM provides a rich framework for exploring the lack of representation of the past and present diaspora. It does so by curating migration narratives, reclaiming diasporic community experiences and illuminating the social and cultural dynamics of retelling national stories. In the COVID-19 era, it became the only means to access the exhibition *The Enigma of Arrival*, while the physical exhibition was no longer available. The global switch to digital media that occurred during the pandemic led to the VMCMM's expansion through additional content as well as redesigned infrastructures.

This shift in focus enabled the virtual museum to become the principal resource highlighting individual stories from the Windrush generation during the project.

The Enigma of Arrival: The Politics and Poetics of Caribbean Migration to Britain

With its arrival from Jamaica to London in 1948, the *HMT Empire Windrush* brought one of the first large groups of post-war West Indian immigrants to the United Kingdom—approximately 600 of the 1,027 total passengers. Post-emancipation migration out of the Caribbean had occurred prior to this timeframe, notably to Britain and Europe in response to the call for recruits in the First World War and to Panama during the construction of the Panama Canal (1903–1914). However, the choice to focus on the Windrush generation came from an identified gap in the collections and narratives of museums across the Caribbean region. Although this theme may have been accessible to diasporic communities in the United Kingdom, UWI's research revealed limited opportunities in Caribbean museums to share the significance of this migration's impact on the economic, social, academic and cultural landscapes of the United Kingdom and the Caribbean. In a fluid, regional space, where so many Caribbean people have come from somewhere else over centuries, whether by force or freely, UWI's research into the collection holdings in Caribbean museums revealed that those institutions had so far made little effort to reflect on past and present experiences of migration and diaspora.

The exhibition *The Enigma of Arrival: The Politics and Poetics of Caribbean Migration to Britain* focused on Caribbean migrations from the 1940s to the 1970s in order to highlight and provide a comparison for the modern migration crisis in Britain. Alongside the VMCMM, it aimed to serve as a medium for informing the general public across the Caribbean and in the United Kingdom about the current state of affairs for Caribbean migrants. Through interpretive panels and interactive audio-visual materials, the audience was introduced to the historical parameters and scope of post-World War II Caribbean migration to Britain and its legacy among later generations, both in the Caribbean and abroad.

The growth of access to technology

Access to technology is at an all-time high and surely set to rise. In 2018, for every 100 people globally, there were 107 mobile and 69.3 mobile broadband subscriptions, compared to 63 and 12 per hundred in 2010. According to ITU (2021) statistics, Barbados saw an increase of 16.66 per cent in internet use in seven years, from 65.1 per cent in 2010 to 81.76 per cent in 2017. Jamaica saw a rise as well, from 26.67 per cent in 2010 to 55.07 per cent in 2017. Greater still, the growth of mobile broadband subscriptions allows access to the internet through social media platforms and search engines. Barbados had significant growth from 2000 at 10.48

subscriptions per 100 people to 124.08 in 2010 and a slight levelling out of 114.74 in 2019. Jamaica saw a similar increase from 13.82 subscriptions per 100 people in 2000 to 113.22 in 2010 and 102.56 in 2019. These figures are consistent with previous experiences working with communities and their museums during the EU-LAC Museums project. In each country and region, the digital infrastructure of mobile phones, computers and digital literacy in communities across Latin America, the Caribbean and Europe was sufficient to facilitate community engagement in digital heritage.

Digital technologies have seen qualitative transformations in recent years in terms of fidelity, usability and presentation. These have allowed digital technologies to represent heritage in a myriad of ways, making mobile and immersive interactions available on commodity devices while promoting accessibility, since potential audiences have both the required digital literacy and the access to devices that are capable of delivering immersive content. This potential coincides with strategies pushing for the creative use of digital technology, such as the UK government report *Culture is Digital* (Department for Digital, Culture, Media & Sport, 2019) and *Digitising Collections: Leveraging Volunteers & Crowdsourcing to Accelerate Digitisation*, which stresses the importance of museums adopting a digital strategy (AXIELL, 2017). The relevance and importance of digitising collections as part of collections management is demonstrated in the European Union (2017) report *Promoting Access to Culture via Digital Means*.

Digital technologies are not only important tools for improving accessibility to museums' resources but can also introduce new methods of gathering digital data. If a museum has the capacity to go digital, it is able to expand from being locally rooted to having a global footprint. Through digital technologies, museums can contribute to the growth of global knowledge, promote social cohesion and inclusion and promote a better understanding of different cultural roots within societies in ways that go beyond direct contact with audiences.

Virtual museum innovations

Virtual museums have typically focused primarily on providing remote access to interpretation of collections held physically by a museum. In recent years, considerable technological progress has been made, allowing virtual museums to advance from simple image galleries, to support 3D and other immersive media and now VR with fully explorable environments. As a result, their scope has undeniably changed, as new media and resources have had to be made compatible with platforms already familiar to heritage professionals to improve interoperability (Fabola et al., 2017).

In order for virtual museums to deliver content, narratives and experiences that adhere to the same principles as their physical counterparts, multiple processes must take place to ensure interconnectivity with known curatorial procedures, schemas and databases. Exhibitions require research into collections as well as interpretation of materials; these processes rely on the capacity and skills of museum staff, volunteers and community members. This is also the case when curating an online exhibition using objects and stories held in a physical museum. Moreover, data and media should link with larger digital libraries for better global accessibility and should be comprehensible to different types of audiences using the content.

Perspectives from a Caribbean round table on virtual museums

The above issues were at the centre of a discussion held during the Museums Association of the Caribbean annual conference in 2018 held within the *Itinerant Identities* conference at UWI Cave Hill Campus in Barbados. A round table discussion was organised during the proceedings, in which participants included community and university museum leaders, as well as digital heritage experts. The members of this 'Caribbean Virtual Museums: Opportunities and Challenges' round table discussed the design, functionalities and specific requirements of future virtual museums for the Caribbean.

The panel was an opportunity to consider how new media might help in the process of learning to create and archive digital heritage, and develop mixed virtual exhibitions in a digital format. The potential for drawing connections and sharing perspectives through digital heritage is perpetually growing. Advances in technology have improved general digital literacies to the point where museums and their audiences can use these technologies for both engagement and dissemination. The panel also addressed common challenges and misconceptions related to digital

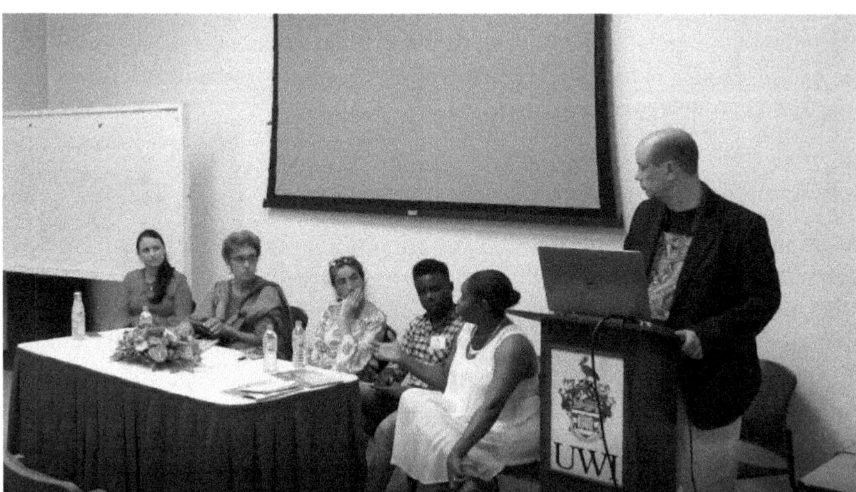

FIGURE 11.1 Panellists in the round table discussion 'Caribbean Virtual Museums: Opportunities and Challenges' during the *Itinerant Identities* conference at UWI Cave Hill campus, Barbados.

heritage and virtual museums and shared solutions with one another. All participants agreed that the rich heritage spread geographically across the Caribbean means that virtual museums have great potential to enable its interpretation and communication.

The panellists began by deliberating and establishing the place of virtual museums within regular museum programming and discussing whether they expand or restrict the museum's ethos. The risk that a virtual museum could overexpose a museum's collections and divert physical audiences exclusively online was raised as a known concern, mostly in organisations that did not already have a prominent online presence. In this regard, the expenses of smaller community museums may not be as large as those of national museums, but these institutions still require patronage. It was agreed that new technology does not threaten to replace a unique in-person museum experience but should run conjointly with existing processes and frameworks to enhance the museum's message. This was said succinctly by anthropologist Karin Weil: '[the museum] wasn't created to bring more people. It was created for identity, memory, human rights, native population [...] it is important to attract more people [...] but also you have to have a balance' (Cassidy, 2018, p. 2). It was agreed that, if community museums are part of their communities, and intended to contribute to their social sustainability, the digital domain can broaden and bolster these aspirations, as well as boost visits to physical exhibitions.

Moreover, as Geology lecturer and museum curator Sherene James-Williamson noted, a virtual museum can change and evolve with both online and physical audiences. James-Williamson observed that '[her] visitors [were] participating more in making [her] virtual museum a virtual museum' (Cassidy, 2018, p. 3). She argued that this could be achieved by allowing visitors 'to take their images [...] and they have their own narrative [...] when visitors come, they also come with information [...]. So I see our online museum environment more interactive and more informative in the next couple years' (Cassidy, 2018, p. 3). Crowdsourcing and co-curation have been a part of museology for the last few decades, but the digital domain can cast a wider net than has been previously accomplished. The internet has facilitated access to a global audience that can contribute to narratives and knowledge, and these contributions could be furthered by a virtual museum. Computer scientist Adeola Fabola confirmed that

> there are so many opportunities here, so many possibilities that our technology would allow, no doubt there would be challenges, but if the interest is there, the tech circles and the heritage experts could work together for a viable solution.
>
> *(Cassidy, 2018, p. 5)*

Within the context of sharing in the digital domain, an issue was raised regarding travelling exhibitions, which is both a distinctly Caribbean challenge and a wider contextual issue. Collections are shared externally through libraries and community centres as well as with other museums throughout the islands, but the process is still difficult. It also presents physical risks to objects that are eliminated through publication as digital models in a virtual museum. Such models could be exhibited with interpretation and metadata, as well as the capacity for viewers to examine the object more closely than they might in person. Rather than printed and sent panels, interpretative material could be included as a digitally packaged exhibition that either remains completely digital or can be printed at the host's discretion. However, it was remarked that converting a physical exhibition to a digital form in a virtual museum is not a 'one size fits all' approach; the resourcing and goals of each exhibition require consideration. Yet, as James-Williamson argued, the addition of an online element to a physical exhibition need not replace it but rather might enhance and augment the experience, leading to different types of engagement and reactions: 'the borrowing and the booking of the exhibition [...] but doing it in an online environment is really really exciting' (Cassidy, 2018, p. 4). Researcher and curator Natalie McGuire could also see the potential for 'these community centres instead of relying on us to bring panels to them, they could download interpretative material provided by institutions and share reach and build agency within their own spaces' (Cassidy, 2018, p. 2).

A common challenge observed by the panellists was the requirement to educate current staff in digital skills and terminology, acquire new digitally conversant staff who could manage a virtual museum and its components and pass down knowledge gained by an individual to other staff. Weil wondered 'if I got out of museums, not working anymore there, who's going to continue, who's going to get training from whom' (Cassidy, 2018, p. 8). The work required to create and add digital content to an online resource, as well as access to technical support when needed, seemed daunting. A few members of the round table had been a part of the EU-LAC Museums project and had experience of helping to generate and manage digital content for a larger project but may not have done so by themselves. A discussion between the museums and technology experts led them to agree that building capacity within museums was valuable, as was utilising the digital literacy of their communities. Tapping into the community as a resource can provide insights, skills and perspectives needed to design a virtual museum that meets the community's needs and interests. This was a frequent observation made during the digital workshops that were run as part of the EU-LAC Museums project and hosted across nine countries. Each participant's skills benefited the digital content created as well as the museum, as new connections were made that could lead to future collaborations.

One aspect of digital collections that was enthusiastically acknowledged was their potential to aid global research. The value of digitised collections is undeniable. Researchers were once required to either travel to an object or request that an item be shipped to them. Barriers that may prohibit or restrict this fundamental part of research can be removed if the object is digitised. Speaking of her museum, James-Williamson commented that

> [it] would benefit a lot, some of our collections people want to study them but Jamaica is quite far away from where the people are. So I'm seeing, in terms of

> a virtual space, making my collections accessible, whether by database initially [...] there are a bit more intricacies to get things going, but to provide an online kind of a catalogue.
>
> *(Cassidy, 2018, p. 3)*

Primary sources were considered. An accurate and accessible digital representation of an object allows research to continue and for comparative analysis and new inquiries to be made. Digital representations allow access for anyone who would not be able to see the physical object in person, which could be as a result of distance, disability or financial strain. Digitisation is then mutually beneficial to the researcher and the museum. Once collections are digitised, their uses are infinite. A digital object is a single file—or, in the case of 3D models, a few file types—that can promote museums, their collections and exhibitions in the form of digital or printed media. It can increase the visibility of the object, museum and associated community. Additionally, digitisation captures an object at a certain point in time, which allows it to be examined as it once was. This is imperative in case of damage or loss. Digitally safeguarding collections is a method for museums to be actively sustainable and to future-proof themselves against unforeseeable incidents.

Participants in the round table also acknowledged the misconception that technology has a high barrier to entry in monetary terms and the skills required for use. When asked how a museum can fit a new technology into its practices, Fabola recommended that museums should rather ask: how can this technology 'supplement how we do things? This technology could walk hand in hand with your existing processes and your existing infrastructures' (Cassidy, 2018, p. 1). Fabola argued that new technologies should not be considered merely for the sake of their novelty or flashiness but must be able to contribute and assist within the museum's existing framework. He also asked that practitioners 'consider the notion of the resource spectrum [...] where as we know there are hundreds of thousands of museums across the world with varying resource availability' (Cassidy, 2018, p. 2). His research suggests that, in order for all museums to take advantage of what emergent technologies can offer, 'it's possible to have some sort of framework where you can use some subtle technology with varying levels of cost and availability irrespective of the resources at the museum's disposal' (Cassidy, 2018, p. 2).

Developments and outcomes in the EU-LAC Museums project demonstrate that even with resource restrictions, the commodity technology held by the community was enough to produce new types of digital content, such as 3D models and 360-degree virtual tours. The threshold to new technologies is lower than commonly thought. Moreover, once established, the use of these technologies can be expanded and adapted in museums' practices.

The question of who virtual museums are for, and what their roles are, was raised as a response regarding resourcing and capacity-building within museums. Uncertainty about staff availability was prominent, especially since a majority of small or community museums have limited resources. Practitioners noted that the role of a virtual museum needs to be resolved before staff allocation can be considered, as the responsibilities of a virtual museum depend on its core function. If it is primarily to increase visitor numbers and promote footfall, then a staff member with tourism and public relation skills could be best placed to manage it. If it is to be a resource for digital collections and exhibitions, however, then a curator may lead its development. A clear understanding of what programme the virtual museum supports can assist in evaluation, which aids in proving its effectiveness to authorities. As Weil suggested, one of the questions to ask before developing a virtual museum is: 'who is going to be in charge of the virtual museum, for what, and maybe [moving forward], you can use it for a lot of things [...] but each one has to be directly in touch with that' (Cassidy, 2018, p. 7).

In the discussion, the possibility of sharing a pan-Caribbean vision for the island nations with the world was an exciting proposal that overrode trepidations. Preserving and sharing knowledge using new technologies and methods was understood as having the potential for sustainability that small museums strive to achieve. Historian and curator Suzanne Francis-Brown observed that the UWI museum's focus on a 'pan-UWI complete Caribbean perspective' is what it was created to pursue (Cassidy, 2018, p. 8). She stated that 'from that perspective of knowledge creation and knowledge sharing, it would be something useful and possible, and perhaps pull in many elements of the university's teaching and learning endeavours, as well the level of students as lecturers' (Cassidy, 2018, p. 8). This involvement would in turn link to capacity-building within the museum, as sustainability for virtual museums depends on the proficiency of a team of staff and volunteers. In cases where a virtual museum relies on only a single person, it must be of the utmost importance to disseminate their knowledge so the management and upkeep of such a resource are continuous.

Summary from the round table discussion

As agreed upon by heritage and technology professionals during the round table, virtual museums are in a unique position to offer local communities, as well as regional and international audiences, access to and inclusion in museum collections and narratives. In a Caribbean context, the functions and capacities of such a resource could solve familiar challenges faced by island museums. The round table discussion, with its variety of panellists, illuminated the possibilities and prerequisites for the success of virtual museums. Drawing on this conversation, we determined that the crucial requirements for a pan-Caribbean virtual museum resource would include:

- community-sourced content and narratives
- access and creation beyond the walls of the museum
- guidance in the use of commodity hardware to create new digital content

A prototype virtual museum, the VMCMM, was developed and trialled at the opening of *The Enigma of Arrival* exhibition, which occurred during the same event as the round table. The VMCMM was created using a practice-based methodology such that user needs and perspectives identified at the exhibition opening were incorporated into the design process The main points made by the panellists were considered and served to shape future iterations of the VMCMM. In turn, this process has informed the design and practicalities of virtual museum resources in subsequent initiatives, including the CINE (2017), CUPIDO (2018) and STRATUS (Northern Periphery and Arctic Programme, 2019) projects.

Engaging with migration and memory

The digital domain enables stories and media to be shared without the constraints presented by the physical world. A transatlantic event such as the Windrush migration includes those from multiple communities in various countries, and its effects resonate with other migration stories all over the globe. It is therefore an ideal topic to explore within the digital domain and to test the technical range in which a virtual museum platform can be developed.

The design for a migration-themed virtual museum would need user journeys to include the following in order to educate the audience:

- an exploration and understanding of the communities involved
- where in the world the event took place
- why it is a significant part of history
- the human connection

These social elements feed into design decisions based on the content available and the best method for dissemination. The virtual museum, as a digital online platform, collates exhibition materials, as well as the primary resources used to create the exhibition, in one location. A comprehensive collection of digitised content is at users' fingertips, with the potential for increased interaction compared to a physical visit. With these intentions, the conclusion that a virtual museum would complement an EU-LAC Museums history-based exhibition was acknowledged and developed.

The mixed-media exhibition for *The Enigma of Arrival* opened at UWI Cave Hill Campus in Barbados in 2018. Curators from the BMHS implemented a community practice approach for co-curating content for the exhibition, a community peer review system and sustainable exchanges of content between the museum and its communities (Cummins et al., 2020).

For the exhibition and the VMCMM, the information panels were converted into interactive digital displays on touchscreens, as well as projections for the physical exhibition. Additional content—such as recorded songs, and historic performances and images—was embedded in the panels, thus enabling a rich

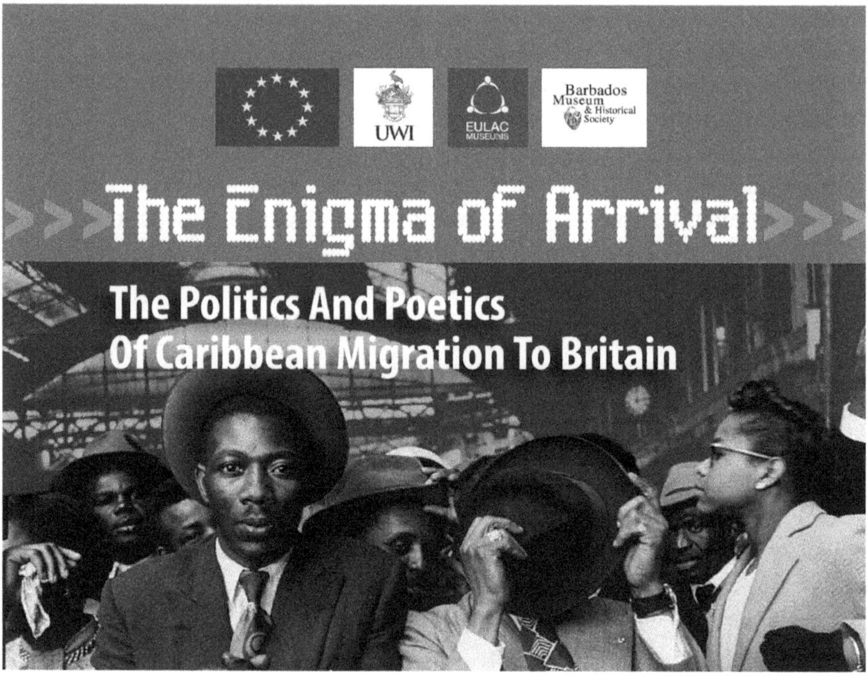

FIGURE 11.2 Digital poster panel from *The Enigma of Arrival* online exhibition, available on the VMCMM.

multimedia environment. Poignant interviews and archival videos were projected on the exhibition's walls. In addition, a mobile application was developed for the exhibition that facilitated interaction with 3D models created from the BMHS and UWI collections; this was available on provided tablets. The process of photogrammetry was shown on the opening night as a sustainable entry into 3D digitisation, and the VMCMM's 3D queue system for remote processing was demonstrated. To complement the curators' intentions for community co-curation of content, a video capture area offered attendees the ability to share their own migration stories and add them to the VMCMM. Participants filled out an upload form with information associated with their stories, including the location of departure and arrival. This information was added to the archival database and then appeared on the mapping interface. Entries to the database in turn contributed to a wiki platform, which built upon and connected multiple contributions. Whether contributing from the physical exhibition or from their own homes, audiences were encouraged to tell their personal stories, adding to the breadth of cultural narratives and co-curated content.

Available online as well as within the exhibition, the VMCMM reaches beyond the physical barriers of the exhibition space and provides a platform for further investigation, research and uploading of personal migration stories. It is designed to be an active resource for content collation and generation, informed by evaluation from the EU-LAC Museums' digital workshops and strong ties between the content curators (BMHS and UWI) and their community.

In the VMCMM, an interactive map is the main navigational tool used to display audience-provided stories as well as spotlighted narratives and notable

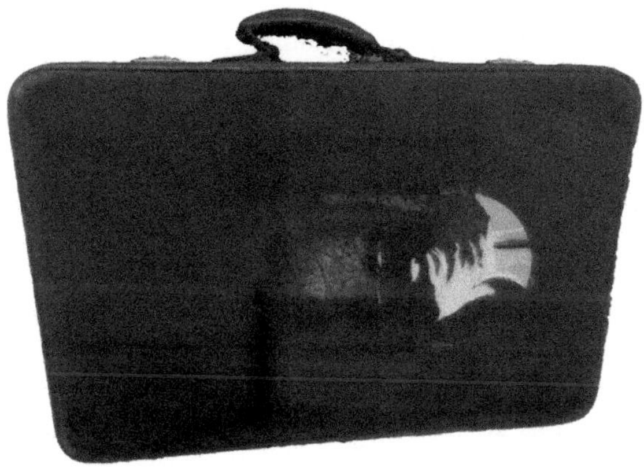

FIGURE 11.3 3D model of a suitcase from the BMHS collections, digitised during their 3D Summer Intensive. The physical object was on display at *The Enigma of Arrival* exhibition at the BMHS in 2019 alongside the 3D model.

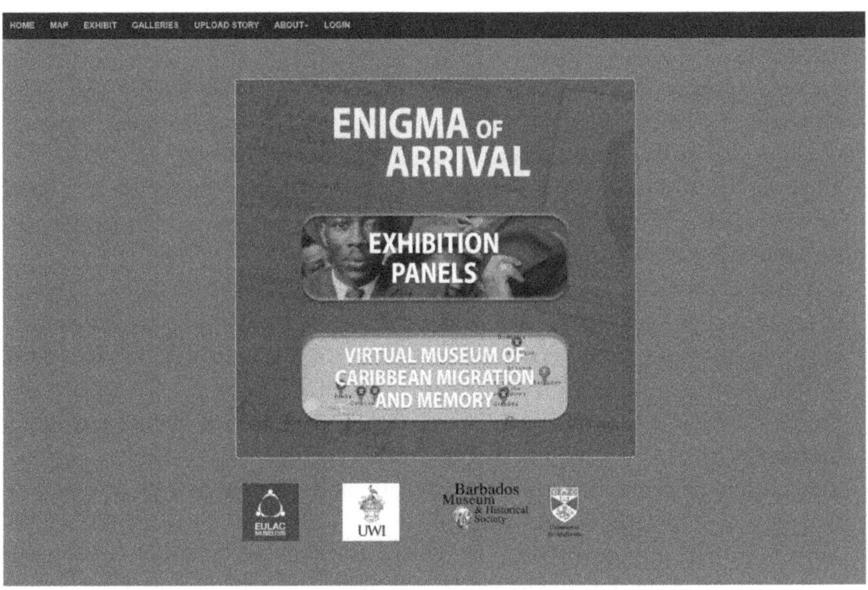

FIGURE 11.4 The homepage of the VMCMM, which allows users to view digital exhibition panels or continue to the map interface.

figures. The map is organised into layers that connect digital resources together: for example, Caribbean carnivals, which have spread globally and are evidence of migration. It also enables users to explore the social media presence and websites of related organisations, making external resources for further research available. Layered within the map are 3D digitised objects that link personal accounts or events with physical objects, most of which originate in BMHS collections. The panels from the physical exhibition, including the embedded media, are available digitally. These provide access to users who could not visit the exhibition and proved the exhibition's resilience during the COVID-19 pandemic.

Virtual museum design and implementation

As the conceptual framework was established, design of the VMCMM was motivated by the state of technology accessible to the heritage sector. The internet, World Wide Web, social media platforms and mobile applications are today part of the everyday practice of museums. There is constant disruption caused by technological developments that are making new functionalities available.

Design priorities

The design of the VMCMM was informed by our previous experience of workshops on creating digital content, from the 'Caribbean Virtual Museums: Opportunities and Challenges' round table discussion, and through practice-based research into developing and using virtual museums. The VMCMM would:

- provide support for emergent media, such as 3D, 360-degree and aerial photography, and integrate it with established support mechanisms for traditional digital media forms, including audio, video, images and text
- allow for the straightforward uploading of personal migration stories, text-based and/or in video format, from the physical exhibition or remotely, with necessary metadata for archiving and analysis
- serve as a platform for the digital panels from the physical exhibition, with embedded media, such as 3D models and videos, associated with the selected themes
- supply a photogrammetry service that automates the creation of photogrammetric models from sets of photographs
- provide a wiki interface that enables community participation in the construction of narratives.
- The design priorities informed the functionality that the VMCMM would require. The VMCMM would need to:
- connect to existing open-source resources to create the functionality needed: leaflet for mapping, Open Street Maps, Omeka, MySQL, MediaWiki, WordPress and the International Image Interoperability Framework (IIIF), which provides support for spherical, 3D and flat media in galleries

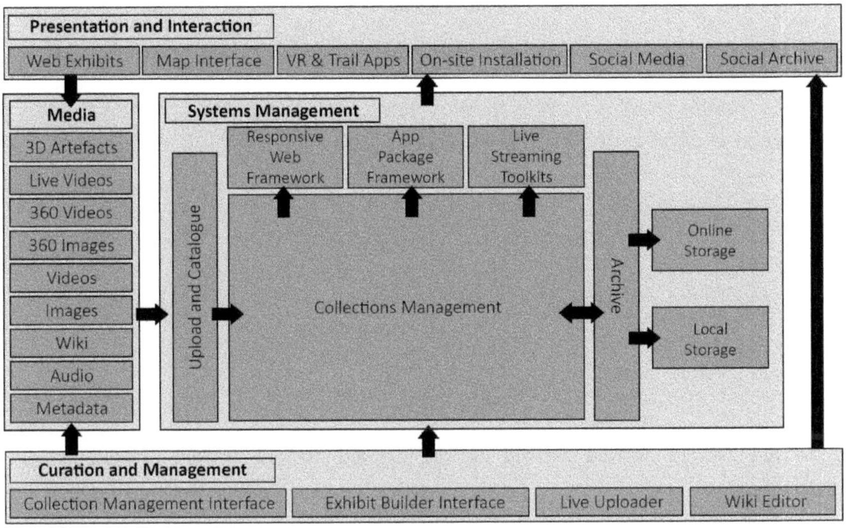

FIGURE 11.5 The architecture of a previously published VMI used for the VMCMM.

- make resources embeddable so that they can be included in web pages and social media and be shared through email and messaging applications.

The VMCMM further developed the Virtual Museum Interface (VMI) framework, which was initially created for the EU-LAC Museums project's virtual museum. The main system components are shown in Figure 11.5. Interaction typically begins with users creating content using the Management Interface. The Management Interface is a web-based form that enables users to upload and modify content, which is stored in the Data Store. Once in the Data Store, the data can be pushed to supported Social Archive platforms and backed up in an Online Store. The data that exists in the Data Store can be used to make exhibits either as VR and mobile applications or museum installations. The content created and curated using the Management Interface and the Exhibit Builder Interface can be accessed using the map-based Web Interface, mobile applications, museum installations and Social Archive sites, each of which draws content from the Data Store. The diagram shows the active nature of the VMI, which supports not only the collaborative creation and management of content but also the presentation and reuse of this content in different ways, such as web-based exhibits, mobile applications, museum installations and social media.

System back end

The back end uses a free and open-source Digital Asset Management System (DAMS) based on Omeka. DAMS provides interoperability with popular web-based

```
function load_data(api_url, api_key, element_type)
{
//api_url e.g. "http://mydomain.org/omeka/api"
//api_key e.g. "al0NgstR1ngwItHnumB3rsanDlett3rs"
//element_type e.g. "items"
$.ajax({
url: api_url+'/'+element_type+'?key='+api_key,
crossDomain: true,
type: 'get',
timeout: 6000,
success: function(json_response)
{
parse_data(json_response, element_type);
},
error: function (xhr, ajaxOptions, thrownError)
{
console.log(thrownError);
}
});
}
```

FIGURE 11.6 API call to retrieve all items from the Omeka database.

systems and frameworks and is supported by a Relational Database Management System, MySQL, which stores the data, and a Representational State Transfer Application Programmable Interface (API), which facilitates retrieving, modifying and adding to the data. Calls to the Omeka API are made using JavaScript (from the online presentation front end) and Python (from the management front end). The code snippet in Figure 11.6 shows a JavaScript function that is used to retrieve items in the Omeka repository using an AJAX call. Upon successful completion of the call, a 'json_response' object is received, which contains the result of the call in JSON format, and is then parsed as required by the 'parse_data' function. If the AJAX call fails or times out, an error is thrown instead. Data management and metadata handling are important features of cataloguing and archiving systems.

Schemas developed using Dublin Core help to provide interoperability between the vocabularies of cataloguing and archiving systems. The Europeana Data Model builds on the Dublin Core schema and is widely adopted across Europe; hence, it boasts a high level of familiarity with heritage practitioners. For this reason, the information provided using the management front end is mapped to UNESCO and Europeana types (which are, in turn, described in Dublin Core terms) so as to facilitate interoperability with existing systems such as Europeana. Classifications cover schema for intangible, tangible movable and tangible immovable heritage, allowing for a varied ingress of metadata.

Management front end

The online front end has two facets: first, a management front end, which is implemented as web-based forms that contributors use to create and manage data; and, second, a presentation front end, which includes a map-based interface, digital galleries and archive searching. The map interface features icon-coded pins that represent entities. Users can click on a pin to reveal a pop-up, which enables further viewing or interaction. Uploaded and archived media appear in IIIF galleries in a Universal Viewer, adapted for 3D models and 360-degree photography.

A web-based archive form enables the general public and heritage practitioners alike to upload files and supply metadata that together represent and describe entities that are presented to users using multiple media types. The associated files and metadata constitute the digital heritage representation mechanisms supported. Three-dimensional artefacts created using photogrammetry and scanning processes are used to disseminate physical heritage with greater digital reach. 360-degree images and videos are used to show landscapes and cityscapes as scenes in immersive virtual tours. Traditional media such as flat images, video and audio capture both tangible and intangible heritage, and associated narratives support heritage interpretation.

The VMCMM facilitates content curation and management through a variety of interfaces. A Live Uploader enables users to transfer digitised content to the Data Store. The content is described using metadata, while narratives are added and

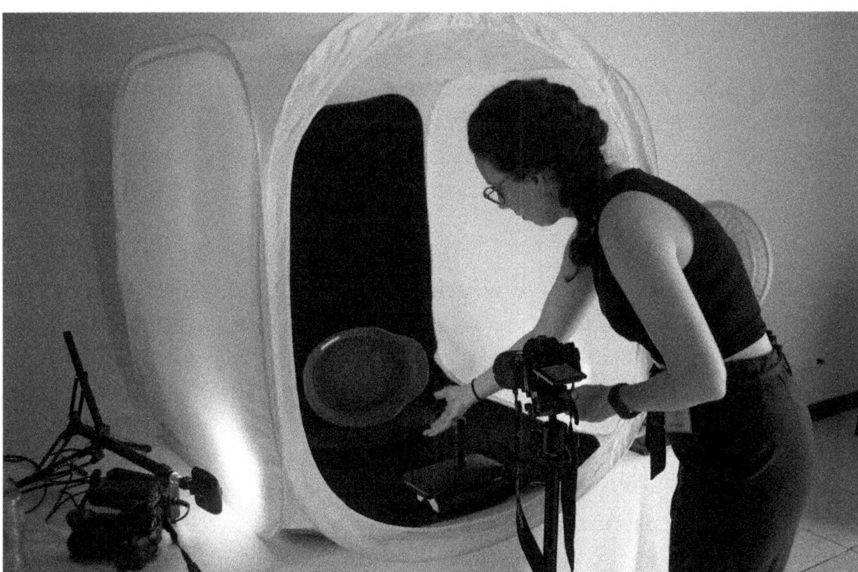

FIGURE 11.7 A demonstration of photogrammetry, used to create 3D models of physical objects in *The Enigma of Arrival* exhibition.

modified using a wiki editor. Collections are created and managed by thematically grouping uploaded items, while an exhibit builder facilitates the creation of exhibits that can be made publicly available. The content that is created using the curation and management interfaces is stored using both local and online storage solutions. Local storage facilitates access to resources despite a lack of internet access, while online storage facilitates interconnectivity and reach. The ability to ensure responsive delivery of resources is a desirable feature of the VMCMM design; it enables the system to serve the appropriate quality of content to users depending on their operating platforms.

The capacity to upload and manage content using the Live Uploader and Management Interface, respectively, and the ability to reuse this content and resources in different ways, emphasises the dynamic nature of the VMCMM. Heritage practitioners and community members can continuously create and digitise content; once uploaded, the same content can be used to build web exhibits and mobile applications and is accessible on a map interface as well as on social media. Metadata added to the content while it is being created or uploaded, or even subsequently, can be changed continuously within the resource throughout its lifespan.

Interactive map

The map-based interface facilitates the geographical visualisation of data, as it represents entities on an interactive map of the world. This enables the visualisation of the spatial and geographical relationships between entities. The interactive map supports:

- zooming and positioning, which enables geographical selection of data. Layers allow the semantic selection and presentation of data.
- categorical visualisation of data, such that users can choose to view only museums, 3D artefacts, tours, images or any of the other entity types available, and any combination of these types.

In addition to the visualisation of content by geographical location and category, the VMI supports descriptive visualisation of content using a wiki system, which collates provided metadata. The wiki is displayed in a panel alongside the map interface and features metadata, descriptive text and data as well as any associated multimedia in the form of 3D artefacts, virtual tours, images or video.

Exploratory visualisation is facilitated using an instant search feature that enables users to query the Data Store for entities that match a given search string. The search results are updated after every key press that modifies the query string, and it is performed on the title and description fields of entities. The instant search is facilitated by locally storing representations of the entities contained in the Data Store so as to preclude the need to repeatedly access the server while searching for entities.

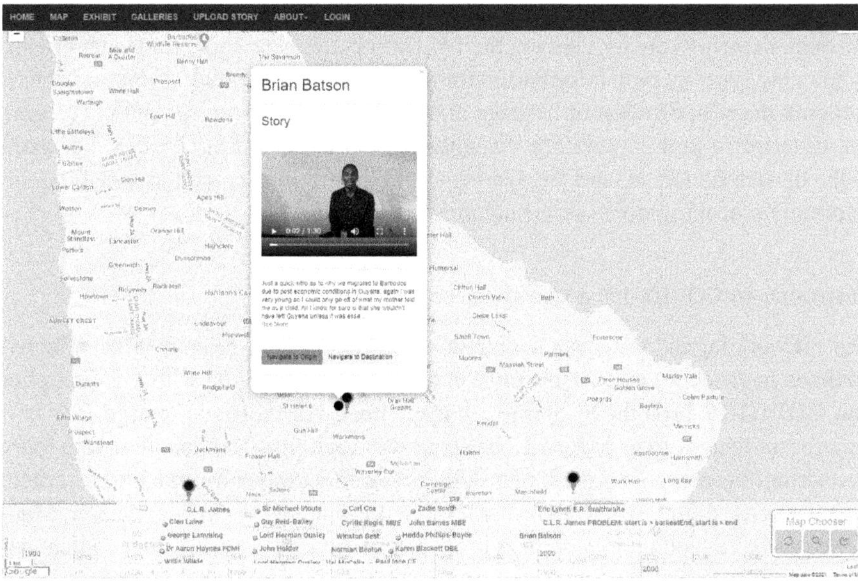

FIGURE 11.8 The video-recorded migration story of Brian Batson in the Destination location on the interactive map interface.

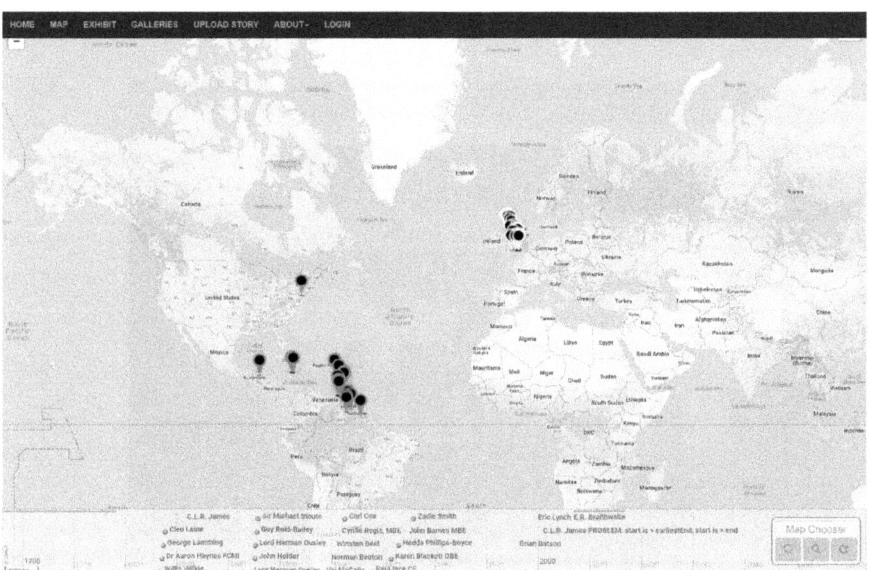

FIGURE 11.9 Mapping interface with an expanded timeline, showing notable figures from the Windrush generation.

Users can access the results geographically by viewing an entity's location on the map, descriptively by viewing the wiki content for the entity and categorically by viewing type-specific information for the entity. These features work together to facilitate the visualisation of heritage content in geographical, semantic, categorical, descriptive and exploratory visualisations, thus providing breadth and depth to the dissemination of, and engagement with, migration, as well as satisfying the information-seeking needs of virtual museum users.

Response to COVID-19: a use case for the VMCMM

The methodologies and system outputs from the project proved to be effective methods in reaching and connecting to audiences when faced with the disruption that COVID-19 brought to heritage engagement. As museums closed, the only way for audiences to engage with heritage was digitally. The use of virtual tours, live virtual events, social media and digitised collections have expanded heritage organisations' digital offerings, whether in a pre-planned or ad hoc manner. Many museums have seen their audiences grow as a result of their additional content and their outreach online.

The topic of Windrush and the VMCMM were part of discussions during several live heritage events broadcast on social media platforms as part of lockdown activities in 2020. As part of the CUPIDO project, alongside the Highland and Islands Enterprise, the curators of *The Enigma of Arrival* joined international conversations around digital themes and participated in discussions about digital methods

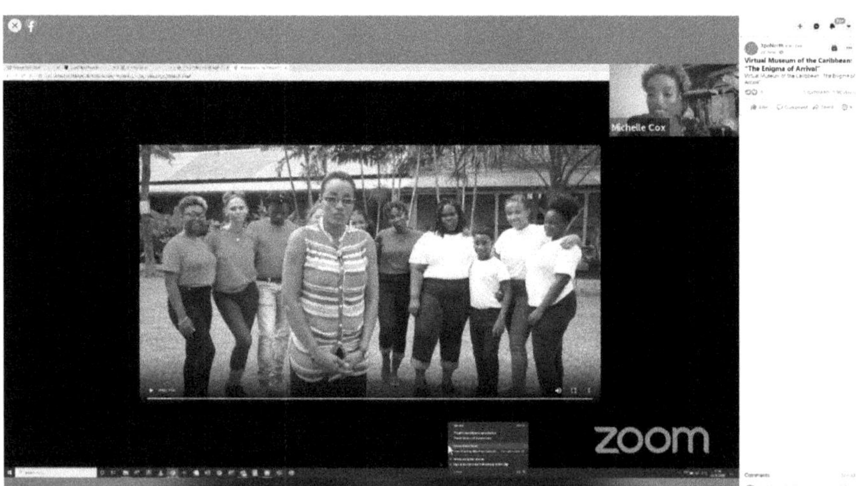

FIGURE 11.10 'Virtual Museum of the Caribbean: The Enigma of Arrival' live Facebook event, showing a video-recorded presentation of the play *Windrush* by members of the Barbados Community College.

for communicating heritage. One live event was solely dedicated to the 72nd anniversary of Windrush. It featured tours of the exhibition panels and the interactive elements of the VMCMM, narratives discussed by the curators, a screening of the Barbados Community College original play *Windrush* and a live reading of poetry shaped by the historical events.

In order to build digital heritage capacity during the period of lockdown, the EU-LAC Museums project's 3D workshops were recreated in the digital realm as a live online webinar series. Topics were expanded to include interactive mapping, the use of social media and virtual museums. The curators of *The Enigma of Arrival* offered their perspectives and knowledge on the topics and gave feedback on the use of digital methods, educational outreach and digital outputs.

Both live event series allowed for the framework and functional features of the VMCMM to be part of the public debate regarding community-based museums and the creative responses to the COVID-19 lockdown. The initial prototype virtual museum, which was launched in 2018 and influenced by the exploratory round table discussion, became a platform for experimentation with digital heritage activity that continued to connect communities to heritage throughout the global lockdown. The use of a virtual museum to supplement live events and webinars can enable heritage organisations to survive and develop resilience in times of unprecedented challenge.

The original plan for *The Enigma of Arrival* exhibition was to include a physical element, such as its travelling to various museums along the geographical migration route of the *Empire Windrush*. The exhibition opened at the UWI Cave Hill Campus in Barbados, the UWI Mona Campus in Jamaica and the BMHS. The BMHS launch was highly successful, and following multiple requests, the exhibition was extended until the end of 2019, doubling its initial duration. The 2nd Generation Barbadian & Friends community group in Birmingham, United Kingdom, alongside Brasshouse Lane Community Centre, hosted the *Enigma* exhibition from December 2019 until January 2020. Through the Centre for Caribbean and Diaspora Studies, the digital poster exhibition was installed at the Goldsmiths University Library in 2019. The display coincided with the celebrations for Black History Month in the United Kingdom.

However, with travel and work restrictions tightening due to COVID-19, the exhibition could not travel further in its physical form. Yet, what seemed like a disappointing turn for the exhibition and curatorial team in fact revealed new ways for interaction with the narratives and digital material. Reading Museum (2020) hosted virtual events for the anniversary of Windrush Day in June 2020. The museum integrated content from the exhibition panels into a virtual display on its website, generated new content in the form of community videos, objects and photographs and developed educational materials. The new content was co-created with a Caribbean community steering group which worked closely with the exhibition team from the BMHS. The digital exhibition poster panels were then used by Vodafone UK through The Black Professional Network for over a month as a

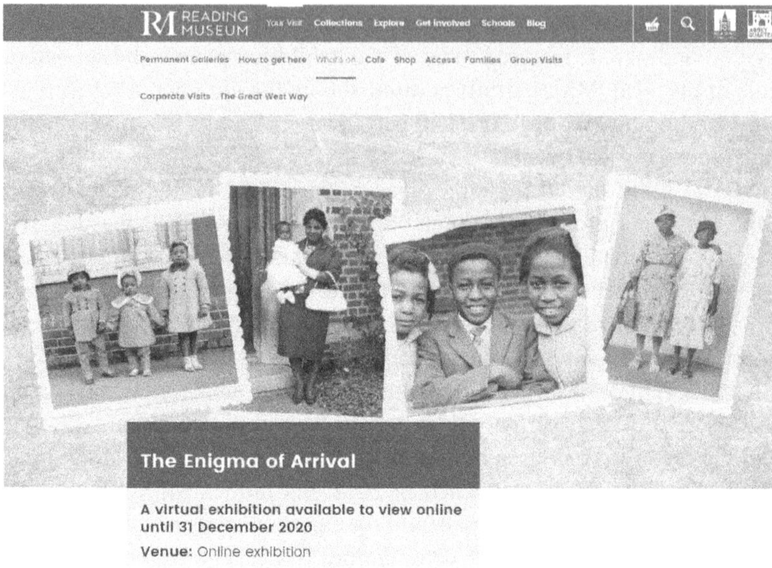

FIGURE 11.11 *The Enigma of Arrival* online exhibition by Reading Museum.

part of their Windrush Day celebrations. The exhibition was free to access for all Vodafone UK employees, of which there are approximately 11,500.

The methods and processes used to convert physical exhibitions to a digital format were employed in other projects after the success of *The Enigma of Arrival* and the VMCMM was acknowledged. For example, an exhibition on Walter Tull, a British footballer who died in the Battle of the Somme, was intended for physical display at the BMHS. Remote work with heritage organisations on the project enabled collaboration through online tutorials to create digital outputs that were then exhibited. In the end, the BMHS (2020) launched a digital exhibition, *Walter Tull: A Strong Heart Beating Loudly*. BMHS hosted a Facebook Live event for the exhibition in conjunction with the Barbados Defence Force and the Tull family on 11 November 2020, Remembrance Day.

Conclusion

The inclusive development of the VMCMM provides a framework to facilitate the development of novel, flexible and creative online exhibitions—in this case, with a focus on Caribbean perspectives on migration and memory. It proved a valuable addition to *The Enigma of Arrival* exhibition as a resource to further explore Windrush stories and histories and a platform for collecting and exhibiting previously undocumented oral histories. In this way, it has helped to expand the world's current knowledge and understanding. The VMCMM is based on the VMI framework developed for the EU-LAC Museums project but included Caribbean-specific features and functions that were developed following the *Itinerant Identities* conference round table discussion. It taps into Caribbean digital skills and literacies while integrating emergent technologies with traditional media and historical records. Although the VMCMM was initially designed to be a companion to *The Enigma of Arrival* physical exhibition, it became the only accessible method for engagement during the global lockdowns as a result of the COVID-19 pandemic. Its unique design and bespoke functionality allowed for broader use and several iterations while the world could only access the content digitally. The framework has since been used in other digitally progressive projects that feature a virtual museum, but the VMCMM stands more particularly as an archetype for migration-themed resources.

Acknowledgements

Thank you to everyone who was mentioned in the above text, all the panellists at the round table, participants in the discussion and everyone involved with *The Enigma of Arrival* exhibition (BMHS, UWI, EU-LAC Museums project and other contributors).

The VMCMM may be found at https://eu-lac.org/vmcarib/

References

AXIELL (2017). *Digitising collections: leveraging volunteers & crowdsourcing to accelerate digitisation.* [Online]. Available at: https://www.axiell.com/app/uploads/2019/04/digitizing-collections.pdf (Accessed: 2 March 2021).

BMHS. (2020). *Walter Tull: a strong heart beating loudly.* [Online]. Available at: http://waltertullexhibition.org/ (Accessed: 2 March 2021).

Cassidy, C. (2018). 'Transcription of "Round table Caribbean virtual museums: opportunities and challenges"'. Zenodo. Available at: https://doi.org/10.5281/zenodo.4573653 (Accessed: 1 March 2021).

Cassidy, C., Fabola, A., Miller, A., Weil, K., Urbina, S., Antas, M. and Cummins, A. (2018). 'Digital pathways in community museums', *Museum International*, 70(1–2), 126–39.

CINE. (2017). *CINE: See the past, imagine the future.* [Online]. Available at: https://cineg.org/ (Accessed 4 March 2021).

Cummins, A., Persaud, V., Hall, K. and McGuire-Batson, N. (2020). *Exhibiting migration and gender. Companion reader: reflections, response & resilience.* [Online]. EU-LAC Museums. Available at: https://eulacmuseums.net/index.php/bibliography-2/details/1/109 (Accessed: 1 March 2021).

CUPIDO. (2018). *CUPIDO.* [Online]. Available at: https://cupidoeu.org/ (Accessed: 4 March 2021).

Department for Digital, Culture, Media & Sport. (2019). *Culture is digital: June 2019 progress report.* [Online]. Available at: https://www.gov.uk/government/publications/culture-is-digital/culture-is-digital-june-2019-progress-report (Accessed: 1 March 2021).

European Union (2017). *Promoting access to culture via digital means. Policies and strategies for audience development.* [Online]. Luxembourg: European Union. Available at: https://op.europa.eu/en/publication-detail/-/publication/7839cb98-651d-11e7-b2f2-01aa75ed71a1 (Accessed: 1 March 2021).

Fabola, A., Kennedy, S., Miller, A., Oliver, I., McCaffery, J., Cassidy, C., Clemens, J. and Vermehren, A. (2017). 'A virtual museum installation for virtual time travel', *Communications in Computer and Information Science*, 725, pp. 255–70.

ITU (2021). *Statistics*. [Online]. Available at: https://www.itu.int/en/ITU-D/Statistics/Pages/stat/default.aspx (Accessed: 1 March 2021).

Northern Periphery and Arctic Programme. (2019). *Project Stratus*. [Online]. Available at: https://www.interreg-npa.eu/projects/funded-projects/project/240/ (Accessed: 4 March 2021).

Reading Museum. (2020). *The enigma of arrival*. [Online]. Available at: https://www.readingmuseum.org.uk/whats-on/enigma-arrival (Accessed: 2 March 2021).

12

ECOMUSEOLOGY IN ARTISTIC PRACTICE

Postcolonial strategies of collective return in Latin America and the Caribbean

Kate Keohane

> Our landscape is its own monument. Its histories are traced on the underside.
>
> (Glissant, 1999, p. 11)

In beginning with the words of the Martinique-born writer, philosopher and theorist Édouard Glissant, I am situating this discussion of participatory art and museological practice within a specific form of postcolonial theory: one founded upon an ecological approach that recentres perception towards the significance of landscape and the ways in which histories are experienced in the present tense. Glissant's theoretical work reorients how we read places and spaces as signs or markers of lived experience and asks us to re-imagine the possibilities of collaborative processes of interrogation. In examining the way that the past is understood or encountered within the lived experience of a place, Glissant asks that we consider the unseen and the ultimately unknowable aspects of landscapes and, in so doing, return to an emphasis upon orality. It is therefore possible to correlate Glissant's postcolonial approach to history-making and collectivity with the field of ecomuseology.

Ecomuseology is a philosophical discipline and a heritage management framework that fundamentally prioritises the needs of local communities. Foregrounding specialised knowledge and engagement—which is in some ways determined by an extended period of time spent in a particular territory—this museological approach preserves stories, traditions and customs. Ecomuseological interpretation is generated by and valued through its emphasis upon the environment and operates in pursuit of the preservation and promotion of cultural heritage (Davis, 2011, p. 5). The word 'environment' refers here to a cultural strategy that responds to the effects of climate change and also relies upon an understanding of ecological connectivity,

DOI: 10.4324/9781003288138-15

along with the multiple factors that generate a particular lived space, including historical, economic, social or political experiences and effects.

The origins of the ecomuseum lie in the 1960s, but the term was first defined in France in the 1970s by the museologists Georges-Henri Rivière and Hugues de Varine. The basic differences between the traditional museum and the ecomuseum are typically illustrated through a pair of formulae developed by de Varine and later added to by René Rivard (1988, pp. 123–4):

> A Traditional Museum = building + heritage + collections + expert staff + public visitors;
>
> whereas,
>
> An Ecomuseum = territory + heritage + memory + population.

Peter Davis (2011) further refined the distinction between the traditional large-scale museum institution and the ecomuseum in *Ecomuseums: A Sense of Place* by stating that the ecomuseum is defined within the context of a designated geographical area or territory. Yet, despite these seemingly self-explanatory defining factors, no two ecomuseums are the same. While theorists have developed various definitions since the term was proposed, with each study adding to the list of key characteristics found in ecomuseums, I suggest that this method might ultimately prove reductive as it does not account for the imbalances of resources and conditions that are intrinsic to the field of ecomuseology (Rivard, 1988; Corsane, 2006; Davis, 2011). The pursuit of an ecomuseum definition is in itself generative—and helpful to cultural institutions for advertisement and funding purposes—but finding a paradigmatic case study or list of attributes only takes analysis so far.

As a shifting concept that was generated to respond to a sense of change within museology for the benefit of local development, the term 'ecomuseum' now demands renewed attention within the context of contemporary debates surrounding decolonisation and ecology. Much like Glissant's theoretical work, which refuses to be reduced to a totalising schema, there can be no unified description of what an ecomuseum looks like or how it can be defined in order to leave space for different interpretations as well as an immediate action in response to changing community needs. Through an emphasis on the adaptability, responsiveness and sustained interrogation embodied by ecomuseums, in this essay, I offer an alternative approach to this field of study. I propose that the ecomuseum is a process or system for working through histories that are embedded within a landscape, and the multiple experiences that contribute to the re-activation of memory and vexed narratives of belonging.

This essay consequently questions the stability of the term 'ecomuseology' through an analysis of three artistic practices centred within specific contexts in Latin America and the Caribbean. I consider initiatives developed by three distinct groups operating within different locations, namely: the collaborative artistic research of Annalee Davis on a former plantation space in Barbados; the community-based theoretical work of Semillero Caribe, which develops strategies to embody theory in relation to specific contexts in Mexico and Cali, Colombia; and the collaborative initiative BetaLocal in Puerto Rico, which attempts to counter damage inflicted during US military occupation. In so doing, I illuminate how collective, ecologically centred artistic practices echo the anticolonial strategies of ecomuseums for responding to landscapes damaged by processes of colonialism, and I further propose alternative methods for ecomuseology itself.

The urgency of recognising the importance of initiatives currently being developed within the field of cultural heritage in Latin America and the circum-Caribbean is associated with a return to the original aims of the ecomuseological project. The early definitions of the ecomuseum are grounded in a specific intellectual context: Rivière believed that museums should do more to place human affairs into broader environmental contexts, while de Varine wished to see museums become democratic institutions, with local communities taking a far more active role in their work at all levels (Hudson, 1992). These standpoints are often associated with the 1972 UNESCO and International Council of Museums Round Table in Santiago de Chile, which declared museums to be 'integral' spaces that hold the potential to play a 'decisive role in the education of the community' (Nascimento Júnior, Trampe and Santos, 2012; Weil and Bize, 2017). Motivated by the Declaration of Santiago de Chile, early theoretical writing emphasised the need to reorient the motivations and focus of museum practice towards communities, particularly those within what has problematically been referred to as the Global South. As Anthony Gardner and Charles Green (2013, p. 444) outline, the term 'Global South' refers to the southern hemisphere and the specific challenges it faces, but

> while historical reflection is central to the South, it does not exclude the significance of constructive initiatives generated out of and in defiance of these histories: that is, the web of potentialities that can connect and be coordinated across the cultures of the South.

The rise of ecomuseology in the 1970s relates to the period that saw the independence of many of these regions, which necessitated a renewed attention to conceptions of the local in order to rebuild and define new forms of cultural heritage and resistance against ever-intensifying climate forces.

The locations in Latin America and the Caribbean considered within this essay are not only some of the places most drastically affected by climate crises but are also home to a wide range of strategies and experimental cultural practices to combat global challenges (Johnson, 2011). While I am hesitant to employ the term 'resilience' due to its problematic overuse, these practices often engage with international audiences while catering to local community needs in order to preserve, promote and propagate cultural heritage and individual memory. To name but a few, initiatives such as Alice Yard in Trinidad; Popop Studios in the Bahamas; Espacio Aglutinador in Cuba; or the Instituto Buena Bista in Curaçao have been pivotal to the development of alternative approaches to creative spaces that involve non-artistic audiences in cultural processes (Wainwright and Zijlmans, 2017; Castellano, 2019; Hadchity, 2020). These cultural collectives promote artistic mobility within the region through opportunities for funding, residencies and networks with other similar initiatives, notably via connection with collaborative platforms like *Tilting Axis* and *Caribbean Linked*. Thus, even as they centre the discussion on a community of cultural workers who have a lived understanding of a space, these initiatives do not deny the possibilities and privileges of international movements and connections. Instead, they employ ecomuseological processes to respond to site-specific needs, while connecting with other people and places through a relational process and practice. Moreover, although they are outside the typical realm of art capital and museumification, these projects share a central goal of alternative forms of community-making and imagined networks. I contend that this previously overlooked approach to 'working through' embodied experience lies at the conceptual heart of the ecomuseological project, which has recently focused its discussions on tourism and income generation (Nitzky, 2012).

The three artistic practices examined in this chapter echo this original purpose and adopt the same forms, thematic focus and intention of what has traditionally been defined as an ecomuseum. Indeed, they each encompass all of Gerard Corsane's (2006, pp. 399–418) 'Twenty-one ecomuseum indicators', which were developed as an evaluative instrument and have subsequently been used to assess the atypical case study of Robben Island Museum in South Africa, among many others. Davis (2011, p. 92) expanded on Corsane's list with five indicators that can be broadly applied:

> First, the adoption of a territory that is not necessarily defined by the conventional boundaries. Second, the adoption of a 'fragmented-site' policy which is linked to in-situ conservation and interpretation. Third, conventional views of site ownership are abandoned; conservation and interpretation of sites via liaison and cooperation. Fourth, the empowerment of local communities; the involvement of local people in museum activities and in the creation of their cultural identity. Fifth and final, the potential for interdisciplinary and holistic interpretation.

The artistic ecomuseological practices considered within this essay are fundamentally united through their pursuit of a 'sense of place' and so could seemingly be defined as ecomuseums according to the pre-existing schema defined by Davis (2011). While this might, in a sense, disprove the value of Corsane's list of attributes and point to the impossibility of developing a unified understanding of what is meant by Davis's 'sense of place', it also points to the expanded creative possibilities embodied by the forms and theoretical underpinning of ecomuseology.

At the heart of the projects by Annalee Davis, Semillero Caribe and BetaLocal is a dialogical art practice (that is, one organised around conversational exchange and interaction) and a sense of embodied theory mediated through an engagement with landscape as both literal place and idea. Although these three case studies are highly unconventional within the field of museology, it is time that the fields of art history and museology work together more closely in order to fully account for the scope and ingenuity of creative labour. The performative approach embodied by these artistic practices allows for greater audience engagement, networking, experimentation with theory and the creation of a space for the sharing of personal experiences.

Furthermore, the literature developed around relational, collaborative and participatory artistic practice shines a generative light on conversations around participation in museums. Art historians Claire Bishop (2006, 2012) and Grant Kester (2011), in particular, have offered an invaluable methodological approach to participatory art. When woven together with existing ecomuseological literature and underexamined site-specific artistic case studies, this approach provides a resource for considering methods of centring local communities within museum and artistic practice. Bishop's (2006, p. 10) contention that participatory art seeks 'to collapse the distinction between performer and audience, professional and amateur, production and reception', with an emphasis upon 'collaboration, and the collective dimension of social experience', seems to speak as closely to the actions that underpin the ecomuseum as it does to contemporary visual art. Kester's (2011, p. 10) question in relation to the 'aesthetic' and 'ethical' criteria for artistic production also bears strongly upon the case studies considered here: 'What forms of knowledge do collaborative, participatory, and socially engaged practices generate?' In these instances, collaborative, community-based artistic strategies actively respond to ecological challenges faced by those working within postcolonial landscapes and facilitate space for transformation and adaptation.

The three art projects founded upon engagements with regions in Latin America and the Caribbean also challenge underlying assumptions around the kinds of landscapes that can be framed or referenced by ecomuseums. Following long debates from the mid-1990s, UNESCO's (n.d.) conception of 'cultural landscapes' was defined as: 'cultivated terraces on lofty mountains, gardens, sacred places [...] [that] testify to the creative genius, social development and the imaginative and spiritual vitality of humanity'. Yet, the European Landscape Convention (2000) expands considerations to all landscapes, emphasising the importance of recognising local communities, and the tension between conservation and progress or transformation and issues of ecology. This approach to landscape therefore allows space for alterations and multiple forms of encounter. In the case studies considered within this chapter, metropolitan, toxic and post-plantation landscapes have as much value and importance as the panoramic, majestic and nostalgic vistas that are typically featured as representative case studies within ecomuseological literature (Borrelli and Davis, 2012, pp. 31–47).

The case studies also work in contradistinction to Peter Davis's (1999, p. 238) argument that 'the one characteristic that appears to be common to ecomuseums is pride in the place they represent'. Instead, the value of collaborative practices of cultural memory work lies in their potential as an experimental method for discussing, sharing and challenging the histories and lived encounters with a space. The individual experience offered by an ecomuseum is consequently less about pride and more about a collaborative, collective effort to discuss, decolonise and develop 'histories [...] traced on the underside' of a landscape, as Glissant suggests (1999, p. 11). The concept of pride, while undeniably an essential motivating factor for underacknowledged cultural efforts, is too ephemeral to operate as a unifying feature and is not necessitated by cultural work of this kind. Rather, much like de Varine and Rivard's formula for defining the difference between the traditional museum and the ecomuseum, the collaborative artworks considered within this essay disrupt museological processes that necessitate walls or collections (which are more associated with notions of pride). They instead rely upon committed collective action and repeated encounters with the specificities of a place, predicated on an awareness of precarity and the complexities of what it is to live and work in a specific location. It is not possible to gain a total or unified understanding of a territory; ecologically based cultural initiatives therefore provide an essential space for participatory preservation and memory work.

We are consequently prompted to question the kind of community that lies at the heart of ecomuseum projects and to ask what it means for something to be community-driven. There are of course communities within communities and there needs to be space within the ecomuseological framework for dissent, disagreement and renegotiation. Moving forward from notions of a unified locality, while retaining an emphasis upon the particular lived experience of a space, the works by Annalee Davis, Semillero Caribe and BetaLocal expand the concept of ecomuseology. They do so by creating space for individualised encounter, transience, connection between territories and alternative forms of sustainability through the documentation and subsequent written, critical and exhibitionary circulation of performances. Although none of the artists acknowledge ecomuseology within their theoretical premises, I suggest that their respective themes of unearthing, embodied theory and walking-as-worlding can also be considered as ecomuseological processes and strategies of resistance. In line with Glissant's theoretical projects, the ecomuseum is therefore not presented as a utopian project dependent upon pride and nostalgia but rather represents an activist response that necessitates a recentring of community, academic and institutional attention.

Unearthing—Annalee Davis and Fresh Milk Barbados

Having grown up on her family's land in the St George parish of Barbados, the work of the artist Annalee Davis focuses on site-specificity and active engagement with figuring the meaning behind the 'sense of a place'. Her house and studio are situated on this inherited land, which used to operate as a sugar plantation and is now a dairy farm. Her father and grandfather were both planters, and her family can trace a paternal line back to 1648, when her earliest recorded ancestor,

Leonard Dowden, came to Barbados as a less than '10-acre man' (Davis, 2019, p. 39). Walkers dairy farm (formerly Willoughby Plantation) at one point held over 300 enslaved labourers, who had been torn from their African homes to cut cane and work the earth. The landscape itself pays testament to this history through the remnants of industry: a crumbling sugar mill and tools for manufacture appear as haunting reminders of the plantation system. Yet the site's rolling bucolic fields and the tropical myths which allow for the circulation of the Barbados landscape in the cultural imaginary (typically framed through white sand beaches and blue skies), occlude memories of suffering and toil and instead develop a picture of a paradise 'outside of time' (Sheller, 2003; Thompson, 2006).

Here, the sense of place associated with the 'histories [...] traced on the underside' of the landscape is connected to hauntings and emotions of discomfort and terror. The complexity arises in the confluence of the plantation as both home-space and site of atrocity. For Davis (2019, p. 15), revisiting this landscape does not imply a pessimistic reaffirmation of the past; rather, it has become a strategy for activating memory and openly invoking difficult experiences and unfinished histories. Davis's work—both artistic and structural—therefore functions in response to the 'plantationocene', a term defined by the feminist theorist Donna Haraway (2015, p. 162) to describe the current geological age. Closely related to the Anthropocene, the plantationocene demarcates the era defined by the irrevocable damage inflicted by humans on the planet during the transatlantic trade in enslaved peoples. As a white Creole woman, Davis uses her practice to work through her positionality within this space, as well as issues of ecology, and the collective practices of honouring sites defined and shaped by enslaved labour.

In actively choosing to live and work on this site over the past 20 years, Davis has developed a series of strategies to memorialise and interrogate the narratives that have contributed to the creation of the territory. She has also facilitated spaces for active discussion by local and international audiences. One of the most obvious ways that she enacts this process is through the creation and development of the art space, Fresh Milk. Since the 1990s, Davis has been campaigning for community needs and the importance of art to notions of national identity and community in Barbados. The Fresh Milk art centre, located on the former plantation, was founded in 2011. It responds to the absence of gallery spaces on the island and the need to create localised systems of support for artists. The name plays upon the current function of the site as a dairy farm; the reference to 'milk' also connotes the turning of blood into a nutritional substance, speaking to themes of healing and transformation. Funded primarily by international residencies, Fresh Milk has become a hub, not only for international researchers, writers and artistic practitioners who pay a fee to live at the site and make use of the Colleen Lewis Reading Room (which houses over 3,500 books related to the cultural heritage of the Caribbean) but also for community members who use the structure as a meeting point for conversations and future planning.

Since its inception, the international residency programme has welcomed individuals drawn to the possibility of researching the specificities of place, the histories of the Caribbean and the need to enact and participate in what Christina Sharpe (2016, pp. 19–20) has termed 'wake work'. In taking up a residency or engaging with Fresh Milk, individuals can engage with artists, students and academics and have the chance to visit other cultural institutions and landscapes on the island. The experience of the space is sometimes described with hesitancy and discomfort by local and international practitioners due to questions of positionality and extractive research practices. Nonetheless, the structures that underpin Fresh Milk inevitably facilitate the development of understandings of the ecology of the space, and the island more broadly, and act as a nexus for vital discussions about what it is to exist within a postcolonial landscape. Fresh Milk therefore stimulates sustainable development on the island by working to recentre the Caribbean as a central node of the contemporary art world or, at the very least, functions as a space for conversations about what it is, in Davis's (2019) words, to be 'committed to a small place'.

Here, it becomes instrumental to compare the Fresh Milk site with a more typical ecomuseological case study. Springvale Eco-Heritage Museum, in the Scotland District of Barbados, is run by Newlands Greenidge and Denyse Ménard-Greenidge. Visitors to this site are provided with tours of the house and the tropical forest that surrounds the building and shown important plants, trees and medicinal herbs that can be found within the territory. Like Davis, Greenidge grew up within the space, and the collection is primarily made up of his family's household contents, including clothing, pots, kitchen utensils, baskets and mahogany furniture, that reflect upon the period of emancipation up to the 1930s (Greenidge and Ménard-Greenidge, 2002). There is also a small library on the site where visitors can learn more about the history of Barbados. The shared emphasis between these entities therefore lies in the preservation of cultural heritage and the need to re-imagine plantation spaces. As in the case of Fresh Milk, however, questions emerge about the kinds of audiences targeted by the Springvale Eco-Heritage Museum. In the earlier years of the initiative, the museum was a valuable resource for educating school groups. However, funding shortages have hampered this kind of interaction, and the majority of visitors are tourists to the island. Both Fresh Milk and Springvale Eco-Heritage Museum depend upon the labour and care of a small group of committed individuals. As such, the sustainability of both initiatives is uncertain.

Fresh Milk works to counter this precarity by capitalising upon the interdisciplinary possibilities of ecomuseological practices, notably within *Unearthing Voices: An Interdisciplinary Archaeology Project* (2014–present). Through a sustained research collaborations with transnational archaeology teams, participants gather at the site to unearth the material heritage of Barbados and to develop historical archives for individuals who were central to the creation of the plantation, and who lived and worked within its bounds. Operating on a yearly basis, groups of students work to excavate materials buried beneath the soil, which frequently include ceramics that were shipped over during the colonial period, as well as tobacco pipes. The initiative gives equal attention to immovable and movable tangible material culture, and to intangible heritage resources through the creation of a discursive space. Any excavated materials are then transferred to the Barbados Museum & Historical Society or are returned to the earth in which they were found. The practice therefore promotes the preservation, conservation and safeguarding of heritage resources in situ, while also connecting in a broader sense to the cultural heritage of the island. As museologist Guisy Pappalardo (2020, p. 7) posits:

> ecomuseums can be an opportunity for strengthening individual-groups-institutions relations, through a voluntary pact. This opens some areas of experimentation in order to understand the implication of the spatial justice framework, with the space at the core of the discourse and institutions as key actors.

Within the context of a site of colonial violence, it is right that the labour of memorialisation and restitution is shared with students, so long as the outcomes of activities ultimately benefit the local community. In an expanded sense, this collaborative work provides the impetus for further study and creative interventions and raises transnational historical consciousness.

The act of unearthing allows for broader conversations and encourages a literal and metaphorical engagement with monuments 'traced on the underside' (Glissant, 1999, p. 11). Building upon the soil-based possibilities, another central strand of Annalee Davis's practice is a knowledge of the curative properties of wild plants that have been passed down orally through generations of Barbadiana (the former indentured and enslaved) who would have grown and harvested wild plants for use in bush teas and bush baths. For example, blue vervain, West Indian bay leaf and cerasee were known respectively to cure insomnia, detoxify the body and cleanse the blood. In sharing the stories and properties of these plants, Davis enacts what she calls 'a botanical uprising', in sharp distinction to the harsh imposition of a monocrop—sugarcane—into the island's landscape for more than three centuries. The fields' subterraneous layers hold an apothecary of seeds that offer a form of transformative remediation to the exhausted topsoil (Davis, 2016). By reactivating these traditions through artistic practice and collaborative projects, this ecomuseological artistic project gains new scope that reaches beyond the walls of the initiative.

Fresh Milk has an architectural structure, but the relatively small studio/gallery/library ultimately serves as a hybrid space for developing relationships with the surrounding landscape. In some ways guided by Davis's individual practice, local artists and international residents are encouraged to create works and conversations about their experience of the place. This might best be characterised by what Davis calls a 'grounding' practice, which involves her taking a daily walk through the geographical territory and watching for changes within the landscape as a form of meditation. This ritual of performing the diffuse border-spaces of her home has also been documented in her painterly projects and performances, notably *Sweeping the Fields* (2016), in which Davis enacts a ritual cleansing of the earth that is then photographed for posterity. In drawing attention to the repeated action of walking a space defined by its colonial histories, Davis highlights the practices that demarcate a territory and the subtle changes and shifts that occur with sustained presence. The work seeks to honour and preserve the cultural heritage of the space: not only through the adaptation of archival material in Davis's re-use of ledger pages from the plantation for paintings and drawings but also in the continuous reference to the colonial craft of Queen Anne's lace and the possibilities for botanical healing. Rather than privileging transnational movements, here an awareness of the space comes laden with issues of genealogy and familial culpability. The work therefore has both a spatial and a temporal aspect, recalling Corsane's (2006) ecomuseological attributes: rather than trying to freeze things in time, it considers both continuity and change. When the subsequent photographic documentation is then circulated internationally, conversations centre on the ecology of the space and what it is to live and work within this specific territory.

While writers like Edward S. Casey (1996, p. 16) have said that a place must be experienced to be understood, with the rise of what has been termed 'the digital community museum' (Cassidy et al., 2018, pp. 126–39), this form of exhibitionary encounter must be taken seriously as a possibility. In some ways, this approach leaves the knowledge of the space to those who have gained understanding through repeated presence, while recognising the privileges involved in art tourism. When circulated internationally, either online or through re-curation, the images that document sustained presence within a space come to represent the generative possibilities of art as a social object from which to discuss shared histories. The research and unearthing processes that underpin Davis's work, which are fundamentally ecomuseological in praxis, become the most important activity in terms of local development, but the international participation and circulation affords further possibilities and sustainable approaches that might reasonably exceed the lifespan of the Fresh Milk initiative.

The collaborative practices foregrounded within the space of Fresh Milk are undeniably affiliated with the forms and aims of the ecomuseum as it was originally conceived (Rivard, 1988). Fresh Milk and Davis's broader practice point to the networks that exist within island spaces to support and generate engagements with cultural heritage. Although the artist's individual work has primarily been shown in international contexts, the ecomuseological actions that underlie her practice respond to a community need. These distinct institutions and individuals might not always see eye to eye, but they form their own form of ecology. The lack of a contemporary art museum in Barbados and the need to continuously reinterpret the meanings embedded within the post-plantation landscape demand the development of alternative spaces for discussion and the preservation of cultural heritage.

Embodied theory—Semillero Caribe in Latin America

Echoing the collaborative projects enacted at Fresh Milk, the work of the artistic collective Semillero Caribe illuminates the possibilities for developing alternative approaches to postcolonial theory within collective practice. It also demonstrates the importance of dialogical and embodied practice as a way of emphasising process rather than product within the aims and framework of ecomuseology. Founded in 2016 by three artists—Madeline Jimenez (Dominican Republic), Ulrik López (Puerto Rico) and Minia Biabiany (Guadeloupe)—the initiative originally responded to the experience of 'being from' the Caribbean, while residing in Mexico City. For its first set of activities, the collective organised a number of workshops that demanded sustained awareness of the body in space, while developing democratic systems for communication and decision-making to answer local community needs. The project is ephemeral by nature: it has since been adapted to produce a publication and a new collaborative practice in Cali, Colombia, and will continue in other locations. Yet, the model offers an ecomuseological educational approach. Indeed, the practice forms the basis of what might be termed a 'diasporic ecomuseum', where the specificities of living in a place are recalled somewhere else.

With a diverse group of participants from multiple walks of life (some who knew each other and others who did not), the workshops focused on creating novel ways of disseminating the multifaceted experience of the Caribbean space within the context of Latin America. This approach was developed by the group through eight multisensory experimental sessions that sought to create alternatives to colonial clichés of Caribbean corporality, exploring climate-related issues, such as humidity or the movements of hurricanes, and developing a postcolonial imaginary within a variety of community contexts. The name Semillero Caribe derives from a Spanish play on words that uses the idea of a Caribbean *semillero* (seedbed) instead of *seminario* (seminar) to reflect the project's generative ecological imperatives, as well as to highlight its non-traditional approach to theory by foregrounding the use of drawing and the body. For example, in the first session at Cráter Invertido, a group of 15 participants from local communities worked on a series of responses that implied the use of the body and its perception, through drawing and dialogues by activating the senses through breathing, shouting, touching, listening and repeating. Theoretical approaches to decolonisation were particularly important to the founding of the project, as were conversations about the specific body-codes at work in Mexico City, which generally function differently from those in the Caribbean. For instance, Biabiany (2020) notably describes the acceptance of sensuality on the street in Mexico City, as well as the extreme experience of the male gaze in public spaces.

As with many cultural initiatives affiliated with ecological aims, this approach was informed by personal knowledge. Biabiany describes how her practice, both individual and collaborative, has been shaped by a specific understanding of the Caribbean informed by the plantation system, present colonisation and the political situation of Guadeloupe, which remains a federation of France. She states that:

> In Guadeloupe, there is a strong identification with France as the governing territory that lies 6,000 kilometers away, but also a different reality, a culture and

> un/conscious resistance strategy challenging French assimilation policies. I see this tension in the perception of territory and the construction of identity as our paradigm.
>
> *(Biabiany, 2017)*

This experience of persistent control by another state can also be witnessed on the home-island of López, as Puerto Rico continues to struggle through its sustained connection to the United States. The Semillero thereby generatively disrupts the structure of the ecomuseum by encouraging consideration of how an embodiment of a specific place can be informed by the actions and needs of another.

Yet, while the structure and programming of the Semillero created space for the retention of different perspectives in the initial workshops, the group sought to draw out similarities within 'Caribbeanness' as both place and idea. Initially, the collective was drawn towards the use of theory to express this affiliation with a particular space but felt discomfort in working with theoretically dense material as artists without extensive research or academic expertise. The activation of the body and the senses became a response to this anxiety, as participants and audiences were encouraged to embody ideas and develop alternative interpretations. These exercises functioned like games, with primary importance placed upon multisensorial interpretation of terms or ideas. Sometimes there would be readings or extended periods of conversation, and at other times, the group would work in silence. Drawing, breathing and moving together, the diverse group of individuals created a shared space for the consideration of a remembered place in the Caribbean in relation to the embodied event space in downtown Mexico. Each activity was linked to one or more concepts by Caribbean authors, for example: Glissant's conception of 'opacity' and 'relation'; the process of internalised colonialism discussed by Frantz Fanon; the sense of chaos defined by Antonio Benítez Rojo; and orality as a form of resistance through the writings of Kamau Brathwaite. In the process of creation, or what the organisers called 'un-learning', the participants worked to relay differences and to figure shared equivalences in a system that sought to avoid colonially implicated strategies of knowledge dissemination.

This avoidance of fixity was echoed in the format of the eight sessions, which took place over the course of a month with the group changing and adapting. Some people arrived only at the end of the workshops, others completed the whole programme and others left in the middle of activities. Each week utilised different, yet thematically related, literary publications as cultural support and to facilitate a form of connection. The Semillero necessitated processes of memory, projection and mindful awareness of the body in space through isolation of the senses. In addition to being one of the tools for the construction of narrative in certain exercises, orality took over from sensation at the end of each session so that responses to the event could be shared. There were no predetermined results or structure;

rather, the needs of the group were answered in real time. This responds closely to Pappalardo's (2020, p. 7) claim that:

> ecomuseums may be educational and transformative processes that start from the collective reconstruction of memory – including tensions, conflicts, contradictions and questions of power – and may evolve in emancipatory paths for liberating the most oppressed individuals of society. This requires interpreting heritage as action, as socially active research.

Rather than directly embodying de Varine's (2006, p. 227) contention that the ecomuseum 'is a two-way medium, where the concrete knowledge and experience of the citizen is exchanged with the more learned scientific or technical knowledge of the specialist, through a jointly built exhibit', here the ecomuseological process allows space for experimental modes of group learning, where all members become experts through an emphasis upon embodiment.

A series of four collaboratively written publications were then produced in 2017–2018 as a form of pedagogical support following these events. These were written in French, English and Spanish to reflect the multilingual reality of the Caribbean region and to expand the reach of the project beyond those who were involved in the Semillero. This emphasis upon accessibility is important for thinking about ecomuseums and how information that is generated through various processes of engagement with a specific territory can be disseminated, documented and incorporated into the formation of the space over time. Critically, in the case of the Semillero, this formulation allowed for adaptation and processes of critical evaluation that encouraged a re-imagination of the aims of the project and a recognition of when the needs of a local community were met.

This evolution is evidenced in the second iteration of the project in 2018, led by Biabiany, where the format was altered to respond to the gender imbalance of the first Semillero. This time, a group of women from Cali, Colombia, engaged in a series of new activities related to five Caribbean women authors. This second manifestation was developed in collaboration with curator Yolanda Chois within the framework of her project *Tópicos entre trópicos* ('Topics among Tropics'). Titled the Semillero Doukou, the event was made possible through the logistical support from Más Arte Más Acción and the cultural division of the Banco de la República in Cali. The reflections and activities, which were intended to create a diasporic bridge between the Caribbean and Black Colombia, focused on the writings of Jamaica Kincaid (Antigua), Nefta Poety (Guadeloupe) and Mary Grueso and Nohelia Mosquera (Chocó, Colombia). This reiteration of the Semillero was founded upon novels rather than theory and actively sought to distinguish itself from logics of intellectual domination in order to create an embodied matrilineal approach to Caribbean studies. Glissant's key theoretical concepts still played an important role in the structure of the sessions, but the terms were removed from their source in order to be re-activated or productively misread, and they were juxtaposed with images taken from the literary texts. Although located in Latin America rather than the Caribbean region, the Semillero actively responded to the processes of imaginative place-creation through collective work. In addition, the artists who initiated the project—who consistently engage with the Caribbean within their individual practice—experimented with relational approaches to correspondences and divergences between the two contexts.

To grant greater authenticity and to encourage alternative iterations, Biabiany invited local researchers to work with local communities in each session. Notable participants included dancer Angélica Nieto with *danza-manglar*; writer Jenny Valencia, author of the urban tale *Shangó y el cerro de las tres cruces de Cali* ('Shangó and the Hill of the Three Crosses of Cali'); artist Carolina Charry with her voice work; and curator Ericka Florez in collaboration with dancer Andrea Bonilla regarding her research around straight and curved lines. Additionally, Otilia Caracas, an author from Valle del Cauca, accompanied the group with her work. The dance of the Bigidi, defined by the choreographer Léna Blou, proved particularly generative (Blou, 2015; Biabiany, 2020). Created by enslaved people, the form of dance was forbidden for many years but remained a core underground practice. The dance involves moving in disrupted lines, in a way that comes close to falling, but which retains a precarious balance. Working to react to living in a place and being unable to see a bright future, but needing to keep moving forward, experiments with the movement become a display of imbalance and the paradoxical logic of irrationality. Rather than becoming a performance of cultural heritage to outsiders, the forms of collectivity, touch, speed, meeting and intimacy made manifest by these forms of dance allows for the generation of new meanings and for associations to be shared within a safe space. The forms of the Bigidi might therefore provide inspiration for understandings of ecomuseology as process or method for working through experiences of dislocation. Unlike the traditional museum, it is vital that the ecomuseum retains space for collapse within its definition, as ecomuseums are by their very nature socially, climatically and financially precarious. As the Bigidi dance suggests, this attribute can be mobilised as a strength, providing a way to respond more quickly to community needs and to move in time with the changing rhythms of the community.

In this way, the second iteration of the Semillero utilised theory within processes of participatory co-creation (Simon, 2010, pp. 268–70). In effect, the sustained incorporation of Glissant's central ideas, rather than being reductive or overly simplified, offered an expansive framework for thinking through other writers and embodied experiences. Beyond the move from theory to practice, the recognition of the importance of art within these conversations also relates to issues of participation, and what Nina Simon (2010, pp. 85–120) describes as the need to move from the 'me to the we' within museum practice: here, the artist and the conception of individualised encounter is retained, while allowing possibilities for imaginative and lived connections. De Varine (2006, p. 228) writes that:

> the ecomuseum must begin as, and remain, an expression of the community, an endogenous product, to be recognized by the community as its own property

> and instrument; so it must speak the language of the community, rooted in the living culture of the people.

By contrast, the work of the Semillero asks what the nature of this community-based language is, and how alternatives might be born. What might be termed 'diasporic ecomuseology' therefore finds its greatest support within the field of visual art.

Walking as worlding—BetaLocal and Puerto Rico

The practices of BetaLocal bring together both the collaborative ecological approach of Annalee Davis and the diasporic landscape embodiment developed by Semillero Caribe. Initiated in Puerto Rico in 2005 by cultural entrepreneur Michelle Marxuach and artists Beatriz Santiago and José 'Tony' Cruz, the non-profit organisation has developed a research and visual arts centre that interweaves residencies and initiatives that present dissenting ways of existing within an island space. The project began by refurbishing a single-floor building owned by Marxuach in Old San Juan to serve as a space for art residencies, exhibitions and meetings. For this purpose, they used the network created in previous years through the organisation of M&M, a biennial event funded by Marxuach. M&M brought a wide range of artists and curators who work in the field of relational aesthetics to Puerto Rico (including Rirkrit Tiravanija, Hou Hanru and Adel Abdessemed) and exhibited the work of world-famous artists, such as Gilbert and George, Marina Abramovic and Vito Acconci (Bourriaud, 1998). After the third edition of M&M, the project was discontinued and transformed into BetaLocal. In contradistinction to M&M, BetaLocal has, since its beginnings, aimed to engage with non-artistic audiences, confronting political causes and fulfilling specific social and cultural needs. Its temporal operation has also differed from the typical biennial system, which relies upon the temporary shipping in of contemporary global art and networking within these systems for the supposed benefit of local practitioners.

Instead, BetaLocal has sought to embody a different, South-centric, model. Initially conceived as an alternative library and archival space that was open to all, it soon evolved in order to integrate other features and programmes. Among those, the first organised was *La Práctica* ('The Practice'), an immersive collective research programme open to artists, researchers and cultural activists. The programme, defined as a 'horizontal, peer-taught' experience, consisted of the annual selection of up to five people with the aim of developing cooperative, practical onsite research. *La Práctica* offers a residency space and encourages the fellowship recipients to devote a number of hours each week to collaborate with different artistic and non-artistic communities. These collaborations result in ongoing joint research initiatives, talks and reading sessions. Under this scheme, the invited individuals are not considered specialists; rather, they are specifically sought out in response to requests made by the various groups who regularly visit BetaLocal. What is important is that all the people who are involved in these actions participate on equal terms in the research process, with the results aimed at addressing and responding to collective expectations and desires. Residents are therefore compelled to decentre themselves and transform their original ideas into cooperative productive action.

The second related initiative that developed chronologically was *La Ivan Illich*, which is founded upon a dissenting approach to educational practice. Conceived as an 'un-learning' project, this curriculum-free educational experience invites anyone who is interested to propose a particular lesson that they would like to receive or teach. BetaLocal provides the space for these exchanges to take place and seeks to satisfy the requests that are received. In the typical ecomuseum, the educational model runs the same risks as the traditional museum of becoming a static preservation or unified vision of vulnerable cultural heritage. Here, however, the skills and needs relate directly to the individuals that interact with the space and initiative and can therefore morph and grow over time. As found in the methodology of Semillero Caribe, this kind of approach will inevitably have its challenges and lead to mistakes, but the expansive potential of the ecomuseological model provides greater fluidity and consequent innovation.

Finally, the most conventional initiative developed by BetaLocal is *The Harbor*, which functions as an art residency for international guests and, as in the case of Fresh Milk, this international residency is premised on the need to interact and actively engage with the specificities of place. Rather than thinking about Peter Davis's 'sense of place', this experience relates more closely to what Bourdieu (2005, pp. 43–9) terms 'the development of reflexive habitus': that is, the potential to reconsider behaviours and develop new practices in relation to an environment constitutive of a society or a group, which can be altered by new actions, education or training.

BetaLocal has resulted in solo projects that reflect upon the specificity of the place, notably by the internationally acclaimed artist Beatriz Santiago Muñoz. However, the organisation also leads a 'walking seminar' that utilises the embodied experience of the landscape as a mode of resistance following the ecological damage caused by US military occupation. Puerto Rico comprises a group of islands that have been a 'possession' of the United States since their invasion in the 1898 Spanish-American War. Following hurricanes Irma and Maria in 2017, Puerto Rico received very limited aid, and questions are still being raised about the US government's lack of engagement and delayed aid response (Reyes Franco, 2018). Until these issues are resolved, or at the very least recognised by those in positions of authority, the situation in Puerto Rico will remain precarious. As in the cases of Fresh Milk and Semillero Caribe, BetaLocal demonstrates that ecomuseology must respond to the condition of being in one place while also being subject to the power or influence of another if it is to adequately engage with the lived experiences of 'post'-colonial spaces. Rather than foregrounding progress, within art-based ecomuseological approaches, continuous return and collective 'working through' are equally important.

BetaLocal's walking practice consequently involves traversing multiple forms of terrain and engaging with oral histories and conversations about the healing properties of plants, often within the grounds of an abandoned military space. The military base originally functioned as a training site and support centre for the control of narcotics traffic. It assisted American invasions of the Dominican Republic in 1965, Grenada in 1983 and Haiti in 1994, but its primary purpose was as a location from which to oversee bombing exercises, or attacks, on nearby Vieques Island. Vieques remained inhabited during this time, resulting in a rise in frequency of cancerous tumours and general harm to the environment and the health of the 9,100 residents due to explosives-related pollutants (Jirau-Colón et al., 2019). In 1999, errant bombs killed a civilian guard during Vieques exercises, sparking a surge in protests, and in 2002, a plane carrying seven airmen crashed in the town of Caguas. The American Navy denied the 1999 charges but decided to close the testing range in 2003. Despite the fact that ambitious plans have been made to turn the empty base into a cruise ship dock, a base for space travel, a commercial airport, tourist resort and industrial park, these proposals have been delayed by environmental studies and a lack of funding, and the space currently remains abandoned.

While BetaLocal requires international residencies for funding purposes, the sustainability of this project, like the work of Annalee Davis and Semillero Caribe, is secured through the re-enactment and documentation of collaborative practices. The resulting materials can then be used to communicate in more expansive networks. Although the spaces in which they are subsequently exhibited tend to be associated most closely with the academic art world, they offer a valuable teaching resource and assist in the process of visualising and recording local community needs and shared issues. Certainly, in some ways, this form of thinking is not revolutionary, since the practices align with ideas surrounding relational aesthetics and participatory art. However, the fields of art history and museology have yet to reconcile their similarities and differences and to work out what each can learn from the other. It is in this regard that initiatives that bridge this divide, including Fresh Milk, Semillero Caribe and BetaLocal, are particularly instructive.

It is consequently important to think about how ideas associated with relational aesthetics and participatory art relate to the development of community-based sustainable tourism (Bowers, 2016, pp. 758–82; Riva, 2017). Incorporating particular ecomuseological principles—such as a holistic approach to interpretation and information sharing, paying equal attention to cultural and natural resources and monitoring the changes to the region over time—can support the three pillars of sustainability (economic, socio-cultural and environmental) within a region, while existing outside of the domain of large-scale museumification (Purvis, Mao and Robinson, 2018).

Conclusion

In highlighting these three case studies, it becomes apparent that through participatory, dialogical practice, artists, activists and audiences from multiple backgrounds can be brought together to respond to the specificities of particular spaces. While the initiatives foreground the importance of ephemerality and transmutability, which seem to go against certain defining principles of the ecomuseological project, these same features might equally be thought of as a way to respond in real time to the needs of local audiences and benefit local communities. Through cultural participation, and an emphasis upon community wellbeing and global sustainability, these creative initiatives avoid didacticism, prioritise collaborative processes of working-through and develop new systems for thinking about museums and 'unlearning' colonially implicated systems of knowledge.

The works of Annalee Davis, Semillero Caribe and BetaLocal raise the question of who ecomuseology is for. The traditional definition of the ecomuseum yields a quick answer: local community audiences. However, this is not always the case, as the experience of sustained presence within a space is not limited to those who dwell in it and communities are themselves fragmented. Rather, ecomuseums almost always have a complex relationship with their perceived internal and external audiences, as well as the broader cultural ecosystem in which they find themselves. In the pursuit of breaking down the binary between local and global, while preserving an emphasis on the needs of those who live in postcolonial spaces, the three projects considered here exemplify the key features of ecomuseums. That is, they allow for change and development; encourage documentation and conservation of cultural heritage by promoting interdisciplinary research at all levels; and utilise a holistic approach to nature and culture to develop a sustainable approach to tourism and international engagement. In this way, the performative aspects of ecomuseology are brought to the fore, particularly the importance of walking, unearthing and orality to collaborative interactions among individuals who are themselves shaped by multiple communities, places and spaces.

Yet, the initiatives also point to the shortcomings of the ecomuseum model. As Borrelli and Davis (2012, p. 43) outline, ecomuseums 'are not a panacea for all environmental concerns. They cannot necessarily resolve, for example, conflicts between conservation and development, or environmental protection and economic interests, or the conflicting interests of communities and ambitious politicians or developers'. By contrast, the kinds of participation developed through experimental artistic practice, networks and interdisciplinary initiatives leave room for unequal relations and alternative hierarchies. They also reject modes of tourism that look for a fixed version of landscape, heritage and culture (Salazar, 2010). Rather, the process of individual encounter is recognised as a powerful factor in shaping new visions for the future of particular spaces, while foregrounding the critical importance of local forms of tangible and intangible heritage.

These art-based ecomuseological initiatives are therefore more sustainable than the types of projects that have traditionally been defined within the bounds of ecomuseology, as they evolve across times and places and gain new meanings in their subsequent documentation and re-display. The artists who choose to develop these practices ground their work in the experience of staying within a particular territory for a prolonged period of time, developing systems that are devised by and for local populations, even as they secure benefits for these communities through international engagement. This approach inevitably relies upon certain privileges, such as access to international movement and the 'global' contemporary art world, but all three initiatives have succeeded in avoiding the constraints and structures of the traditional museum, by developing alternative systems of value and pointing to the importance and possibilities of staying within or returning to a place.

The actions of Annalee Davis, Semillero Caribe and BetaLocal ask us to collectively reflect on the way that history is traced on the underside of the landscape. They further prompt us to be actively involved in the process of recognising and meeting the socio-ecologic challenges experienced within contemporary postcolonial spaces. The most valuable function of the ecomuseum is therefore not to encourage pride in a specific territory—although such museums might depend upon hopeful visions for the future—but instead to continuously develop processes for seeing the same space differently. The practices involved in the creation, development, enactment and afterlives of the above initiatives and, by association the ecomuseological school of thought, therefore become less about progress than the experience of the effects of the past on the present and the need for a continuous return to places, spaces and ideas. In recognising complexity and conflict, as well as issues of power, precarity and belonging within the discourse surrounding ecomuseums, space is left for multiple future interpretations, collaborations and engagements with and within a specific environment.

References

Biabiany, M. (2017). 'In conversation with Minia Biabiany: poetics of immanence', *C&*, 6 June. [Online]. Available at: http://contemporaryand.com/fr/magazines/in-between-places/ (Accessed: 5 February 2021).

Biabiany, M. (2020). Private communication with Kate Keohane, 8 May.

Bishop, C. (ed.) (2006). *Participation*. London and Cambridge, MA: MIT Press.

Bishop, C. (2012). *Artificial hells: participatory art and the politics of spectatorship*. London and New York: Verso Books.

Blou, L. (2015). *Le Bigidi: une parole de l'être!* Tedx Talks. [Online video]. Available at: https://www.youtube.com/watch?v=u8Oojo5pJqg (Accessed: 26 March 2021).

Borrelli, N. and Davis, P. (2012). 'How culture shapes nature: reflections on ecomuseum practices', *Nature and Culture*, 7(1), pp. 31–47.

Bourdieu, P. (2005). 'Habitus', in Hillier, J. and Rooksby, E. (eds.) *Habitus: a sense of place*. Aldershot: Ashgate, pp. 43–9.

Bourriaud, N. (1998). *Relational aesthetics*. Paris: Les Presses du Réel.

Bowers, J. (2016). 'Developing sustainable tourism through ecomuseology: a case study in the Rupununi region of Guyana', *Journal of Sustainable Tourism*, 24(5), pp. 758–82.

Casey, E. (1996). 'How to get from space to place and back again in a fairly short stretch of time: phenomenological prolegomena', in Field, S. and Basso, K. (eds.) *Senses of place*. Santa Fe, CA: School of American Research Press, pp. 13–53.

Cassidy, C.A., Fabola, A., Miller, A., Weil, K., Urbina, S., Antas, M. and Cummins, A. (2018). 'Digital pathways in community museums', *Museum International*, 70(1–2), pp. 126–39.

Castellano, C.G. (2019). *Beyond representation in contemporary Caribbean art: space, politics, and the public sphere*. New Brunswick, NJ: Rutgers University Press.

Corsane, G. (2006). 'Using ecomuseum indicators to evaluate the Robben Island Museum and World Heritage Site', *Landscape Research*, 31(4), pp. 399–418.

Davis, A. (2016). *The dark domain*. [Online]. Available at: https://annaleedavis.com/archive/the-dark-domain (Accessed: 19 March 2021).

Davis, A. (2019). *On being committed to a small place*. Costa Rica: TEOR-éTica.

Davis, P., (1999) Ecomuseums: A Sense of Place, London & New York: Leicester University Press

Davis, P. (2011). *Ecomuseums: a sense of place*. London and New York: Continuum.

Gardner, A. and Green, C. (2013). 'Biennials of the south on the edges of the global', *Third Text*, 27(4), pp. 442–55.

Glissant, E. (1999). *Caribbean discourse: selected essays*. Charlottesville: University Press of Virginia.

Greenidge, N. and Ménard-Greenidge, D. (2002). *Springvale Eco Heritage Museum: a visitor's guide*. Barbados: Springvale Plantation.

Hadchity, T.K. (2020). *The making of a Caribbean avant-garde: postmodernism as post-nationalism*. Indiana: Purdue University Press.

Haraway, D. (2015). 'Anthropocene, capitalocene, plantationocene, Chthulucene: making kin', *Environmental Humanities*, 6, pp. 159–65.

Hudson, K. (1992). 'The dream and the reality', *Museums Journal*, 92(4), pp. 27–31.

Jirau-Colón, H., Cosme, a., Marcial-Vega, V. and Jiménez-Vélez, B. (2019) 'Toxic metals depuration profiles from a population adjacent to a military target range (Vieques) and Main Island Puerto Rico', *International Journal of Environmental Research and Public Health*, 17(1), p. 264.

Johnson, S. (2011). *Climate and catastrophe in Cuba and the Atlantic world in the age of revolution*. Chapel Hill: University of North Carolina Press.

Kester, G. (2011). *The one and the many*. Durham, NC: Duke University Press.

Nascimento Junior, J. do, Trampe, A. and Santos, P.A. dos (eds.) (2012). *Mesa redonda sobre la importancia y el desarrollo de los museos en el mundo contemporáneo, Resoluciones, 1972*, Vol. 1. Brasilia: Ibram.

Nitzky, W. (2012). 'Mediating heritage preservation and rural development: ecomuseum development in China,' *Urban Anthropology and Studies of Cultural Systems and World Economic Development*, 41(2), pp. 367–417.

Pappalardo, G. (2020). 'Community-based processes for revitalizing heritage: questioning justice in the experimental practice of ecomuseums,' *Sustainability*, 12(21), 9270. [Online]. Available at: https://doi.org/10.3390/su12219270 (Accessed: 12 March 2021).

Purvis, B., Mao, Y. and Robinson, D. (2018). 'Three pillars of sustainability: in search of conceptual origins', *Sustainability Science*, 14, pp. 681–95.

Reyes Franco, M. (2018). 'The visitor economy regime', *Independent Curators International Research*. [Online]. Available at: https://curatorsintl.org/research/the-visitor-economy-regime (Accessed: 12 March 2021).

Riva, R. (ed.) (2017). *Ecomuseums and cultural landscapes*. Santarcangelo di Romagna: Maggioli Editore.

Rivard, R. (1988). 'Museums and ecomuseums – questions and answers', in Gjestrum, J.A. and Maure, M. (eds.) *Okomuseumsboka – identitet, okologi, deltakelse*. Tromso: ICOM, pp. 123–8.

Salazar, N. (2010). 'Tourism and cosmopolitanism: a view from below', *International Journal of Tourism Anthropology*, 1(1), pp. 55–69.

Sharpe, C. (2016). *In the wake: on blackness and being*. Durham, NC: Duke University Press.

Sheller, M. (2003). *Consuming the Caribbean: from Arawaks to Zombies*. London and New York: Routledge.

Simon, N. (2010). *The participatory museum*. Santa Cruz: Museum 2.0.

Thompson, K. (2006). *An eye for the tropics*. Durham, NC: Duke University Press.

UNESCO (n.d.). *Cultural landscapes.* [Online]. Available at: https://whc.unesco.org/en/culturallandscape/ (Accessed: 24 February 2021).

Varine, H. de (2006). 'Ecomuseology and sustainable development', *Museums and Social Issues*, 1(2), pp. 225–31.

Wainwright, L. and Zijlmans, K. (eds.) (2017). *Sustainable art communities: contemporary creativity and policy in the transnational Caribbean*. Manchester: Manchester University Press.

Weil, K. and Bize, C. (2017). *An analysis of the 'Round Table of Santiago' of 1972, the social role of museums, and its relevance to sustainability in the present.* [Online]. Available at: https://eulacmuseums.net/index.php/bibliography/details/1/93 (Accessed: 9 March 2021).

13

EXHIBITION-MAKING AS STORYTELLING

The 14th FEMSA Biennial in Michoacán Mexico

Ana S. González Rueda and David A. J. Murrieta Flores

In her manifesto *How to Make Art at the End of the World*, artist Natalie Loveless (2019, p. 21) reflects on using Thomas King's *The Truth About Stories* in her teaching to conceive of stories as 'material-semiotic events that configure worlds' and affect how we see the world and act within it:

> Stories are wondrous in their capacity to reorganize our approaches to our social material worlds; they are dangerous for their capacity to produce themselves as compelling objects of belief [...] the telling of stories is a political performative. A world-making, knowledge-making practice.
>
> *(Loveless, 2019, p. 21)*

Loveless (2019, p. 21) stresses the need to think about the stories '*we are crafted out of* as well as which we participate in crafting'. This chapter investigates Mexican modern and contemporary art's implication in both national history and the stories that challenge and unsettle established narratives. It concentrates on the decentralising curatorial proposition of *Inestimable azar* ('Inestimable chance'), the 14th Fomento Económico Mexicano S.A.B. de C.V.(FEMSA) Biennial (February 2020–February 2021), based in the Mexican state of Michoacán and directed by Daniel Garza Usabiaga. The biennial's exhibition programme centred on a series of 'museological interventions', such as site-specific commissions that responded to the cultural heritage of the venues—especially various understudied mural paintings from the 1930s and 1940s in the cities of Morelia and Pátzcuaro (Bienal FEMSA, n.d.). We approach the interaction between the biennial and the venues' murals through storytelling to address the kinds of worlds and knowledges it constructed. Our analysis engages with the wider call for a postcolonial reinvention of the museum (Chambers et al., 2014; Simpson, 2001; Von Oswald and

DOI: 10.4324/9781003288138-16

Tinius, 2020). Introducing her volume on museums and indigenous perspectives, historian Susan Sleeper-Smith (2009, p. 2) considers museums as powerful rhetorical devices intentionally built to tell stories and maintains that the public museum became the site for 'official and formal versions of the past'. Traditionally, the history told by objects and their curatorial and interpretive contextualisation followed an evolutionary narrative that contrasted the 'primitiveness' of indigenous cultures with the 'progress' of Western societies, a story that served to justify the violent colonial imposition of 'civilised order' across the world (Sleeper-Smith, 2009, p. 2). Sleeper-Smith (2009, p. 4) underlines indigenous peoples' demands to 'deconstruct the colonization narrative from the viewpoint of the oppressed [and introduce] a multiplicity of voices, a variety of narratives, and the use of museums as tools of revitalization'. In this case, the biennial introduced a collaborative framework that sought to facilitate the participation of Michoacán's indigenous communities.

The FEMSA Biennial was founded in 1992. Its eighth edition introduced a new itinerant format that aims to engage with the local heritage and artistic production of the host state (Bienal FEMSA, n.d.). Funded by the FEMSA Foundation, the biennial is part of a more extensive cultural programme through which the multinational corporation professes its commitment to support the sustainable development of the communities where it operates (FEMSA Foundation, 2021).[1] Philosopher and political theorist Oliver Marchart (2014, p. 264) has argued that biennials from the periphery contribute to the decentring of the West, especially regarding issues 'around the legitimacy and status of non-Western art'. He observes that, while biennials are often instrumental in enhancing the public impression of a particular city and are an asset to the tourist industry, they crucially assist in 'constructing local, national and continental identities' (Marchart, 2014, p. 264). Marchart cites the 1989 Havana Biennial as a key reference that introduced a model less focused on spectacle and more concerned with a specific discursive interest. Its theme, 'Tradition and Contemporaneity', addressed 'anticolonial politics and non-Western modernities' (Marchart, 2014, p. 271). The 1989 Havana Biennial also posited the now widely accepted idea that biennials should interact with their host locations and not, as Marchart (2014, p. 273) describes, 'simply descend like a UFO'. This landmark exhibition anticipated the current curatorial interest in participatory and critical education strategies. For Marchart (2014, p. 273), the idea of decentring biennials refers to an effort 'to shift the canon and to open the field for dissident practices and discourses'.

The widespread influence of the 1989 Havana Biennial is evident in the case of *Inestimable azar,* particularly in its decentralising, discursive approach and interest in collaborating with local, indigenous craft makers. However, art and cultural theorist Panos Kompatsiaris (2020) underlines the ambivalences of contemporary art biennials as proclaimed sites of resistance that ultimately rely on the dominant neoliberal order they seek to contest. In this sense, it is important to consider that FEMSA, the biennial's sponsor, is a multinational corporation, owner of the world's largest Coca-Cola bottling franchise and the largest convenience store chain in Mexico. Over the years, it has been the target of criticism for its appropriation of public resources such as water, often to the detriment of indigenous communities (Pearson, 2017; Franco, 2020); its consistent opposition to public policies that address health issues from which it profits (De Alba, 2020); its hostile labour practices across the country (Lobo, 2019) and the extensive environmental impact of its operations (Peredo, 2011, 2015). To make matters worse, FEMSA's extractivism has often been abetted and protected by the Mexican government (see Ramírez Miranda, Cruz Altamirano and Marcial Cerqueda, 2015; Pacheco-Vega, 2015). It is crucial, then, to consider how the biennial worked within the local heritage sites and arts scene, reflecting wider cultural struggles about identity where the state is no longer the dominant actor.

The role of the 14th FEMSA Biennial is particularly significant considering the dire state of cultural institutions in Mexico at the time of writing. The government, led by President Andrés Manuel López Obrador, has chosen to concentrate on a single mega-project: 'Chapultepec Park: Nature and Culture', directed by renowned artist Gabriel Orozco, which took 12 per cent of the federal culture budget for 2020 with a total estimated cost of £368 million (Cepeda, 2020). In light of budget cuts to culture in the public sector and concomitant mass layoffs, critics have pointed to the project's reinforcement of a central site in a wealthy area, while peripheral institutions struggle to survive the effects of the COVID-19 pandemic. The government's attention to this single enterprise demonstrates its adherence to neoliberal logic: the grand scheme serves as a distraction from the deep-seated issues and prevalent precarity within the cultural sector (Cepeda, 2020).

This chapter considers the critical agency of the 14th FEMSA Biennial and its intervention in museum knowledge production processes at a moment of crisis. We begin by considering the biennial's curatorial proposition: the setting up of a dialogue between contemporary art and Michoacán's local heritage, an encounter that was conceived as a historiographical revision that might challenge official national history. The second section provides a historical contextualisation of the post-Revolutionary, state-led cultural and educational campaign in Mexico during the first half of the 20th century. In particular, we examine the didactic role of muralism, as well as the redefinition of crafts within the narrative of modernisation. The third section concentrates on the biennial's re-reading of the local modernist heritage, especially regarding issues of Mexican identity. As an example of the biennial's storytelling, we discuss Graciela Speranza's participation in the public programme and her analysis of Juan O'Gorman's mural *Historia de Michoacán* (1942). Finally, we present two of the artistic commissions and their different approaches to local craft: Marco Rountree's imaginative rethinking of modernist aesthetics and Adela Goldbard's collaboration with Arantepacua's Communal Indigenous Council (2019–2021) on a craftivist project.

An encounter between the modern and the contemporary

As explicitly stated by artistic director Daniel Garza Usabiaga and public programme curator Esteban King, *Inestimable azar* was intended as a platform for dialogue between the 20th-century Mexican avant-garde, whose works feature prominently in the buildings and institutions housing the biennial, and contemporary artists (Miércoles de SOMA, 2021). The title itself alludes to the complex geographical network of the avant-garde in Europe and the Americas, being a reference to the 1938 manifesto written by André Breton, Diego Rivera and Leon Trotsky, which resulted from a series of discussions held in the town of Pátzcuaro, Michoacán, that year (Tarcus, 2019). The manifesto conceived of chance encounters as a way to know the world better (or change it); the biennial's title included this reference to centre the state of Michoacán as a meeting point between local, regional, national and international cultural currents.

Acting within the discourses of the art world while also expanding into the related terrain of history and art history, the biennial's curatorial line sought to break with the unitary quality of Mexican art historiography, which has tended to homogenise the country's artistic developments through national perspectives (see Bienal FEMSA, n.d.). Its inaugural conference bore a provocative title, *Adiós historia oficial* ('Goodbye official history'), suggesting that the nationalist narrative that characterises Mexican art and education would be left behind. This claim on the history of the nation was based on an interpretation of the biennial's simultaneously local and global focus as offering an alternative to official nation-making narratives. The dialogue between the 20th-century avant-garde and contemporary artists was thus framed as an opportunity to renew the historical links between both: first, through the involvement of local artists, artisans and curators in the biennial's development, and, later, by the inclusion of international audiences in the digital instances of the programme due to the COVID-19 pandemic.

The preceding iteration of the FEMSA Biennial (2018), directed by curator Willy Kautz, had already revised its structure in order to encompass activities and proposals beyond the usual exhibition-centred format. It took place in the state of Zacatecas, which has even less involvement in the contemporary Mexican art world than Michoacán. Entitled *Nunca fuimos contemporáneos* ('We were never contemporary'), Kautz's biennial integrated heterogeneous works within a programme designed to involve public spaces and institutions peripheral to the art world.[2] Garza Usabiaga and King's continuation and expansion of the FEMSA Biennial's possibilities two years later reflected an even greater interest in the interactions between the local and the global, positioning these as integral to an understanding of art's role in Mexican history and identity. The role of a robust pedagogical programme articulated through free conferences, workshops, networking events and academic activities continued the precedent set by the 13th Biennial in terms of knowledge production. However, where the latter's questions and critical standpoint were related to generic concepts of modernity and contemporaneity, the 14th iteration focused on particularly Mexican approaches to history.

The staged encounter between modern and contemporary Mexican art and the biennial's production of knowledge through artistic, representational and educational means—whether presence-based or digital—reveals similar patterns to avant-garde muralism. The common starting point is history, and, in this sense, the re-evaluation of historical discourses usually belonging to the nation represents the primary site of struggle where the biennial inserted itself as a contender. The Mexican muralists' context, shaped by the Revolution that broke out in 1910, was thoroughly permeated by discussions about history. The murals themselves were conceived by intellectuals as part of a vast educational programme first designed in 1921 after the creation of the Secretaría de Educación Pública (Public Education Ministry, hereafter referred to as SEP). As will be discussed below, the educational logic of muralism, which stakes a claim on historical discourse and even becomes indistinguishable from it, is the basis from which the biennial's proposals emerged. In this, they mirrored statements of a historical nature that touch upon issues of identity and nationality.

Muralism's teaching of national history

The creation of the SEP in 1921 was guided by an educational ethos that transcended the more technical aims of progressive instruction and acculturation that were typical of prevailing positivist views at the end of the 19th century (see Vázquez de Knauth, 1970, p. 81). In the hands of the Revolutionary intellectual José Vasconcelos as Education Secretary, the institution's aim was social betterment and national harmony understood as a path towards the redemptive 'light' of civilisation (Garciadiego Dantan, 2015, p. 34). Vasconcelos's project had three major components: the school, the library and the arts. All were organically related in a way that is best expressed through the free textbooks created in the 1920s and distributed nationally to schools and libraries, which were illustrated by artists aligned with the values of the Revolution. Each of the three components was represented by professionals and students (teachers, librarians, artists), who were recruited by the SEP to go on 'missions' across the country to improve literacy, and for the practical education of peasants and the working class. One of Vasconcelos's ambitions was to 'decentralise culture' through the creation of arts and teaching centres even in the most geographically challenging parts of the country (Garciadiego Dantan, 2015, p. 44). Education understood in this way was part of the implementation of social justice. Harmony, under the nation's banner, was seen as the necessary endpoint of the creation of Mexican citizens (Garciadiego Dantan, 2015, p. 49). Thus, nation-building became a primary concern for the Mexican system of education throughout the post-Revolutionary period, up to and beyond the Lázaro Cárdenas presidency (1934–1940), when practically all the murals that feature in the biennial were commissioned and completed.

The federal scope of the SEP meant that schools proliferated under the mantle of the state, which extended the nationalisation of the population of the Mexican territory even in places where governments previous to the Revolution had little outreach. The process involved a complex conceptualisation of Mexican identity derived from the new Revolutionary values and interpretations of history in which indigenous populations, local traditions and cosmopolitan or nationalist outlooks played crucial roles. The colonial process of 'Mexicanising' indigenous peoples was one of the main threads of the educational system, with various positions vying for hegemony throughout the 20th century. In the 1920s and 1930s, there were two prevailing views: the first advocated erasing indigenous identities in favour of a Mexican one centred on the figure of the *mestizo* (mixed-race person); the second sought to produce a dialectical dynamic in which indigeneity and *mestizo* identity would be synthesised into a new Mexican identity, 'elevating' the indigenous in the *mestizo* and the *mestizo* in the indigenous. Thus, political and educational processes became deeply intertwined, leading Mexican intellectuals of the period to conceive of education as an organic remedy to all the ailments of society. Following this model, schools would transform 'not only the individual, but the entire social medium comprehending the entire community' (Bruno-Jofré and Martínez Valle, 2009, p. 49). With history as the core discursive node, the homogenisation of the country (culturally, but also politically, economically and socially) as a task to be realised by educative means implied a broad array of informal pedagogical tools, such as art, public rituals and ceremonies that would engage entire communities (Bruno-Jofré and Martínez Valle, 2009, p. 49). The school would be integral to the social life of populations, and its jurisdiction would extend beyond traditional teaching facilities. Among the vehicles of nation-building were mural commissions and vast art-historical, anthropological and archaeological projects, such as the creation of the Instituto Nacional de Antropología e Historia (National Institute of Anthropology and History) in 1939.

In this sense, it is significant that the SEP was the centre of both educational and cultural projects. For instance, in 1937 the Cárdenas government created the Departamento Autónomo de Prensa y Publicidad (Autonomous Department of Publicity and Press, hereafter referred to as DAPP), which was explicitly and expressly charged with propaganda matters both within and outside the country. The DAPP conceived of education as a tool of propaganda (Cruz Porchini, 2014, p. 243), and, as a result, its functions came directly into conflict with the interests of the SEP. For instance, SEP mural projects that post-Revolutionary governments understood to be crucial in the creation of a national imaginary were seen by the administrators of the DAPP as interior propaganda efforts (see Dümmer Scheel, 2018). Consequently, the DAPP began to model their posters on murals commissioned by the SEP.[3] The DAPP's interpretation was made possible by the SEP's nationalist programmes themselves, which also included posters, and which were often designed to convince the people of the benefits and rights they had—perhaps indirectly—gained from the Revolution. Nonetheless, the Cárdenas presidency represented a culmination of the educational projects born from the Revolution. Deliberations on national identity and the historical foundations of the country were concentrated in the SEP, and the borders between history, memory and the nation became porous (Farías Mackey, 2010, pp. 261–2). Within Cárdenas's doctrine of a 'socialist education' (Bruno-Jofré and Martínez Valle, 2009, p. 48), history would be a crucial subject, used to show how Mexican society is, how it had been and how it *should* be—in other words, the role of history in socialist education would be one of socialisation (Vázquez de Knauth, 1970, p. 80; Buenfil Burgos, 2004, p. 48). This conception is similar to the muralist avant-garde's sustained development of historical subjects as the revolutionary key to activate the Mexican masses.

The creation of the DAPP was but one instance of the Cárdenas government's comprehensive nation-making strategy and the wide-ranging nature of its approach to education, demonstrated by its adoption of a visual communication system in which murals shared the same discursive space as propaganda due to a common historical subject matter. Turned into vehicles of education, murals proliferated across the country in SEP projects that targeted less wealthy regions outside the metropolitan centres, such as Michoacán, Veracruz, Guerrero and Sinaloa (Cruz Porchini, 2014, p. 16). Born in Michoacán and a former governor of the state (1928–1932), Cárdenas embraced regional projects and even experimented with Pátzcuaro, one of the 14th FEMSA Biennial's seats, as a modernist tourist attraction (Jolly, 2018, loc 10.68). The realisation of Vasconcelos's 'decentralisation of culture' by the Cárdenas government is well represented by its numerous mural commissions, distributed across towns and cities of the state of Michoacán, including Pátzcuaro. As suggested by art historian Jacqueline Jolly (2018, loc 7.17), the development of Pátzcuaro under Cárdenas offers 'two competing ways of imagining the region vis-à-vis the nation crystallized': first, that Mexico 'was the sum of its regions, each with distinctive cultures, products, and landscapes to contribute to the national whole'; and, second, that 'the regional could embody the national', so that 'the local might serve as a microcosm of the nation'. This tension is reflected by the mixed themes of mural commissions in the period, which oscillate between local, national and even international historical topics (as in Phillip Guston and Reuben Kadish's *The struggle against terrorism*, also known as *The struggle against terror and fascism*) (see Boime, 2008). As vehicles of nation-building, murals attempted to situate Mexicans in various *present* contexts grounded upon history, moving between the local and the global.

Conceived as part of the educational system, muralism generally produced historical discourses without recourse to the conventionally text-based, academic processes of history-making. Its bases were developed throughout the Revolutionary and post-Revolutionary periods by the artistic avant-garde in relation to both political and aesthetic issues that were expressed, for the most part, in representations of history. Against Romantic conceptions of the individual genius (although ultimately reaffirming them through an image of heroic participation in public life), the muralists privileged collective work, regarding art as a public endeavour that would take place outside and beyond the art world and its established institutions (Jaimes, 2012, p. 19). They proposed a break with the conventions of looking at artworks within museums; their intent was to dissolve the limits between art and life, critiquing the existing institution of art and explicitly freeing the spectator from art world constraints and conventions, such as the exhibition space or the individualised aesthetic experience. As art historian Renato González Mello (2008, p. 15) argues, the muralists conceived of their practice as an ethical imperative, meaning that it had to act beyond artistic concerns to impact social life. Moreover, their ethos coincided with the educational project of Vasconcelos's SEP, since they also understood art as a privileged form of knowledge, even more apt than social and natural sciences, which ignored the spiritual dimensions of human existence in its path towards *illumination* (González Mello, 2008, p. 88). The murals' representations of history turned public spaces into national sites where the nation's memory and the identity of Mexicans were at stake. The murals' historical discourse, however, put the intricacies of historical processes aside, limiting its knowledge and its world-making to the nation construed as a homogeneous—or at the very least homogenising—entity.

Muralism's revolutionary origins resulted in an art *for* the masses, emphasising its public nature as the rejection of traditional art world institutions.[4] Its social themes highlighted historical content as a way to impact reality by interpellating viewers as Mexicans. The aesthetics and politics of murals taught viewers through visual and narrative means *what* Mexico was, *who* a Mexican was and *how* they came to be such. One of the critical elements of the educational aspect of muralism was its capacity to generate popular historical knowledge while ignoring historiographical debates. Its appeal to a strained heterogeneous identity was rooted in the 'inclusion and appropriation of a glorified and nationalized indigenous culture' (Oesterreich, 2018, p. 5). The muralists' representations of the Mexican people as the primary subject of history entailed defining who exactly said 'people' were, a process articulated around the 'elevation' of the category of popular art and its highest expression: the *artesanía* (craftwork). This view was promoted by the first Exposición de Arte Popular (Exhibition of Popular Art) in 1921, which was commissioned by President Álvaro Obregón and coincided with the SEP's creation. The category of popular art as redefined by artists and intellectuals of the period implied the transformation of everyday objects, which is to say craftworks from all over the country, into *works of art*. In other words, the heterogeneity of objects from daily life across the Mexican territory—mostly indigenous in origin—was reduced to a homogeneous category of (high, culturally acceptable) popular art that was necessarily tied to a unifying image of the nation.

Additionally, popular art was seen as a result of the revolutionary process and its path towards further emancipation. As such, it connoted a newly achieved modernity in terms comparable to the claims of the muralist avant-garde (see Subirats, 2018, p. 119). This modernity-born-from-revolution was not exclusive to the major cultural centres of the country and could be found everywhere, including the periphery. As art historian Miriam Oesterreich (2018, p. 9) argues:

> the early staging of the indigenous as representing the national and the traditional as epitomizing the modern, like the *Exhibition of Popular Arts* in 1921, can be interpreted as manifestations of the national for Mexicans themselves, as an aesthetic strategy to unify diverse social strata into a single national identity by means of cultural politics and touristic development of the provinces.

As mentioned in the previous section, the Cárdenas government actually developed many of these cultural discourses that had been first advanced—but never realised—in the 1920s, and that privileged 'the provinces'. Alongside its mural commission projects, it founded the Museo local de artes e industrias populares (Local Museum of Popular Arts and Industries) in Pátzcuaro in 1938, which was among the first of its kind in the country.[5] The formation of an image of the nation that was both homogeneous and heterogeneous, *mestizo* and indigenous, suggested a complicated relationship between the country's centres and its margins (Jolly, 2018, loc 10.73). The Cárdenas regime, in its effort to decentralise culture, essentially conceived of Mexico as the sum of its peripheries.

Through government policies in which politics and aesthetics were entangled, murals and schools went hand in hand when it came to the production of knowledge about the Mexican self and its context. While it would be difficult to claim that murals themselves *were* schools, they did perform functions that had previously been the preserve of the school environment, developing beyond the art world and considerably affecting the everyday lives of Mexicans. First, by 1934, the rate of illiteracy was very high: only one in six adults could read (Lira García, 2014, p. 132). Second, the influence of the Catholic Church in education was a threat to the state's secularism, as well as its anchoring of the Mexican identity in revolutionary social values. In this context, the state's expropriation and occupation of church buildings that were central to smaller cities and towns, as well as its missionary ethos of school-building, necessitated more than textual tools of education. Murals played the role of not only establishing the state's presence within essential buildings but also teaching Mexicans about their history in a purely visual, aesthetic manner that did not need them to be able to read. Since every Mexican must visit public buildings, whether for bureaucratic or educational purposes, murals became one of the centrepieces of daily public life in the country. Considering all of the above, the history that murals taught—with very few exceptions—could be summarised as follows: first, an indigenous golden past is ruptured by the barbaric Spanish conquest, followed by three centuries of resistance and torture (often at the hands of the Church)[6]; then, a new libertarian rupture occurs in the form of the country's independence (where the heroes of the motherland are born), followed by a century of struggle against foreign interventions and imperialism; the last rupture is the Revolution, when justice is done for indigenous peoples and the working class comes into being (along with its new heroes), and whose future is bright with further emancipatory potential. This is the core of the 'official history' that the FEMSA Biennial explicitly rejected through new approaches to storytelling. However, it faced several difficulties in its attempt to overcome the foundations of Mexican identity with which it engaged.

The biennial's reinterpretation of local modernist heritage

On the one hand, the biennial's organisers explicitly utilised terms reminiscent of the artistic avant-garde and focused on the historical connections between modern and contemporary art, such as in their proposal to create an 'anti-manual' about the 'encounters and crossings between art, curating and pedagogy' (Bienal FEMSA, n.d.). Articulated around axes that include 'realisms', 'artistic integrations' between, for instance, painting and architecture created by muralists and their state patrons, and 'traditional artistic practices', the biennial's curatorial approach was firmly based on modernist concerns about the relationship between art and life. As King's statements show, the organisers sought to think of the biennial beyond the art world, conceiving of it as a space that could have an impact outside the exhibition space. It would be a place for art, as well as for knowledge exchange and community-making (Miércoles de SOMA, 2021).

On the other hand, the organisers also engaged with historical issues about Mexican identity and questions of indigeneity, cultural centralisation and the revision of the category of popular arts. In Garza Usabiaga's words,

> This year's public programme is bringing into discussion the historiographic task of re-reading our local cultural patrimony. Due to its rich artistic history, the state of Michoacán presents interesting examples of where the local intersects with the national and the global.
>
> *(Garzon, 2021)*

The biennial's artistic and curatorial commissions reflect both King's and Garza Usabiaga's statements, promoting local curators and institutions under the international framework of the biennial, but also giving great importance to artists whose projects involved Michoacán communities through collaboration with their workshops and artisans. While some of these projects will be discussed further below, our interest at this point is to suggest that the attempted destabilisation of artistic categories, such as popular arts, and the rethinking of established narratives and historical canons mirror modernist educational approaches to the same issues.

Muralism's elaboration of historical discourse was well supported by the Mexican post-Revolutionary state. It was used by governments throughout the 20th century to turn the country's history into a series of static myths and images (official history). However, the muralists themselves were continually at odds with the state's attempts to co-opt their works. Art historians and scholars such as

Subirats (2018) and Jaimes (2012) are among the most recent authors to argue that the relationship between the muralist avant-garde and the governments that would often sponsor them was not free of conflict and contradiction. This is a relevant factor when considering the likewise contradictory developments of the Mexican educational system, which involved murals, public rituals, ceremonies and traditional educational institutions. This system produced a shift in art-historical hierarchies related to artworks and popular craftsmanship; developed historical discourses that privileged the marginalised and peripheral; attempted to connect the local and the national with the international; and attempted to impact the everyday lives of Mexicans through knowledge production.

At times, the biennial's pedagogical discourse becomes indistinguishable from its modernist counterpart. Discussing the commissioned curatorial work of Erandi Dávalos in an interview with SOMA, Garza Usabiaga insisted on the biennial's role as a platform for the recognition of artisanal work, stating that the intent was to 'bring these artists out into the light' ('*sacar a estos artistas a la luz*') (Miércoles de SOMA, 2021). This act of 'elevation' mirrors those made by post-Revolutionary intellectuals. It was supported by the accompanying programme's various conferences about culture during the Cárdenas period, monographic talks on Juan O'Gorman—author of *Historia de Michoacán* (History of Michoacán), one of the biennial's modernist centrepieces—and the relation between art and propaganda in the 1930s. The programme with which the event staked its historiographical claim closely followed the authoritative methods of modernist knowledge production, undertaken through a diversification (perhaps even 'regionalisation') of means: workshops, talks, events and exhibitions in public spaces across cultural centres in Michoacán that put artisans, artists and spectators in dialogue with one another about historical issues of identity and the nation. It is significant, in this sense, that both Garza Usabiaga and King are agents from the art world, like most muralists were, and that they are both from Mexico City. Additionally, all of the website's materials are in only Spanish and English (there are four indigenous languages in Michoacán alone), and the talks were delivered solely in Spanish.

What makes the biennial distinct, first, was its private, corporate origin, since it was able to enter the struggles of history-making without the burden of nation-building that characterised the educational core of the muralist avant-garde. The post-Revolutionary state saw education as a path towards modernisation and citizenship, national *illumination* and emancipation, creating a 'regionalist' aesthetic in which murals played the role of monumentalising the state's appropriation of various indigeneities for identity purposes. The biennial's programmes critiqued the consequences of this process, which are especially relevant in a context where the current Mexican government, led by President Andrés Manuel López Obrador, has fallen back on historical appeals to unity in national identity. Thus, instead of appropriating marginalised voices in historical representations, the biennial attempted to let them speak out through its critique of official history and by involving local communities and artisans throughout its development. Instead of knowledge driven by homogenisation, the plurality sought by the biennial produced heterogenisation and the possibility of new narratives about the local, national and international. The intention was to present Mexican identity as a site of encounter, a history in the making, instead of a settled imaginary or an immovable past. Second, then, the biennial framed its public programme in response to the COVID-19 pandemic, facilitating dialogue through its own infrastructure, no longer the state's. The state's articulations of history were not mediated conversations in which several points of view could be articulated. By contrast, the biennial promoted a view of history in which the periphery *should* have a voice, amplified through the private and global nature of its own structure, and best seen in the artistic commissions. In this regard, it is important to mention that the López Obrador government, self-proclaimed as leftist and thus supposedly committed to further representation of the country's peripheries, has so far mostly reproduced the strategies of appropriation that characterised nationalist discourses throughout the 20th century. The government has even made itself a protagonist in Mexican history in a process it has called 'the fourth transformation' (an allusion to what it considers to be Mexico's epoch-building historical events), offering its own interpretation of its place in official history. Furthermore, its claims to having single-handedly ended the neoliberal fragmentation of cultural institutions is hardly a reality; the biennial, in this context, only reaffirmed the market's incisive participation in struggles over historical discourses.

During her lecture for the biennial's public programme, Argentine critic Graciela Speranza (XIV Bienal FEMSA, 2020) presented a detailed reading of O'Gorman's mural, *Historia de Michoacán* (1942), whose narrative follows the pattern outlined above.[7] At the top, O'Gorman depicted the Purépecha cosmogony and scenes of indigenous life before the conquest, including dances, rituals, violent confrontations with the Aztecs, villages, temples and pyramids. The Spanish colonisers advance at the centre of the composition, followed by evangelisation scenes and the main heroic characters of the Mexican Independence and Revolution struggles. Speranza observed that while the mural synthesises centuries of history, O'Gorman has attended to every detail: each water ripple, each feather on the Purépecha headpieces, the manes of the conquerors' horses, the thread of a fishing net, the ribbon bows tying a weaver's braids, the folds on José María Morelos's headscarf and so on. Speranza argued that, through these details, the artist achieved 'a referential illusion, an effect of reality that vivifies the history lesson' (XIV Bienal FEMSA, 2020). However, she noted that O'Gorman also resorted to surreal figures for the darkest moments of the story. The mummy at the centre and the hand-headed monster with serpent arms at the right anticipate the artist's post-apocalyptic later work. O'Gorman completed the lesson with his didactic use of text. A dog carries a sign with an ironic commentary: '*conque así es la famosa civilización humana*' ('so this is the famous human civilisation'). At the bottom left, the artist's self-portrait holds a written statement that refers to the resistance of the oppressed peoples and their latent strength, which will someday produce extraordinary art and culture 'like a gigantic erupting volcano'. In alignment with the biennial's curatorial premise, Speranza's analysis of O'Gorman's realist, surrealist and didactic storytelling strategies offered a historiographical revision, a retelling.

By considering how we might look at the mural today, Speranza's lecture also contributed to the biennial's intention to draw connections between the modern and the contemporary. Her talk compared the 'excess' of muralist figuration with the digital overload of the 21st century. For instance, she discussed Trevor Paglen's *From 'Apple' to 'Anomaly' (Pictures and Labels)* (2019–2020), a mosaic of thousands of images that problematises machine-learned categories. Speranza also mentioned Carlos Huffman's painting *El Juegador* (2013) and its meticulous depiction of realist and surrealist figures: the fern leaves among the cables, techno-garbage, old printers and routers that allude to a dystopian future. She drew further connections between O'Gorman's late, post-apocalyptic, 'anti-architectural' work and contemporary artists' responses to the Anthropocene, such as Adrián Villar Rojas's monumental, futuristic, clay and cement ruins in *The Murderer of Your Heritage,* the 2011 Argentine pavilion at the Venice Biennale and Pierre Huyghe's *After ALife Ahead* (2017), an evolving ecosystem installed in an abandoned ice rink that brought together organic, inorganic and augmented reality components. In this sense, she presented O'Gorman's surrealist visions as prophetic.

Most importantly, Speranza reflected on the non-anthropocentric Purépecha worldview and the blurring of the boundaries between humans and animals suggested by the masks portrayed in the mural. For the critic, these scenes suggest a more equitable relationship with nature. Her most compelling insight, which offered a radical reinterpretation of the mural, is borrowed from philosopher Déborah Danowski and anthropologist Eduardo Viveiros de Castro (2017, p. 104):

> indigenous people have something to teach us when it comes to apocalypses, losses of the world, demographic catastrophes, and ends of History [...]: for the native people of the Americas, *the end of the world already happened* – five centuries ago. To be exact, it began on October 12, 1492.

While their thinking informs Speranza's rediscovery of O'Gorman's mural, it can also assist our analysis of the rest of the biennial, especially at the time of the COVID-19 crisis. Danowski and Viveiros de Castro argue that the American genocide of the 16th and 17th centuries has been the largest demographic catastrophe in history, even taking into account the current and future threats of nuclear war and global warming. They observe that survivors found themselves as '*humans without world* [...]. They carried on in *another world,* a world of others, their invaders and overlords' (pp. 105–6). In what follows, we discuss two of the biennial's artistic commissions and examine how exhibitions may investigate and present 'the many worlds in the World' (p. 120).

Contemporary art and craft retellings

Marco Rountree

Marco Rountree's untitled installation drew from local heritage to investigate progressive historical narratives and trouble the notion of a national modernist aesthetic. The dogs from O'Gorman's mural reappear in clay as one of three components. Rountree commissioned local craftsman, Juan Carlos Marín, to reproduce them life-sized, in the same colours and positions as the mural: one is standing and the other is sitting holding the sign with its mouth. Detached from the context of the painting, their unimpressed, sceptical remark 'so this is the famous human civilisation' is open to new interpretations. The artist also had four wooden columns with fish motifs carved locally. These reference the *troje*, the region's traditional housing which consists of various structures surrounding an *ekuaro*, a central area demarcated by greenery, low walls and different units that constitute the interior living space of extended families. Spatially, the *troje* is formed by a square or rectangular room, a raised platform used to store corn and seeds and a porch at the front with decorated wooden columns (Ettinger, 2015, pp. 71–2). Finally, Rountree's installation includes a mural made of colourful seeds, another reference to

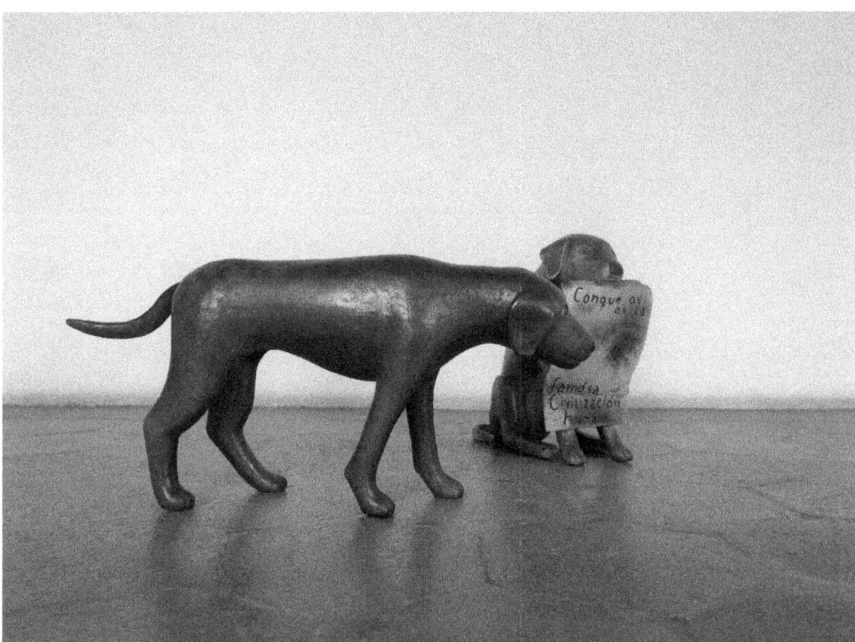

FIGURE 13.1 Marco Rountree, *Untitled* (2020). Wood, ceramic and seeds. Installation detail. 14th FEMSA Biennial, Centro Cultural Clavijero, Morelia, Mexico. Courtesy of the artist.

the *troje*. However, the mural was also inspired by a common craft activity for children in Mexico (Rountree, 2021). Most people who grew up in Mexico remember arranging shapes with beans and pasting them on paper. Rountree's sculptures, columns and murals revisit a fundamental principle of modern art and architecture in Mexico, and one of the biennial's central lines of research: the integration of the visual arts. The artist has developed an imaginative approach to Mexican modernism, as seen in previous works, such as *Xitle Volcano School of Sciences and Artisanry* (2019), a series of interventions of the Museo Anahuacalli, built by Diego Rivera to house his collection of pre-Hispanic figurines and opened posthumously in 1964 (Museo Anahuacalli, n.d.).

Both Rountree's project at Anahuacalli and his installation at the biennial reveal his interest in artist Adolfo Best Maugard's drawing teaching method which, according to art historian and curator Karen Cordero Reiman (2010, p. 45), was formulated to create decorative images 'endowed with a Mexican national character'. Cordero observes that the state endorsed Best Maugard's method as part of its cultural programme and school curriculum in the early 1920s. His method proposed a basic vocabulary for a national art, based on elements allegedly taken from pre-Hispanic art. Best Maugard put forward seven primary elements found in

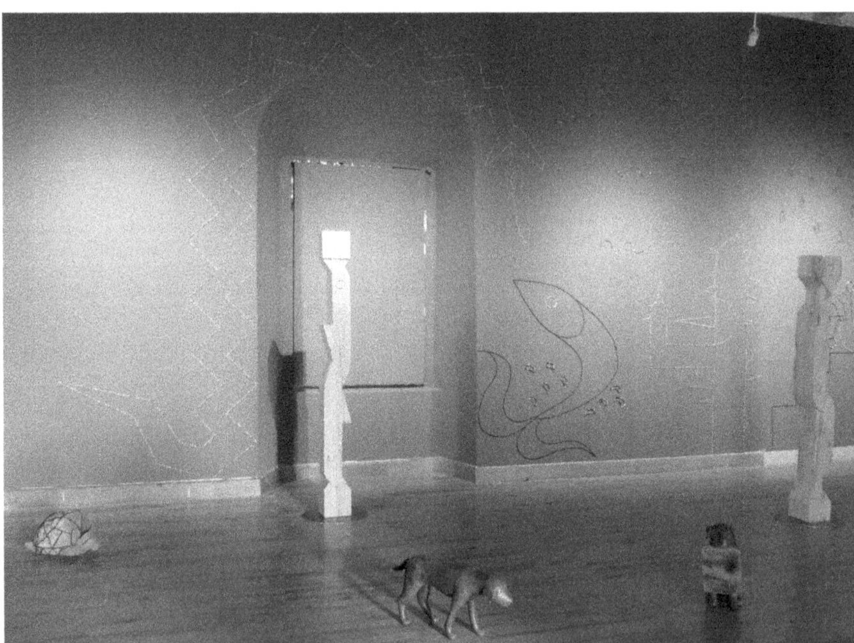

FIGURE 13.2 Marco Rountree, *Untitled* (2020). Wood, ceramic and seeds. Installation detail. 14th FEMSA Biennial, Centro Cultural Clavijero, Morelia, Mexico. Courtesy of the artist.

different combinations across the 'primitive' art of all nations: the spiral, the circle, the half circle, the S motif, the curved line, the zig-zag line and the straight line (Cordero Reiman, 2010). Rountree's mural starts from this basis; there are water drops, a pot of flowers, a fish, a snake, and a tree and its roots, all highly schematised. The tree's trunk and branches are made up of straight lines of black beans; more colours are used for the blooms represented in circles. The roots below are also drawn in straight lines, mostly in white and pink. A snake surrounding an arch on the wall is mainly drawn using a zig-zag line, and Rountree has even left some free-floating spirals. During our conversation, the artist emphasised the importance of the line in his work (Rountree, 2021). His reference to Best Maugard is particularly telling in the context of the biennial. Cordero Reiman underlines the significant influence of the method in introducing and popularising modernist aesthetics in Mexico. She notes that it encouraged the adoption of rural material culture as a model rather than a subject of contemporary painting. It provoked a generational shift towards a more abstract use of line and colour, reinforced the compositional role of drawing in the canvas and extended the stylisation of figurative motifs. According to Cordero Reiman (2010, p. 55), Best Maugard was driven by the need to produce a national art 'on a grassroots level', as part of the public education programme. Together with the columns which stand for the local, traditional way of life and O'Gorman's dogs, which are unconvinced of the enlightening narrative, Rountree's seed mural interrogates the nation-building, unifying didacticism of avant-garde muralism, making space for other stories.

The installation invites renewed scrutiny of the 'modernism of *artesanía*' (Montgomery, 2014, p. 233). Art historian Harper Montgomery (2014, p. 235) has problematised the post-Revolutionary conception of indigenous artisans as 'natural, innate creators' integral to Mexico's modernisation. Her study delves into the conflicting discourses surrounding popular art. She notes that before crafts were commercialised, they were displayed as a resilient system of production resistant to capitalist markets. Montgomery pays particular attention to Dr Atl's commentary on the volume accompanying the 1921 Exposición de Arte Popular. There, the artist considered the popular market as a 'site of socialist integration' that demonstrated communal self-reliance and sustainability as an alternative social and economic system, which was resistant to US industrialisation (Montgomery, 2014, p. 240). Dr Atl's ideological reflections emphasised the rural communities' connection to the land, based on the traditional standing of minerals, earth and clay as communal property in Mexico. However, Montgomery draws attention to the more problematic implications of Dr Atl's vision: the idea that indigenous labour was driven by an instinctive, creative drive rooted in 'race', and his notion of *habilidad manual indígena* (indigenous manual skill) as the basis of a mythical work ethic. These informed immigration policy and discourse that posited Mexican labourers as exceptionally skilled and an asset to the US economy. Accordingly, Montgomery notes the contrast during the 1930s between the Mexicans crossing the border to work in the US, and the US citizens travelling to Mexico to vacation.

In this context, crafts played a significant part in supporting tourism and presenting Mexico as a colourful, peaceful and non-industrial retreat. Montgomery reflects on the 'deeply problematic myth of the Mexican as a "naturally" able worker' (2014, p. 247). For all the utopian values embedded in *artes populares*, however, artisans remain vulnerable to the forces of globalisation, especially considering the increased privatisation of cultural tourism in Mexico, in which the biennial is implicated (see Coffey, 2010). We must underline at this point the paradoxical role of biennials as both critical agents and sites of spectacle, and the political ambiguity that legitimises these events within global neoliberal culture (Green and Gardner, 2016; Kompatsiaris, 2017). In this case, Rountree's installation made a subversive historiographic intervention by offering an alternative retelling. At the same time, the biennial's emphasis on crafts as representative of resilient, anti-capitalist ways of life provided a unique selling point, even if visits were hampered by the COVID-19 pandemic's travel restrictions.

Arantepacua's Communal Indigenous Council 2019–2021 and Adela Goldbard

In contrast to Rountree's more conceptual, individual approach, Adela Goldbard chose to collaborate with the Purépecha community of Arantepacua on a craftivist project. The co-authored installation centred on the events of 5 April 2017, when more than 300 members of the Michoacán police and army forces suppressed and attacked the community using police cars, trucks, helicopters and a 'rhinoceros' armoured tank. Four community members were killed and another nine were detained (Goldbard, 2021). The previous day, a delegation from Arantepacua had attended a meeting with officials in Morelia (the state's capital) to discuss a land ownership issue with the neighbouring village of Capácuaro. Far from being resolved, the conflict escalated, and the Arantepacua community organised a protest, including road blockages, which prompted the police operation (Ureste, 2020). After the traumatic event, the community decided to reject and effectively expel political parties and the local police. They sought justice by exercising their right to self-determination as an indigenous community (United Nations General Assembly, 2007; Aparicio Wilhelmi, 2009). For instance, they established a communal patrol called *kuaricha* and formed a horizontally structured communal council made up of four women and four men, which acts as the local authority (Ureste, 2020). The state of Michoacán officially recognised their decision to self-govern in 2018. The neighbouring communities of Comachuén, Sevina and Nahuatzén are also struggling to reclaim their right to self-determination and resist impoverishment, political persecution, harassment and criminalisation. Arantepacua continues to demand that those responsible for the 5 April operation are brought to justice (Ureste, 2020).

Goldbard (2021) first approached Arantepacua's Communal Indigenous Council and met with a relative of one of the victims. The artist recalls that during these initial meetings they 'discussed the importance of making their struggle visible through collaborative and creative work' in opposition to the dominant official narratives, and in support of their legal fight for justice. Her proposal involved a *trueque de saberes* (exchange of knowledges) with other nearby communities, based on an understanding of traditional craft and first-hand narratives as forms of resistance. As the central component, Goldbard (2021) proposed the fabrication and destruction of a real-scale papier-mâché rhinoceros, which was later commissioned in Cherán. Schoolteacher, Juana Morales, suggested the addition of cross-stitched embroideries made by craftswomen from Arantepacua and Turícuaro. The embroideries were based on photographs and video stills sourced from the council's archive and Auani Pascual's documentation. Three hundred clay *diablitos* (little devils) and over 70 wooden police cars, trucks and vans were made in Ocumicho and Pichátaro. Finally, Goldbard commissioned traditional songs known as *pirekuas* that narrate significant events for the community. An edited version of the artist's interviews with several community members was integrated into the installation at the Centro Cultural Clavijero (Clavijero Cultural Centre) in Morelia. These interviews were essential to the artist as first-hand accounts from the community.

Goldbard (2021) stressed that every decision was made in agreement with the council as the local authority. She reflects that the process was not easy; trust was gradually built through dialogue, and short- and long-term goals negotiated, including several commitments on her side, such as facilitating workshops for children and showing the installation in Tijuana and Chicago (which have large Purépecha populations). Council members, Juana Morales and Valentín Jimenez, acted as co-producers, facilitating the collaboration among neighbouring communities and initiating the *trueque de saberes*. Scholar Mary Loveday-Edwards suggests that, in this kind of approach, the artist assumes the role of 'co-learner, facilitator, [or] social transformer' (cited in Robertson and Vinebaum, 2016, p. 6). Goldbard (2021) sees herself 'as a weaver, a producer and a catalyst'. Her role consisted of 'intertwining' the shared narratives and bringing various components together. Ultimately, she aimed to 'reconstruct and preserve the collective memory of Arantepacua [...] and purge some of the harm inflicted on the community by the bloody events of April 5th, 2017' (Goldbard, 2021). In this regard, the project raises questions regarding authorship and a potentially uneven collaborative relationship between the artist and the community. In some respects, Goldbard's contemporary approach reproduces the modernists' intention to preserve endangered indigenous cultures. At the same time, however, the community members involved also recognised the value of the project for their own purposes.

Artist, critic and curator Nicole Burisch (2016) has examined the recent attention to craft within politically engaged, collaborative and performative projects. She notes that, in these cases, the centrality of the crafted object shifts to become a record, a prop or a tool, and sometimes the object disappears completely. Considering that performance art's transition from objects to actions has been historically interpreted as a political stance against commodification, Burisch (2016, p. 59) argues that the analysis of dematerialised craft practices must attend to 'the role of gestures, actions, and encounters'. As the central component of Goldbard's *Kurhirani no ambakiti*, the life-size rhinoceros stood in for the police's armoured tank, embodying the harm inflicted on the community (Goldbard, 2021). On 4 December 2020, the effigy was carried in a procession that followed the same route as the yearly remembrance procession for the victims of the 5 April raid. The papier-mâché rhinoceros arrived at Arantepacua's central square to be burnt and destroyed with fireworks, while local musicians performed the commissioned *pirekuas* that narrated the events of 2017 and the community's subsequent struggle for self-governance (Goldbard, 2021). The rhinoceros's head was cut and displayed as a trophy at the exhibition. According to Goldbard (2021), 'the aesthetic violence of this action [was] intended as a purging'. It sought to destabilise the politics of memory, dismantle 'oppressor/ oppressed' dichotomies and assist in healing collective trauma. The artist reflected that the project's title, translated as 'burning the devil: since that's the only way they listen to us' suggests that violence is sometimes the only means left for oppressed populations to contest the violence inflicted on them, and that, in fact, it offers a radical approach to storytelling. While there is an undeniable gap between the performative action and its documentation, the audio conversations with community members and the video of the rhinoceros's procession and burning, which was presented as part of the installation, offer a glimpse into the resistant potential of the project, which lies in the community's sense of ownership over it.

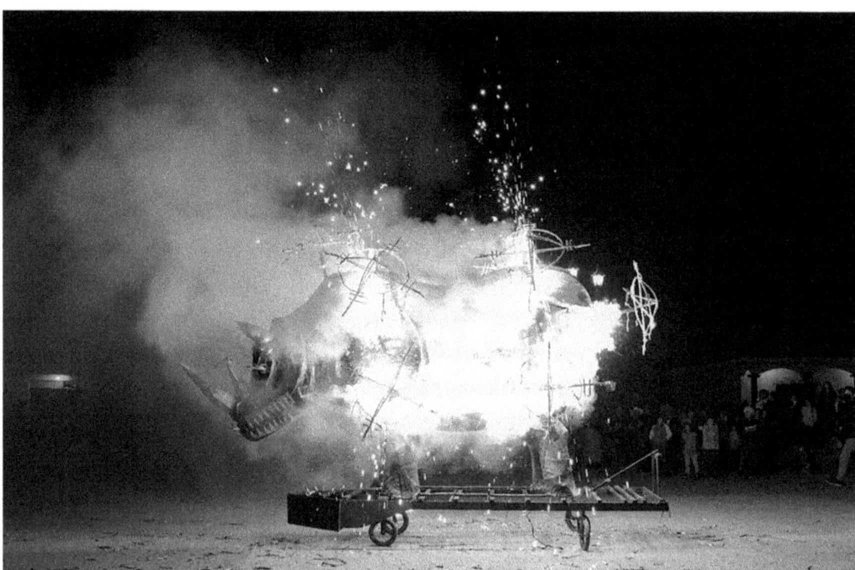

FIGURE 13.3 Adela Goldbard, *Kurhirani no ambakiti (quemar al diablo): porque solo así nos escuchan* (2020). Video still. 14th FEMSA Biennial. Courtesy of the artist.

Based on the communal archive, the cross-stitched embroideries present a visual counter-account of the event. Scholar and curator Ellyn Walker (2021, pp. 303–6) has studied how diverse communities across the Americas use embroidery 'as a site of resistance and re-imagination [...] to expose histories of gendered, colonial and state-sanctioned violence, and create models of feminist making, community-building and Indigenous resurgence'. In this case, the archival photographs and video stills became 'pixelated', tactile images (Goldbard, 2021). As Julia Bryan-Wilson (2017, p. 7) claims in her seminal study on art and textile politics, 'to textile politics is to give texture to politics, to refuse easy binaries, to acknowledge complications'. Not only do the stitches 'insist on the women's survival', but embroidery also supports healing and decolonising processes, demanding truth-telling and accountability (Walker, 2021, p. 308). In Goldbard's installation, the stitched police barricades and approach, along with the community's losses, defence and protest, present the counter-narrative that has been suppressed by the state. In this sense, the textiles perform 'a vital act of memory work, allowing others to bear witness' (Walker, 2021, p. 313). *Kurhirani no ambakiti* demonstrates how craft activism can support self-determination, autonomy and cultural memory, as well as their interconnections with global citizenship and justice (Black and Burisch, 2021, p. 56). In the context of the biennial, the commission's curatorial framework encouraged and facilitated the collaborative process, allowing the Arantepacua community to craft their retelling as part of their ongoing fight for justice.

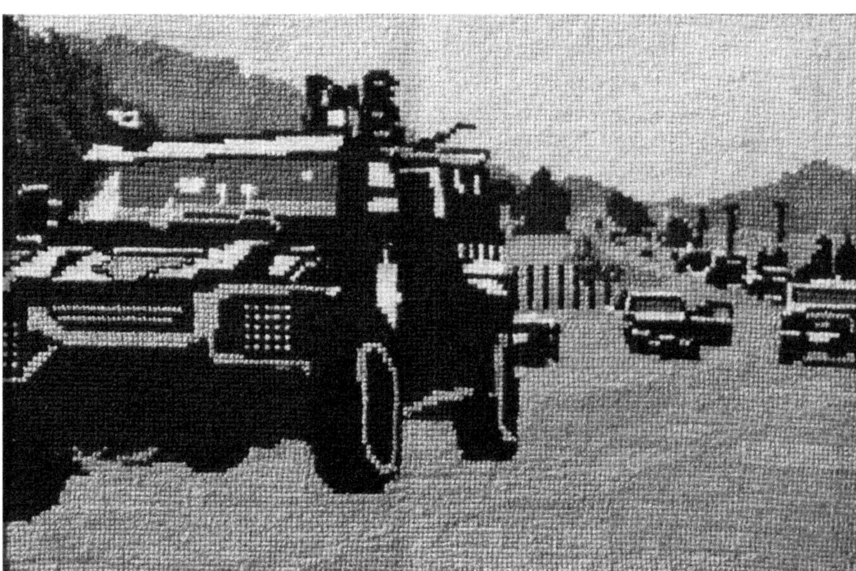

FIGURE 13.4 Adela Goldbard, *Kurhirani no ambakiti (quemar al diablo): porque solo así nos escuchan* (2020). Installation detail. 14th FEMSA Biennial. Courtesy of the artist.

The more than 300 clay *diablitos*, hand-painted with police and military uniforms, and more than 70 wooden police cars and trucks point to the excessive use of force by the Michoacán police (Goldbard, 2021). The *pirekuas*, mostly sung in Purépecha, narrate the events of 5 April, remember the deceased and praise the strength and resilience of the community (Goldbard, 2021): '*Arantepacua vive y seguirá viviendo; hoy se escucha su voz*' ('Arantepacua lives and will keep living; its voice is heard today'). Overall, the installation presents the tensions that lie in the distinction between art and craft. As Bryan-Wilson (2017, p. 6) suggests, the dynamics between 'fine art/non-fine art [bring] to the fore extraordinarily fraught questions about race, cultural appropriation, valuation and class disparity'. However, while the line drawn between art and craft has emphasised the latter's functionality or use-value, analysis and interpretation of art in the 21st century tends to explore the collapse of such boundaries (Bryan-Wilson, 2017, p. 13). Goldbard (2021) notes that indigenous communities' artistic practice preserves and complicates oral memory, expresses identity and connects people to their territory, tradition and culture—all of which are urgently needed in a world in crisis. Similarly, in their introduction to *The New Politics of the Handmade*, editors Anthea Black and Nicole Burisch (2021, p. 31) stress the need for 're-articulating craft as a world-making and geographically specific aesthetic practice that connects to the land'. Black and Burisch reflect that while craft alone might not overturn colonial frameworks, it can offer alternative ways of knowing and imagining that contribute to

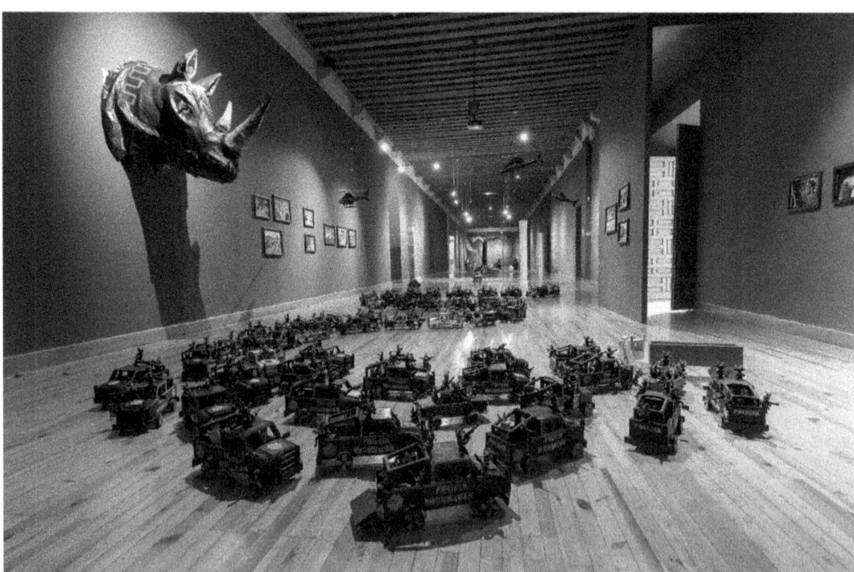

FIGURE 13.5 Adela Goldbard, *Kurhirani no ambakiti (quemar al diablo): porque solo así nos escuchan* (2020). Installation view. 14th FEMSA Biennial, Centro Cultural Clavijero, Morelia, Mexico. Courtesy of the artist.

cultural transformation. For Goldbard (2021), the biennial platform sustained some of the main aims of her project: making visible the attack of 5 April and giving voice to the community's struggle. Nevertheless, she stressed that, more than vindicating popular art, decolonising contemporary art requires moving away from its commodification, and abolishing or renewing its 'alienating and stagnant institutions' (Goldbard, 2021). Embroiled in these complex politics, the meeting between art and craft staged by Goldbard's project can be seen as what Bryan-Wilson (2017, p. 19) calls 'forms of making side by side' that offer no straightforward conclusions. These practices are best approached with a 'both-at once' or 'both/and' logic (Bryan-Wilson, 2017, p. 36): art and craft, authored and collaborative, action and object, local and global, aesthetic and political. While, during the 20th century, tradition was retrieved as part of the country's modernist project (ultimately reinforcing binary distinctions), contemporary practices concentrate on blurring their limits. In this case, by working at the seams of these boundaries, the project presents the community's claim over their history.

We have necessarily focused on only 2 out of 24 artistic commissions and 5 local exhibitions organised by the biennial. Our analysis is inevitably limited to the works that, in our view, best reflected the biennial's curatorial proposition and its emphasis on reconsidering official narratives and promoting co-creative processes of knowledge production. To briefly cite one more example, Costa Rican artist Carlos Fernández's site-specific installation *Continua despensa de saberes*—comprised of a series of ten abstract paintings and three photographs—responded to the 16th-century fresco paintings depicting botany lessons on the walls of the Old Jesuit School in Pátzcuaro.[8] Fernández (2021) regards his paintings as a 'live register' that incorporates graphics from the agronomy classes he teaches and the virtual dinners he hosted during lockdown, in which he performed a monologue tracing food products and capitalist trade networks. Fieldworkers, cooking processes and local markets are layered onto the canvas. As curator Gabriela Saenz observes, Fernández's work unveils traditional, more sustainable agricultural practices (Fernández, 2021). Overall, the biennial's decentralising, revisionist approach presented situated artmaking at the end of the world. Danowski and Viveiros de Castro (2017, p. 5) describe the Anthropocene as a 'passive present' or a present 'without a view'. We are living through a 'shared catastrophe' that we can no longer revert, which makes its mitigation more urgent (Chakrabarty, 2009, p. 218). Crucially, as philosopher Bruno Latour (2017, p. 90) sustains, the events we need to cope with lie largely in the past rather than the future.

In the context of precarious cultural labour and contested narratives about Mexican identity, the 14th FEMSA Biennial offered a decentralising, revisionist perspective on the role of art in history-making. The biennial's discursive approach presented a historiographical intervention that questioned homogenising national narratives and re-examined, in particular, the concept of *artesanía* and its part in the post-Revolutionary definition of Mexican identity. Our analysis considered both the role of muralism within a larger cultural and educational programme during the 20th century and the biennial's revision and challenge to official historical discourse. Through its public and exhibition programmes, the biennial facilitated contemporary art retellings in close collaboration with the local indigenous communities. While the corporate framework sustaining these commissions raises concerns regarding their critical and political potential, the curatorial proposition brought a crucial issue to the fore: the pressing concern about whom history speaks for and the possibility of communities crafting their counter-stories in response to their erasure. By approaching the biennial through storytelling, this chapter proposes that large-scale exhibitions may be knowledge- and world-making practices that potentially reflect the many worlds in the world.

Notes

1 FEMSA Foundation works in three main areas: water sanitation and security, early childhood development and a cultural programme that promotes Latin American modern and contemporary art.
2 The title of the 2018 FEMSA Biennial referenced Bruno Latour's epistemological critique in *We Have Never Been Modern*, first published in French in 1991.
3 Another instance of friction between the SEP and the DAPP surrounding notions of propaganda was the film production programme that various post-Revolutionary governments, including that of Cárdenas, had implemented as part of their plans for education. The programme aligned with the 'Mexicanisation' project and the formation of a national imaginary, originally developed by the SEP (see Aobites and Loyo, 2010, p. 246). However, by 1938, it had been taken over by the DAPP, which understood it as less a cultural issue than one of interior propaganda in which a good amount of documentary films promoting the works of the Cárdenas administration be funded and created in a very short time (see Dümmer-Scheel, 2018, p. 294).
4 Despite muralism's focus on the masses, it was simultaneously for elite 'initiates', as González Mello (2008) demonstrates in his detailed reading of the masonic and occult elements of murals by Rivera and Orozco from the 1920s to the 1940s.
5 As the museum's website states, its purpose was to 'assert the economic and aesthetic value of products by Purépecha people' native to the state of Michoacán (INAH, 2020). The first Museo de Arte Popular in Mexico City opened in 2006.
6 So ingrained was this kind of reading of murals that Philip Guston and Reuben Kadish's work in Morelia about fascism and racism in the US came to be popularly known as *The Inquisition*, with its hooded figures understood as representations of Spanish Catholic torturers.
7 A detailed visual analysis of the mural is available from Canal Crefal (2018). *Conoce el Mural de Juan O'Gorman.* 24 October. [Online video]: https://youtu.be/4j79zVo5f_o
8 The school was founded in 1574. It belonged to the Jesuits until 1767. It subsequently held diverse functions until around 1960, when it was abandoned. The building was restored between 1990 and 1994 and is now a dependency of Michoacan's Culture Ministry (Sistema de Información Cultural, 2017).

References

Aboites, L. and Loyo, E. (2010). 'La construcción del nuevo estado, 1920–1945', in Gómez Ruiz, F. (ed.) *Historia general de México ilustrada*. Mexico City: El Colegio de México, pp. 196–259.

Aparicio Wilhelmi, M. (2009). 'La libre determinación y la autonomía de los pueblos indígenas. El caso de México', *Boletín Mexicano de Derecho Comparado*, 124, pp. 13–38. [Online]. Available at: https://revistas.juridicas.unam.mx/index.php/derecho-comparado/article/view/4087/5262 (Accessed: 4 May 2021).

Bienal FEMSA (n.d.). *What is the FEMSA biennial?* [Online]. Available at: https://www.bienalfemsa.com/en/xiv-inestimable-azar/acerca-de/que-es (Accessed: 15 March 2021).

Black, A. and Burisch, N. (eds.) (2021). *The new politics of the handmade: craft, art and design.* London: Bloomsbury.

Boime, A. (2008). 'Breaking open the wall: the Morelia mural of Guston, Kadish and Langsner', *The Burlington Magazine*, 150(1264), pp. 452–9. [Online]. Available at: http://www.jstor.org/stable/40479800 (Accessed: 18 June 2021).

Bruno-Jofré, R. and Martínez Valle, C. (2009). 'Ruralizando a Dewey: el amigo Americano, la colonización interna y la Escuela de la acción en el México posrevolucionario (1921–1940)', *Encuentros sobre Educación*, 10, pp. 43–64.

Bryan-Wilson, J. (2017). *Fray: art and textile politics.* Chicago and London: The University of Chicago Press.

Buenfil Burgos, R.N. (2004). *Argumentación y poder: la mística de la revolución mexicana rectificada.* Mexico City: Plaza y Valdés.

Burisch, N. (2016). 'From objects to actions and back again: the politics of dematerialized craft and performance documentation', *TEXTILE*, 14(1), pp. 54–73.

Cepeda, G. (2020). '"Billion-peso tarp": the case of Mexico City's Chapultepec Park', *Art Review*, 27 October. [Online]. Available at: https://artreview.com/billion-peso-tarp-the-case-of-mexico-city-chapultepec-park/ (Accessed: 15 March 2021).

Chakrabarty, D. (2009). 'The climate of history: four theses', *Critical Inquiry,* 35(2), pp. 197–222.

Chambers, I., De Angelis, A., Ianniciello, C., Orabona, M. and Quadraro, M. (eds.) (2014). *The postcolonial museum: the arts of memory and the pressures of history.* Reprint. London and New York: Routledge, 2016.

Coffey, M. (2010). 'Marketing Mexico's great masters: folk art tourism and the neoliberal politics of exhibition', in Berger, D. and Wood, A.G. (eds.) *Holiday in Mexico: critical reflections on tourism and tourist encounters.* Durham, NC and London: Duke University Press, pp. 265–94.

Cordero Reiman, K. (2010). 'The Best Maugard drawing method: a common ground for modern Mexicanist aesthetics', *The Journal of Decorative and Propaganda Arts*, 26, pp. 44–79. [Online]. Available at: http://www.jstor.org/stable/41432953 (Accessed: 15 March 2021).

Cruz Porchini, D. (2014). *Proyectos culturales y visuales en México a finales del cardenismo (1937–1940).* PhD thesis. UNAM. [Online]. Available at: http://132.248.9.195/ptd2014/marzo/0709778/0709778.pdf (Accessed: 22 January 2021).

Danowski, D. and Viveiros de Castro, E. (2017). *The ends of the world.* Translated by R. Nunes. Cambridge: Polity Press.

De Alba, J.I. (2020). 'Femsa: el negocio de la enfermedad y la cura', *Pie de Página*, 12 June. [Online]. Available at: https://piedepagina.mx/femsa-el-negocio-de-la-enfermedad-y-la-cura/ (Accessed: 12 March 2021).

Dümmer Scheel, S. (2018). '¿De quién es la diplomacia pública? El rol del Departamento Autónomo de Prensa y Publicidad (DAPP) en la propaganda exterior cardenista', *Estudios de Historia Moderna y Contemporánea de México*, 55, pp. 279–312.

Ettinger, C.R. (2015). 'Perspectivas sobre la conservación de la troje purépecha. Entre la conservación ideal y la realidad', in Ojeda Dávila, L., Mijangos Díaz, E. and Mercado

López, E. (eds.) *Cultura, sociedad y políticas públicas. Pasado y presente del patrimonio cultural en Michoacán*. Morelia: Universidad Michoacana de San Nicolás de Hidalgo, pp. 69–87.

Farías Mackey, M.G. (2010). 'Cárdenas, el indigenista', in León y González, S. (ed.) *El cardenismo, 1932–1940*. Mexico City: Fondo de Cultura Económica, pp. 258–324.

FEMSA Foundation (2021). *Our commitment*. [Online]. Available at: https://femsafoundation.org/ (Accessed: 15 March 2021).

Fernández, C. (2021). Interview by Ana S. Gonzalez Rueda [Skype], 20 January.

Franco, L. (2020). 'Los dueños del agua: trasnacionales acaparan reservas, mientras México avanza al temido "Día Cero"', *Contralínea*, 6 January. [Online]. Available at: https://www.contralinea.com.mx/archivo-revista/2020/01/06/los-duenos-del-agua-trasnacionales-acaparan-reservas-mientras-mexico-avanza-al-temido-dia-cero/ (Accessed: 11 March 2021).

Garciadiego Dantan, J. (2015). *Autores, editoriales, instituciones y libros: estudios de historia intelectual*. Mexico City: El Colegio de México.

Garzon, S. (2021). 'Mexico's FEMSA Biennial finds ways to decentre', *Ocula Magazine*, 7 January. [Online]. Available at: https://ocula.com/magazine/conversations/14th-femsa-biennial-in-mexico/ (Accessed: 21 January 2021).

Goldbard, A. (2021). Email to Ana S. Gonzalez Rueda, 3 February.

González Mello, R. (2008). *La máquina de pintar: Rivera, Orozco y la invención de un lenguaje*. Mexico: UNAM.

Green, C. and Gardner A. (2016). *Biennials, triennials and documenta: the exhibitions that created contemporary art*. Malden, MA: Wiley-Blackwell.

INAH (2020). *Museo local de artes e industrias populares*. [Online]. Available at: https://www.inah.gob.mx/red-de-museos/299-museo-local-de-artes-e-industrias-populares-de-patzcuaro (Accessed: 21 January 2021).

Jaimes, H. (2012). *Filosofía del muralismo mexicano: Orozco, Rivera y Siqueiros*. Mexico City: Plaza y Valdés.

Jolly, J. (2018). *Creating Pátzcuaro, creating Mexico: art, tourism, and nation building under Lázaro Cárdenas*. Austin: University of Texas Press. EPUB edition.

Kompatsiaris, P. (2017). *The politics of contemporary art biennials: spectacles of critique, theory and art*. New York and London: Routledge.

Kompatsiaris, P. (2020). 'Curating resistances: ambivalences and potentials of contemporary art biennials', *OnCurating*, 46, pp. 313–22. [Online]. Available at: https://on-curating.org/issue-46.html (Accessed: 15 March 2021).

Latour, B. (2017). *Facing Gaia: eight lectures on the new climatic regime*. Translated by C. Porter. Cambridge: Polity Press.

Lira García, A.A. (2014). 'La alfabetización en México: campañas y cartillas, 1921–1944', *Traslaciones: Revista Latinoamericana de Lectura y Escritura*, 1(2), pp. 126–49.

Lobo, A. (2019). 'Trabajadores en huelga de la Coca-Cola se rebelan contra sindicato en Ciudad de México', *World Socialist Web*, 28 June. [Online]. Available at: https://www.wsws.org/es/articles/2019/06/28/coca-j28.html (Accessed: 11 March 2021).

Loveless, N. (2019). *How to make art at the end of the world: a manifesto for research-creation*. Durham: Duke University Press.

Marchart, O. (2014). 'The globalization of art and the 'Biennials of Resistance': a history of the biennials from the periphery', *World Art*, 4(2), pp. 263–76.

Miércoles de SOMA (2021). *XIV Bienal FEMSA*. 14 January. [Online video]. Available at: https://youtu.be/zAQy2ECXvnM (Accessed: 25 January 2021).

Montgomery, H. (2014). 'From aesthetics to work: displaying Indian labor as modernist form in Mexico City and New York', *Modernism / modernity*, 21(1), pp. 231–51.

Museo Anahuacalli (n.d.). *Xitle Volcano School of Sciences and Artisanry*. [Online]. Available at: http://museoanahuacalli.org.mx/en/xitle-volcano-school/ (Accessed: 15 March 2021).
Oesterreich, M. (2018). 'The display of the "indigenous" – collecting and exhibiting "indigenous" artifacts in Mexico, 1920–1940', *Artelogie*, 12.
Pacheco-Vega, R. (2015). 'Agua embotellada en México: de la privatización del suministro a la mercantilización de los recursos hídricos', *Espiral*, 22(63), pp. 221–63.
Pearson, T. (2017). 'A US charity helps big businesses take indigenous people's water in Mexico', *Open Democracy*, 19 September. [Online]. Available at: https://www.opendemocracy.net/en/democraciaabierta/us-charity-helps-big-businesses-take-indigenous-peoples-water-in-me/ (Accessed: 15 March 2021).
Peredo, X. (2011). 'Cuando te tomes una Indio recordarás esto', *Animal Político*, 17 August. [Online]. Available at: https://www.animalpolitico.com/el-blog-de-anca/cuando-te-tomes-una-indio-recordaras-esto/ (Accessed: 11 March 2021).
Peredo, X. (2015). 'El estadio BBVA Bancomer, monumento al despojo ambiental', *Horizontal*, 29 July. [Online]. Available at: https://horizontal.mx/el-estadio-bbva-bancomer-monumento-al-despojo-ambiental/ (Accessed: 11 March 2021).
Ramírez Miranda, C., Cruz Altamirano, L. and Marcial Cerqueda, V. (2015). 'Luchas por el territorio y soberanía alimentaria en el Istmo oaxaqueño, México', *Eutopía*, 8, pp. 29–44.
Robertson, K. and Vinebaum, L. (2016). 'Crafting community', *TEXTILE*, 14(1), pp. 2–13.
Rountree, M. (2021). Interview by Ana S. Gonzalez Rueda [Skype], 14 January.
Simpson, M.G. (2001). *Making representations: museums in the post-colonial era*. Rev. edn. London and New York: Routledge.
Sistema de Información Cultural (2017). *Centro Cultural Antiguo Colegio Jesuita*. [Online]. Available at: https://sic.cultura.gob.mx/ficha.php?table=centro_cultural&table_id=259 (Accessed: 11 March 2021).
Sleeper-Smith, S. (ed.) (2009). *Contesting knowledge: museums and indigenous perspectives*. Lincoln and London: University of Nebraska Press.
Subirats, E. (2018). *El muralismo mexicano: mito y esclarecimiento*. Mexico City: Fondo de Cultura Económica.
Tarcus, H. (2019). 'Una voz libertaria en la medianoche del siglo: el "Manifiesto por un arte revolucionario independiente"', *Nueva Sociedad*, 283. [Online]. Available at: https://nuso.org/articulo/una-voz-libertaria-en-la-medianoche-del-siglo/ (Accessed: 22 January 2021).
United Nations General Assembly (2007). *United Nations declaration on the rights of indigenous peoples*. Resolution A/RES/61/295. [Online]. Available at: https://www.un.org/development/desa/indigenouspeoples/wp-content/uploads/sites/19/2018/11/UNDRIP_E_web.pdf (Accessed: 4 May 2021).
Ureste, M. (2020). 'Arantepacua: la comunidad indígena que expulsó a los partidos y creó su propia policía', *Animal Político*, 25 February. [Online]. Available at: https://www.animalpolitico.com/2020/02/arantepacua-expulso-partidos-politicos-policia/ (Accessed: 15 March 2021).
Vázquez de Knauth, J. (1970). *Nacionalismo y educación en México*. Mexico City: El Colegio de México.
Von Oswald, M. and Tinius, J. (eds.) (2020). *Across anthropology: troubling colonial legacies, museums, and the curatorial*. Leuven: Leuven University Press.
Walker, E. (2021). 'The sovereign stitch: rereading embroidery as a critical feminist decolonial text', in Black, A. and Burisch, N. (eds.) *The new politics of the handmade: craft, art and design*. London: Bloomsbury, pp. 303–32.
XIV Bienal FEMSA (2020). *Programa público: real, surreal, virtual. De la desmesura muralista a la sobrecarga digital*. 25 November. [Online video]. Available at: https://youtu.be/taPODqu6UqM (Accessed: 15 March 2021).

14

THE *ARRIVANTS* EXHIBITION

Art, migration, museums and resurrections

Allison Thompson

Introduction

For more than 25 years, since the quincentennial of Columbus's arrival in the Caribbean, there has been a growing interest in exhibitions that 'explore' Caribbean art. The majority of these shows have been presented in European or North American institutions, with only a few having reached the Caribbean.[1] What is required to construct a more equitable global platform from which to articulate a discourse on contemporary Caribbean visual practice? What is required to curate an exhibition of Caribbean art first and foremost from a Caribbean perspective and for a Caribbean audience—an exhibition that is regional in its focus and its staging, able to travel first through the Caribbean but also internationally, as a counterpoint to those exhibitions that have gained wider international exposure in the past?

The exhibition *Arrivants: Art and Migration in the Anglophone Caribbean World* contemplated these questions in the early stages of its conception. Taking its title and its focus on 'the journey' from Kamau Brathwaite's *The Arrivants* trilogy, this exhibition, which included the work of 25 artists from the Caribbean and the wider diaspora, explored the diasporic nature of Caribbean society as documented and interrogated through its artistic production.[2] Planned as part of the Horizon 2020 EU-LAC Museums and Migration project, the exhibition, which opened at the Barbados Museum and Historical Society in November 2018, sought to investigate the impact of migration and gender, and the resulting cultural diasporas, on the field of contemporary visual art and on curatorship in particular. This exhibition, curated by myself and Veerle Poupeye, cast its gaze on the issues represented from within the Caribbean itself, taking into consideration how such projects are negotiated in the Caribbean context.[3]

DOI: 10.4324/9781003288138-17

Kamau Brathwaite's *Arrivants: A New World Trilogy* (1973), consists of three long poems—*Rights of Passage* (1967), *Masks* (1968) and *Islands* (1969)—that each deal in their different ways with journeys and a 'rediscovery' of Africa (Brown, 1995). Brathwaite, who originally left Barbados in 1949 to study history at the University of Cambridge, subsequently travelled to Ghana in 1955 to work as an education officer. While there, he witnessed Kwame Nkrumah's rise to power and the emergence of Ghana as the first African state to gain independence from Britain, two events that impacted his ideas about Caribbean culture.

Brathwaite later said of the transformative experience of his time in Africa:

> Slowly, slowly, ever so slowly, I came to a sense of identification of myself with these people, my living diviners. I came to connect my history with theirs, the bridge of my mind now linking Atlantic and ancestor, homeland and heartland [...]. And I came home to find that I had not really left. That it was still Africa; Africa in the Caribbean.
>
> *(Coombs, 1974, as cited in Morris, 1995, p. 118)*

Brathwaite returned to the Caribbean in 1962 and the following year accepted a teaching post at the University of the West Indies Mona Campus in Jamaica, but a few years later, he returned to the UK to pursue a PhD at the University of Sussex. It was at this time that he, along with John La Rose and Andrew Salkey, founded the Caribbean Artists Movement (CAM), an important collective of Caribbean writers and artists. This fertile period spent across three continents formed the grounding of the *Arrivants* book, which has been described as 'a major document of African reconnection [that] [...] charts a set of overlapping psychic journeys to, from and within the New World and Africa, acknowledging achievement and some painful realities, examining self and community, past and present' (Morris, 1995, p. 129).

Migration, both voluntary and forced, is and always has been central to the story of the Caribbean; as such, it is deeply embedded in the psyche of Caribbean people and has shaped their identities and experiences, whether they are migrants themselves or not. From the arrival of European expeditions, dating from the end of the 15th century, and the extermination of the indigenous population, the Caribbean was repopulated first by Europeans as well as through forced migration from Africa and subsequent waves of migration from Asia and North and South America. The 20th century also witnessed significant movement out of the Caribbean to diasporic centres such as London, New York, Miami and Toronto. There has also been ongoing migration into and throughout the region. The art selected for this exhibition focused on the social and cultural impact of these migratory patterns and histories, their political significance, as well as acts of defiance and resistance and the implications for individual and collective identities. The work of Kamau Brathwaite provided inspiration on how traditional practices, models and languages might be reconsidered, altered or creolised.

Both curators of the *Arrivants* exhibition, Veerle Poupeye and myself, migrated to the Caribbean in the 1980s from Belgium and Canada, respectively. We both have had long careers as educators, curators and writers in the Caribbean, but we remain in some sense 'outsiders'. This perspective of living between two places, of belonging in some way to a somewhere else, is the experience of the migrant. Indeed, many of the key texts on Caribbean identity and culture have been impacted by the writer's experience of distance or remove, strangeness, of being outside or away.

The Caribbean, while often superficially characterised as an undifferentiated region, is in fact a complex and highly varied space in terms of geographies, histories, ethnicities, religious or spiritual practices, languages and more. For the *Arrivants* exhibition, it was decided to give particular focus to the Anglophone Caribbean from the early 20th century to the present day and to the cultural impact of migration from and to the UK, North America and Europe, as well as movement within the Caribbean and Central American region. Implicit in this are the earlier histories of forced and voluntary migration that have shaped the Caribbean as we know it today and the manner in which these have shaped the identities and experiences of Caribbean peoples, whether they are themselves migrants or not. Most of all, the exhibition focused on the social and cultural impacts of these migratory movements, their political significance, the histories of defiance and resistance, and their implications for individual and collective identities. While the decision to focus on the Anglophone Caribbean was the result of certain practical considerations—notably the relatively small scale of the project, the ease of accessibility to work and the ability to provide more focus on a restricted scope of research—this brings with it several shortcomings and biases. It continues to give priority to the divisions within the Caribbean based on the history of colonisation, most notably evident in the divisions of language today. In reality, people within the Caribbean have moved across these barriers for multiple reasons, including education, employment and family. While Barbados is unique in its uninterrupted history of colonisation under the British, many Caribbean countries experienced periods of control under differing imperial powers, with present-day cultures that reflect this.

Many of the artists included in the exhibition have lived in multiple locations and have reflected on this experience in their work; as well, the broader effects of diaspora, displacement and migration are key themes in the work of many artists in the Caribbean and its diaspora alike. Given the recent migration crises throughout the world, and particularly the questions about the immigration status of members of the Windrush generation in the UK, the subject has taken on particular potency in this moment.

A second but important consideration in conceiving this exhibition pertained to the representation of Caribbean art in survey and thematic shows, most of which have been initiated, funded and toured by major institutions in metropolitan centres and most of which have never even been shown in the Caribbean—a major imbalance in the representation of Caribbean art that needs to be addressed. Was it possible, the organisers wondered, to rethink the exhibition format to respond to the specific needs and aspirations of exhibition-making in the region, to speak first to a Caribbean audience, but then also present an inflected voice to a global audience? Caribbean writers throughout the 20th century have had a profound influence on cultural and postcolonial studies—Aimé Césaire, Frantz Fanon, Edouard Glissant, Kamau Brathwaite, George Lamming, Wilson Harris, Derek Walcott, Sylvia Winter, Stuart Hall and the list goes on. These critical thinkers are known and cited globally to describe the social transformations over the last century. But notwithstanding the broad application and relevance of their radical ideas, the kernels of this thought were rooted within the unique space described as the Caribbean.

Noted European curator Hans Ulrich Obrist (2014, p. 14) has famously remarked on his reliance on Caribbean theorists, especially Édouard Glissant in his attempt to formulate an approach of globality in his projects. But as art educator and curator Nanne Buurman has pointed out in her essay 'The Blind Spot of Global Art? Hans Ulrich Obrist's *Ways of Curating*', the proliferation of contemporary art that is circulated as 'global art' in biennials and art fairs worldwide raises ethical questions about the tropes of globality and practices that presume to transcend territorial borders and the inequitable access to resources and visibility. Themes of migration, in particular, have been identified as one of the major conundrums of the current global condition, but as Buurman indicates, there is a difference between voluntary and forced migration (Buurman, 2018, pp. 301–22). How can curators in the Caribbean address these power imbalances as they relate to access to resources and visibility? And how do we rethink the exhibition model to resituate the dialogue of a Caribbean contemporary and its relationship to the global?

In their book *Situating Global Art*, Buurman and her co-authors address the structural conditions of exclusion and systemic discriminations caused by the logic of national and regional canons, art history's 'colonial unconscious' (Dornhof et al., 2018, p. 11). They call for efforts to decolonise art historical knowledge and replace binary epistemological models with more relational approaches that focus on 'contacts, flows and circulations, as well as global relations of production' (Dornhof et al., 2018, p. 12). The traditional Eurocentric or Western narrative of a linear, chronological historiography of progress has been widely criticised for failing to acknowledge the existence of multiple modernities as well as the heterogeneity of coexisting contemporary art practices. The authors' description of a more fluid and non-linear model echoes Kamau Brathwaite's concept of Tidalectics, his riposte to the fixed, assured, earthbound reasonings of European thought embodied in Hegel's dialectic and inspired instead by the fluctuating tides, rhythmic waves and itinerancy of the ocean in modelling a different interpretative approach.

Museums and their practices of exhibition display are rooted in a European tradition of 19th-century empire building through amassing colonial collections of looted objects and staging 'great exhibitions' and human zoos, and they are implicitly linked with the agendas of nation-building. However, Dornhof et al. argue that the exhibition format, with its ability to present a variety of images and objects from diverse contexts simultaneously, has the capacity to function as a counter-model and a 'critical corrective' to linear historiographies and diachronic narratives of progress—more so than the linearity of written text. (Of course, Brathwaite's poetry confounded this very presumption of linearity!)

The emergence of so-called global art as represented, for example, in the proliferation and expansion of biennials can mask deep inequities and biases—the radically varying social, political and economic conditions that impact art production, distribution and reception worldwide. Dornhof et al. argue for a perspective that acknowledges 'the inherent transculturality of artistic practices and artefacts', in an effort to account for their 'dynamic cross-cultural constellations, migrations and transformations, locations and dislocations' (2018, p. 17).

The concept of transculturation, first articulated by Cuban anthropologist Fernando Ortiz Fernández in the 1940s, has more recently been revived for its potential to address some of the current imbalances. Ortiz developed the term in his classic text *Cuban Counterpoint: Tobacco and Sugar* (1940), as an alternative to acculturation, explaining that when a dominant culture imposes its ideas and practices on another, both are transformed through the multidirectional reciprocity of the exchange (Ortiz, 1995). The German philosopher Wolfgang Welsch introduced the term 'transculturality' in the 1990s (seemingly without acknowledging Ortiz's precedent half a century earlier) as an alternative to *multiculturality* or *interculturality*, which portray cultures as homogenous and monolithic (like distinct spheres), since contemporary societies are characterised by greater inner differentiation and complexity as a result of multiple cultural influences, whether they experience migration or not (Welsch, 1999). For Welsch, the transcultural was equated with the cosmopolitan or syncretic and could serve as a political and ethical corrective to ethnocentrism and xenophobia.

More recently, art historian Monica Juneja has identified the limitation of Welsch's description, which assumes that border crossings and cultural mixing were unique attributes of modernity. She argues that transculturation denotes a dynamic process of transformation that unfolds through extended contacts and relationships between cultures. She reminds us that the emergence of the discursive category of 'culture' within the social sciences is tied to the idea of the modern nation, premised on the belief that identifiable groups were 'ethnically bound, internally cohesive and linguistically homogeneous spheres' (Juneja and Kravagna, 2013, p. 25). But this seemingly stable conception was challenged or threatened by contradictory trends generated by mobility and extended contacts that have characterised societies for centuries. The terms *transculturation* and *transculturality* are explicit critiques of this notion, for the prefix 'trans' enables emancipation from the concept: 'Transculturality is about spatial mobility, circulation or flows, an insight drawn from studies of globalisation, but is neither synonymous with nor reducible to these' (Juneja and Kravagna, 2013, p. 25).

Importantly, Juneja argues for using transculturality as an analytical mode rather than a theoretical given. It is necessary to acknowledge a range of possible transactions rather than fixed dichotomies or polar positions. While concepts such as syncretism, creolisation and hybridity, which have emerged in those regions that had close ties with colonialism, notably the Caribbean and Latin America, are related, they have become 'globalised' in their usage but also diluted from their original meaning and should not be conflated with transculturation:

> it [transculturation] rather operates on a different register and highlights the procedural character of a broad variety of phenomena, including flows, entanglements, and other forms of circulation, and confronts us with the challenge of finding a precise language to capture the morphology of the relationships built into these phenomena.
>
> *(Juneja and Kravagna, 2013, p. 29)*

Decolonisation is identified as a transcultural project. The model of colonialism describes the transmission of culture from metropolitan centres to colonised peripheries where it is uncritically absorbed as a result of asymmetries of power. But this model, even in cases where colonies have reconfigured the culture exported to them, is distorted by its construction as the coloniser–colony binary. A transcultural view refutes the myth of the dominant, fixed and immutable culture of the coloniser and instead allows us to locate these processes in a global context that transcends this opposition and views cultural phenomena as multi-sited interactions.

Relying on the adaptive concept of transculturation as articulated by Juneja, Dornhof et al. propose focusing on 'transcultural topologies' of global art—made up of institutions, actors and specific art practices as well as historiography and curation—that serve as 'nodal points' in networks that transcend bounded or geographical categories of nation, region or city.

> Focusing on the transcultural topologies of global art thus permits the study of relational processes of circulation and exchange while also calling into question the idea of ethno-cultural locality as a nostalgic marker of authenticity as well as celebrations of multicultural plurality that disregard ongoing inequalities in capitalist and (neo) colonial power relations.
>
> *(Dornhof* et al.*, 2018, p. 18)*

If transculturation, and by extension Tidalectics, can function as an analytical mode, the challenge is to conceive how this manifested in curatorial practice, in the face of the gross inequities in the distribution of resources, influence and power.

Responding to earlier exhibitions of Caribbean art, the *Arrivants* project intended to situate this discussion within the context of the Caribbean, beginning with Barbados. While the exhibition focused on the Anglophone Caribbean, it included artists from the diaspora, several of whom were exhibiting in the region for the first time. The original exhibition proposal emphasised the intention to cast its gaze on the issues represented from within the Caribbean itself. And while diaspora artists were included, there was a deliberate intention to represent the diversity of artists and artworks that originate in the Caribbean, as this is often also a deficiency in externally curated exhibitions. As a project, *Arrivants* also reflected on the processes involved in art exhibition-making in the Caribbean, the challenges as well as the opportunities for new thinking and innovative approaches and the need for capacity development. A blog was established on which members of the curatorial team could share their thoughts and reflections, along with photographic and video documentation of the installation process and the exhibition, as well as short interviews with participants.[4]

Two broad themes were eventually identified, which helped to frame our thinking and choice of works: place/displacement and diasporic subjectivities.

Place/displacement

A key issue in diasporic experiences is the connection to place—both in terms of the imaginaries that surround the original homeland and the sense of connection, or lack thereof, to the place of arrival and settlement—and at times perilous and alienating process of moving from place to place, whether by force or by choice. These are common preoccupations in the work of artists who are themselves migrants, whose subjectivities are shaped by various diasporas and who are part of the cosmopolitan societies of the Caribbean as frequent travellers. Such a sense of displacement also occurs in the context of Caribbean tourism, which can be seen as the flipside of migration, which generates a largely fictional sense of place that is rooted in nostalgia, stereotype and exploitation.

Diasporic subjectivities

Between the diasporic origins of the Caribbean and the continued transnational movements of Caribbean people, identities are constantly renegotiated, with regard to notions of 'home' and responses to life in the diaspora, where cultures collide as much as they do in the Caribbean itself. This shapes the experiences and definitions of self, community, family, race, social status and gender and sexuality, in both positive and negative ways.

The exhibition venue

Unlike neighbouring countries, such as Guyana, Jamaica, the Bahamas or Cayman Islands, Barbados does not have a national gallery, despite decades of calls for one from the local art community. Nor does it have a purpose-built museum for contemporary art such as those of Martinique, the Dominican Republic or Puerto Rico. Thus, it was largely out of necessity that the decision was taken to locate the exhibition at the Barbados Museum and Historical Society. As such, the *Arrivants* exhibition was conceived as an intervention into the historically charged environment of the museum, which is located in a 19th-century military prison, located within the environs of the historic Garrison Savannah area.[5] The works were installed in two small temporary gallery spaces (the Aall Gallery and a former storage room, converted in the 1990s and referred to as the Exhibition Gallery), as well as the Cunard Gallery, which houses a permanent collection of historical prints. Additional works were installed outside the museum by the entrance, in the interior walkways, in a prison cell and in one of the period rooms of the Warmington Gallery.

Outside the museum, works by Eddie Chambers and Hew Locke were situated on either side of the large entrance. Chambers was born in Wolverhampton, UK to parents who had emigrated from Jamaica as part of what became known as the 'Windrush generation', a term referring to West Indian people who were invited by the colonial British government to help rebuild Britain in the aftermath of the Second World War.[6] The 1948 British Nationality Act gave all subjects of the British Empire citizenship and the right of settlement in the UK. Despite this, the new Caribbean arrivants encountered intense racism and a colour bar that excluded them from housing, many types of employment and social spaces such as pubs and restaurants. The British Parliament subsequently passed immigration laws in 1962, 1968 and 1971 that radically curtailed immigration to Britain from the Commonwealth, bringing the Windrush era to a close. The term has recently received renewed attention in the wake of the Windrush scandal, during which many of these early West Indian settlers found their citizenship called into question by the authorities. Hundreds of citizens were detained, deported and denied legal rights as a result of a 2012 government policy to create a 'hostile environment' for immigrants. The Windrush report, which investigated the scandal, ultimately presented evidence that harmful immigration policies were the result of the public's and officials' poor understanding of Britain's colonial history.

This lived experience of the hostile and racist environment in which West Indians lived, one that led to riots in the late 1950s and escalated to militarised police aggression in the 1970s, informs the work of Eddie Chambers.[7] In *Untitled* (1994), which was included in the *Arrivants* exhibition, Chambers reconfigures the Union Jack, changing the colours to red, gold and green, symbolic of Rastafari culture. The work was first produced as part of a project in which artists were invited to design flags for the Liverpool Town Hall. Chambers explained:

> Growing up Black, in Britain, in the 1970s, it seemed to me that I did not really have a flag. I had never been to Jamaica, so I did not feel that the flag of that country was mine. Racism and a certainly alienation from the British nation state meant that I did not see the British flag as being mine either. What I did start to feel, by my mid-teens, was a strong pride in my Afrocentric ancestry, history and heritage, engendered, in no small part, by the teachings of Rastafari, and reggae music. My flag, made in the mid-1990s by a flag maker, was my attempt to create a Black British ensign, that took account of the influence of Rastafari on the making of Black Britain. With its Rastafari colours of red, gold and green, for me at least, this was a flag that I could finally identify with.
>
> *(Chambers, 2018).*

Veerle Poupeye has identified in Chambers's work a 'subversive inversion of images and symbols' similar to the mocking tactics traditionally used in Caribbean carnivals, such as Jamaica's Jonkonnu, observing that Chambers imposed a 'black' identity on a quintessentially 'white' symbol (Poupeye, 2022, p. 19).

There are protocols in place that govern the display of national symbols, and when Chambers's flag was first exhibited at the Liverpool Town Hall, it was removed after only one day. Poupeye describes that, in an ironic twist, Chambers's flag, which was deemed unsuitable for display, was nevertheless ceremoniously folded and formally returned to him. This history gave us pause as we pondered what procedures to put in place to fly Chambers's flag on the flagpole outside the Barbados Museum, next to the Barbados flag. The imposition of a black identity on a symbol of Britishness in a formerly colonised island where the vast majority of the population are descendants of African slaves resonates in a very different way than it does in the UK. We contemplated alternative options for displaying Chambers's work if the museum was instructed to remove it, but that never happened. The only incident occurred the night of the exhibition opening when a museum guard lowered Chambers's flag along with the Barbados flag at sunset, as protocol dictated. We had to explain that this was a work of art and the flag needed to be quickly raised again before the guests arrived.

Also installed at the front of the museum, next to Chambers's flag, was a new work commissioned from Guyanese British artist Hew Locke, which also manipulated iconic yet contested national imagery. Locke has been investigating the history and symbolism of public monuments for over 15 years, altering their images with paint and collage. In response to the many recent controversies surrounding statues commemorating war heroes and political figures, and in questioning the notion of designating public and national heroes, Locke has created interventions in the form of altered photographs, which he describes as 'mindful vandalism'.[8] Images of statues of Christopher Columbus, Edward Colston and Peter Stuyvesant have been garishly draped in cheap gold chains, medals and other adornments—the excessive wealth and booty harvested by the enterprise of empire.

For the *Arrivants* exhibition, Locke was invited to address the statue of Lord Horatio Nelson, which had been located in the centre of Bridgetown since the early 19th century. The bronze statue was originally commissioned by public subscription from Sir Richard Westmacott, following Admiral Nelson's death in battle at Cape Trafalgar in 1805, and was erected at the top of Broad Street in front of the Parliament Buildings in 1813, in what was then named Trafalgar Square. The statue pre-dates Nelson's column in London's Trafalgar Square by almost three decades. The Barbados monument became a source of controversy in the post-independence era, particularly after the site was renamed 'National Heroes Square'. This move alone necessitated a rethinking of the historical narratives surrounding Nelson's legacy as the protector of Great Britain's hold on the Caribbean island.

In *Nelson, Bridgetown* (2018), Locke's treatment of Lord Nelson differs from many of his other digital manipulations of monuments. Locke describes the earlier images as more graphic while this one is more impressionistic. Rather than being draped with the trappings of his conquests, Nelson is surrounded by images that seem to emerge as ghostly apparitions, an aura of the violence that characterised the colonial era. Skulls and bodies appear from the patinated bronze of Nelson's figure like silent witnesses and victims. His military jacket is transformed into the Union Jack, emphasising Nelson's role as the embodiment of the empire's naval power. The graphic diagram of bound human cargo aboard the slave ship Brooke is printed across the plinth of the statue below Nelson's name, unmistakably linking his exploits with the British transatlantic slave trade. The larger-than-life digital image of Locke's altered Nelson was the first to be presented in the form of a two-dimensional cut-out figure as a simulated public monument. 'History is messy', Locke states. 'But there are means of reconciling with it' (Locke, 2018).[9]

Locke's intervention at the Barbados Museum was a temporary one, and at the end of the exhibition, the work, which was digitally printed on foam-core was, at the artist's instructions, destroyed with a sledgehammer. In retrospect, this was an uncanny foreshadowing of events around the world, when public monuments were pulled down, notably that of Edward Colston in Bristol. In November 2020, on the occasion of Barbados's 54th anniversary of independence, the government formally removed the bronze statue of Nelson from his plinth in Bridgetown.

Stepping through the large fortified doors of the Barbados Museum, the first work viewers encountered was Kelley-Ann Lindo's *Sending Love Inna Barrel* (2018). Four cardboard shipping barrels were suspended from the ceiling, end to end, forming a long tunnel. Chairs were placed at either end, and visitors were invited to engage in conversation through the long echoing chamber. The work is a response to the feelings of abandonment experienced by Caribbean children left behind when their parents migrated overseas. It was not uncommon for parents to leave children with family members or neighbours for years while they struggled to secure housing, employment and some financial security before bringing their children to join them. In the meantime, parents would send clothing and foodstuff in large cardboard shipping barrels, which were eagerly received, not only for the goods inside but also as a longed-for connection with mothers or fathers. The term 'barrel children' is used to refer to this phenomenon, acknowledging the traumatic repercussions this has had. Lindo has commented:

> Traumatic memories are forever susceptible to change, each time there are attempts to recollect it, and it is that fragility I have explored, through the use and manipulation of fragile materials. My ongoing body of work seeks to establish a conversation around the dynamics surrounding the 'barrel children' syndrome within the Caribbean culture – a term referring to children who have been left behind by one or both parents who have migrated.
>
> *(Lindo, 2018)*

The barrel is a recurring image in a number of works in the exhibition, symbolic of a nomadic existence as well as the movement of both peoples and goods that has governed the region's history. For *Reparation* (2003), Guyanese artist Philip Moore repurposed the cardboard container used by West Indians to import consumer goods from England and the United States, painting the surface with intricate patterns, sweeping brushstrokes and sequin-like dabs of paint. Prominent among the images is a crowned, two-faced colossus, his armour itself the profile of another face with a row of all-seeing eyes. The heart-lined strip that wraps around the barrel is highlighted by rows of holes punctured through the cardboard surface and repeated in patterns across the rest of the object. These are illuminated by a string of Christmas lights inside the barrel, evocative of a metropolitan city at night; metal rings at the top indicate that the barrel could be hung like a chandelier or beacon. The title, *Reparation*, refers to the paltry compensation handed to slaves at the moment of Emancipation as well as more recent calls for economic restitution to be paid to descendants of slaves by those who profited from their enforced and unpaid labour.

Barrels also appear repeatedly in the epic painted series by Guyanese artist Stanley Greaves, *There is a Meeting Here Tonight*; however, here, these are the steel drums used to export oil. In *The Annunciation* (1993), a man stands inside the drum, which is transported on a dolly, pushed by the woman beside him, an agricultural worker identified by her stalk of sugar cane. While the specific meaning of the figures, objects and relationships in these surreal works is evasive, the barrels are pervasive, signalling the extraction of resources and circulation of capital that continues to influence the political instability in the region.

Kishan Munroe, an interdisciplinary artist from the Bahamas, produced a singularly remarkable work for the *Arrivants* exhibition. Munroe, whose practice is rooted in extensive historical research, employs documentary practice to promote engagement with underexplored narratives of the African diaspora. *Drifter in Residence* (2018) was a live expedition and video installation in which the artist undertook what he described as an 'artist's residency at sea' (Munroe, 2018). Based on extensive research and training in survival techniques, Munroe constructed a raft—kept afloat on a platform of barrels—for an expedition on the Atlantic Ocean at the peak of hurricane season. In a statement, the artist explained:

> This 'introspective/retrospective' pilgrimage marks a ten-year milestone in my professional artistic journey, actively engaged in extensive anthropological investigations through cultural immersion. With this phase of the project I literally plunge into the foreign world of the ocean, seeking to tap into the narratives, realities and histories of man's precarious relationship with the sea and his never-ending quest for 'home'.
>
> *(Munroe, 2018)*

For the *Arrivants* exhibition, nine video screens were installed in the Exhibition Gallery in three rows of three. The outer eight screens showed images of the research and construction process as the artist conceived of and built the raft, while the central screen presented a live feed of the artist drifting at sea, conveying the isolation and vulnerability of this singular yet heroic figure adrift in the expanse of the ocean.

The sea was another recurring theme. Nadia Huggins presented two works from her *Transformations* series (2014–2016), a group of diptychs that explore the relationship between the artist and the marine ecosystem. The artist pairs cropped photographic self-portraits with marine organisms resulting in a new and hybrid relationship. Huggins explains:

> In the sea, as a woman who identifies as other, my body becomes displaced from my everyday experiences. Gender, race, and class are dissolved because there are no social and political constructs to restrain and dictate my identity. These constructs have no place or value in that environment.
>
> *(Huggins, 2018)*

Cosmo Whyte's *The Expat* (2017) provided a poetic corollary to Huggins's images—a self-portrait of the artist seen from behind, his damp back dotted with clumps of barnacles, a performative display that presents bodies, and bodies of water, as liminal.

Whyte's *In the Belly of the Whale* (2018), responded to the physical space and history of the Barbados Museum, specifically a jail cell which has been preserved as a record of the building's original function as a military prison. With the cell door shut tight, viewers had to peer through the small opening to glimpse the spartan and cramped environment which Whyte transformed with veils and soft lighting. Artefacts from the museum's collection—an African drum, large ceramic jugs, a funerary urn and a ceramic replica of an Ife head—were surrounded by dozens of empty and discarded rum bottles, a rusty machete and an old shoe; a clash of historical artefacts and detritus of a colonial past.

Veronica Ryan also created an installation that responded to the Museum's existing exhibition spaces with work that like Whyte's, challenged traditional notions of the artefact. *Shack Shack* (2018) consists of a series of small mixed media sculptures displayed in a pair of glass cabinets built into the thick coral stone walls of the Aall Gallery. The title references the long wooden seed pods produced by the flamboyant tree and woman's tongue tree (Albizia Lebbeck) that appear in the cabinets, bound together with twine. Ryan combines natural, organic elements with colourful nylon fishing line, woven plastic bags and crumpled soft-drink bottles. Wrapped, bound, nestled and confined, these various materials—natural and manmade—revel in their contorted eccentricities. Some of the objects are stitched into pillows, concealed within crocheted covers or displayed on doilies. The singular objects spaced out on the shelves exist as separate islands, curiosities that are paradoxically familiar and strange. They teeter between museum artefact and domestic ornament, preserved in an anthropological display case or the cabinet of a West Indian front room.

Ryan's focus on the rich and varied—and increasingly imperilled—natural life of the region resonated with the paintings of Lynn Parotti and Winston Kellman. Parotti's *Microatoll I* and *II* (2016), part of her *Bahama Land* series, provide a microscopic view of a thriving reef head with corals as seen through the crystal Caribbean water, an environment of extraordinary life forms and beauty. But Parotti introduces foreign elements that suggest an imperilled ecosystem.

Winston Kellman's seascapes are recorded from the ocean's edge, looking out across the rugged east coast of Barbados from a small fishing village known as Bathsheba, which lies on the brink of the 'Black Atlantic'. Paul Gilroy uses this phrase to describe an African diasporic model of modernity understood from a transnational and intercultural perspective that acknowledges the centrality of the experience of slavery as central to plantation economies and imperial capitalism (Gilroy, 1993, p. 15). Every day, Kellman paints the same yet changing view; Paul Gilroy's 'changing same', a phrase that emphasises the connectedness between different moments that allows us to identify systems of cultural exchange and continuity across time (Gilroy, 1993). The artist completes each work in one sitting, recording a site which is layered with a history of trauma that is both personal and collective, but also capturing the ephemeral atmospheric sensations of that unique encounter. Collectively—and ultimately this is how the works need to be understood—the *Bathsheba Series* forms a diary or journal that records the intersection of place and memory.

Marianne Keating brings a unique perspective to the selection of works in the Arrivants exhibition. Born in Ireland and based in London, Keating has created *Landlessness* (2017), a two-channel video installation, filmed on location in Ireland and Jamaica, that interrogates the largely undocumented migration of Irish indentured labourers to the Caribbean during the early 19th century. Based on records found in the National Archives in Ireland, England and Jamaica, Keating presents conversations and recovered textual traces, which previously had been consigned to disappear within the archives.

Simon Tatum's response to the theme of migration was *Tropical Forms* (2018), a collection of organically shaped paintings that function as organisms capable of adapting to the various environments and exhibition spaces that a nomadic existence takes them to. The wooden crate in which they were shipped lies at the base of the installation of forms—a mixture of plants and human limbs—that expand upwards and across the walls, claiming space and visibility.

Caroline Holder's *Homeland Insecurity* (2006) is a 24-piece ceramic dinner set decorated with intricate sgraffito images and text that reference the heightened state of fear, suspicion and paranoia that have been fostered in New York after the collapse of the World Trade Center on September 11, 2001. As a Barbadian artist based in New York, Holder recorded the atmosphere of paranoia and fear that pervaded the city:

> I was in a prime position as an 'outsider' to observe the enormous psychological toll; the world had become less safe and we less certain of our place in it. Fear

> cast a thin layer of dust over the city. I found it particularly poignant to see the burning towers recurring in the drawings of my young students two years after the tragedy occurred. These, alongside the ubiquitous advertising campaign, 'If you see something, say something,' and police presence everywhere compelled me to develop this installation.
>
> *(Holder, 2018)*

The clay pieces are inscribed with images of planes flying into burning buildings, neighbours peeking through their curtains, phones being tapped, with accompanying messages: 'Yo mamma is a terrorist [...] So turn the bitch in'. For the *Arrivants* exhibition, the work, which is in the Barbados National Collection, was displayed in the 'dining room' of the Warmington Galleries, period rooms that are on permanent display at the museum, viewed through openings, as if peering through a window. The pairing of the ostentatious interior décor of an 18th-century colonial 'great house' with pointed commentary on contemporary xenophobia poses opportunities to contemplate the long and ongoing trajectory of historical contestations and power imbalances within the intimate sphere of domestic rituals.

On a wall facing the Warmington Galleries, Leasho Johnson's expansive mural *Land of Big Hood and Water* (2018)—the third incarnation of a 'guerrilla' street art action originally located on Hope Road in an upscale part of Kingston, Jamaica—transformed the serene ambience of the museum's tree-lined upper courtyard into a raucous party. Vinyl cut-out figures frolic with abandon in a sea of vibrant red. These modified figures pose in a variety of contorted dance hall-inspired postures. Like Tatum's *Tropical Forms*, Johnson presents a hybrid melding of human and plant forms. The title parodies Jamaica's informal motto, 'Land of Wood and Water', 'hood' being an American term for ghetto, but also a colloquial Jamaican term for penis. These humorous characters seem to mock tourist expectations of a hyper-sexualised excess but equally evoke how racist stereotypes are rooted in histories of violence and exploitation.

The imagery in Johnson's mural shared some links with Sheena Rose's work. *'This Strange Land' Sketchbooks #anotherconfession* (2018) was the title given to a display of six small and well-worn drawing books. *'This Strange Land'* is, for the artist, Barbados. Through a vast compilation of line drawings, the artist explores her own feelings of anxiety and alienation in the island of her birth. In depictions of herself as a half-submerged island monster, her sense of self becomes synonymous with place—an identity that both embraces and breaks apart the tropes of the Caribbean as an act of rebellious self-exploration and self-realisation. The intimate scale of the notebooks perhaps makes their diaristic self-confessions less confrontational but no less provocative.

As objects, the sketchbooks fit well in the museum's Cunard Gallery, named for Sir Edward Cunard, a member of the eponymous shipping line dynasty and a donor to the Barbados Museum and Historical Society. He was one of several wealthy British visitors who built villas on the Barbados West Coast in the mid-20th century. Encouraged by Neville Connell, the director of the museum, Cunard collected colonial Caribbean prints, which were bequeathed to the museum upon his death in 1962. The collection included iconic early images of the Caribbean by artists including Agostino Brunias, Isaac Mendes Belisario and Lieutenant J.M. Carter. These works form the core of the historic print collection permanently on view in the Cunard Gallery. As Veerle Poupeye noted in the *Arrivants* blog, these prints, produced largely by itinerant and military artists, present the Caribbean as seen through the eyes and the world view of the planter and colonial administration classes, and form an important and multilayered visual archive of life in the colonies during the 18th and 19th centuries.

For the exhibition, we retained a selection of the historical prints and paintings in the Cunard Gallery to operate in contrast to, and in tension with, the modern and contemporary works. These acted as interventions into the historical narratives, in an effort to subvert these colonial perspectives while commenting on the social and cultural contradictions of postcolonial Caribbean life. The room was dominated by Ras Ishi Butcher's epic painting *400 Years: New World Order* (1994), which extended across the length of the end wall in the rectangular gallery, directly facing the panoramic mid-18th-century painting *Governor Robinson Going to Church.* Butcher's *400 Years* presents a revised historical overview of Europe's tragic and violent encounter with the Caribbean, including Columbus's voyages, transatlantic slavery and the proto-industrial plantation system. The large figure of the overseer surveys the patchworked fields, with smaller vignettes recording scenes of colonial domination and death. The overseer's thorned and undulating whip snakes across the diptych. It is an epic retelling of transatlantic encounter that led to the long and brutal institution of plantation slavery. Referring to this work, Richard Powell states, 'one notices certain technical procedures and recurring motifs that, apart from distinguishing him as a remarkable witness with something special and powerful to say, individuate Ras Ishi as a dedicated and cerebral painter' (Powell, 2010, pp. 19–20).

Also located in the Cunard Gallery was a work by Ewan Atkinson, created for the *Arrivants* exhibition to be displayed in the 1950s-styled cases that house a selection of historical maps. *Peregrination, A Playable Reproduction* (2018) is a Victorian-styled board game that was presented as if 'in play', laid out with accompanying game pieces, a pair of dice and a stack of 91 buttons as 'counters'. As described in the accompanying instructions, the supposedly serendipitous game pieces have been taken from the pockets of the imaginary players: a wooden shoe, a Vape mat (mosquito repellent), a dried passion fruit, a peanut-candy wrapper, the key from a can of corned beef, a plastic toy and a commemorative pin.

Using a language both visual and written that mimics colonial texts and parlour games, Atkinson fashions this faux artefact as a storytelling device. The full title, as inscribed on the board, is *A New Neighbourhood Amusement: Peregrination! An instructional game of chance replete with folly and adversity for the benefit of recent arrivants*. Like an ancient map, the game is presented as a well-worn and much-used document, complete with illusionistic fold marks, rips and stains. The route or journey to be pursued through the game is mapped out as a spiral moving through 50 'symbolic figures' or stations that represent locations, artefacts and residents or characters found in 'the Neighbourhood', an imaginary community of misfits and migrants that has encompassed Atkinson's production for the past 16 years. The two text panels that frame the central image provide a description of the 50 figures along with the meticulously articulated rules of the game. Nevertheless, ambiguity recurs and randomness seems to rule in the end. The often unreasonable and dehumanising bureaucratic hurdles, the waiting in line, the prospect of being sent back, the loopholes and penalties are familiar pitfalls for all who travel or migrate. In the era of visa lotteries, Atkinson's surreal world is all too familiar.

Francis Griffith's painting *A History of Time* (*c.* 1966) provides an interesting comparison, as it can also be understood as a map of sorts, a diagrammatic representation or symbolic depiction that charts not only space but also time, a constructed world view that compresses biblical stories, historical events and recent international happenings into an architecturally ordered framework, tying these disparate elements into a seemingly preordained expository presentation.

The story of Francis Griffith's life is an extraordinary one that involved migration and travels across the world and provided him with rich experiences and mystical revelations that were recorded in his paintings. Griffith, who was born in Barbados, became a seaman with the British Merchant Marines and worked as a gunner with the British Royal Navy during World War II and, later on, the docks repairing and painting ships. In the 1950s, Griffith continued to work in Cardiff in construction and manufacturing, during which time he studied welding and technical drawing. By the early 1960s, Griffith returned to sea life, travelling, by his own account, to 76 countries. He was most impacted by visits to Africa and the Middle East, where, through mystical interventions, he was given the name 'Son et Luimere' (*sic*), which he translated as 'Son of Light'. It was at this time that Griffith began painting as a way to record not only the places he had visited but also important world leaders and significant dates, mapping out a complex web of indecipherable connections and meanings.

A History of Time is one of Griffith's earliest known paintings as well as the largest and most ambitious. The majority of the painting is taken up by an elaborately articulated banquet hall where crowds of Arab men and women have come to honour the Queen of Sheba and King Solomon, who appear twice at the top and bottom of the composition. Rows of elegantly arched windows and richly attired guests are separated by mountain ranges and rivers with travellers arriving by boat and on camel. The centre of the composition is dominated by a large circular fountain with a six-petaled structure that functions like a large compass—a magnetic, orienting force. Along the top border of the composition, Moses presents the tablets with the Ten Commandments. On either side, horse-drawn chariots bear the flags of the various nations engaged in World War II: Australia, the United States and Greece on the left, and Pakistan and the USSR on the right. Through the central north–south axis of the painting is a meeting of powers, both ancient and modern, overseen by God's laws, imposing structure and order over the history of time as understood and divined by the artist.

The theme of diasporic subjectivities finds particular expression in a series of portraits located throughout the exhibition. *The Poet* (1947), one of the earliest known works by Karl Broodhagen, is a portrait of writer George Lamming when he was only 20 years old, six years before the publication of his acclaimed debut novel, *In the Castle of My Skin*. This sensitive terracotta bust reflects Broodhagen's lifelong interest in portraiture, and specifically the representation of Caribbean people. The sculpture was made the year Broodhagen began teaching at Combermere school, where Lamming had studied under Frank Collymore.[10] Lamming's later writings, such as *The Emigrants* (1954) and *The Pleasure of Exile* (1960), focused on the migrant's journey and the alienation and displacement caused by colonialism. The portrait busts and paintings by Broodhagen provide their own subtle insights into the diversity, complexity and richness of the Caribbean experience.

Paul Dash's *Self-Portrait* (1979) appears as a remarkably intimate and honest confrontation with self, as the artist, palette and brush in hand, faces the viewer with an intense and steady gaze. By the end of the 1970s, Dash had been actively involved with CAM. His acquaintance with other Caribbean artists and writers may have influenced his determination to focus on his own identity as an Afro-Caribbean man. He later explained:

> At that time I had not painted a full-on portrait of a black sitter and hadn't seen many portraits of black people in the flesh; paintings in which there was a black presence yes, but few portraits in which artists struggled to say something specific about such sitters. Rembrandt, Pieter Paul Rubens, Marie Benoist, Augustus John and others had made wonderful paintings of black subjects but I hadn't yet seen them in a gallery setting or had the opportunity to study such works in depth.
>
> *(Dash, 2018)*

Dash, who later had an influential career as an art educator at Goldsmiths, University of London, had to wait until after his retirement to gain long-overdue recognition for his painting.[11]

In Sheena Rose's photographic double self-portrait *Flowers and Pearls, Gorgeous* (2018), the artist is presented as a famous personality, wearing dark glasses and literally coated in glitter. Rose has developed a rich cast of personae over the years that she performs through photography and videography, both as a means to explore multiple dimensions of her own personality and to allow herself to live other lives. She writes:

> Sometimes, I feel so disconnected from my home Barbados, that I feel like an outsider and it doesn't feel like an ordinary space; its more than that. There is

magic, spirits, beauty and mystery in this strange land; a quiet magical space. Perhaps an exaggerated space that draws the viewers to be very curious of space and surroundings.

(Rose, 2018)

Keith Piper's *Ghosting the Archive* (2005) presents the largely forgotten studio contents of British commercial portrait photographer Ernest Dyche, who recorded the likenesses of residents of the inner-city area of South Birmingham during the 1950s, 1960s and early 1970s. Piper came across boxes of glass negatives during an artist residency undertaken in the archival spaces at the Library of Birmingham in 2005. Having grown up in the same area, Piper was inspired to 'reactivate' this record of an evolving inner-city community composed largely of immigrant families. In the single-screen video presented as a slide show, the artist stands behind the camera in the library storage room and extends his hands into the frame, wearing white archival gloves and holding up photographic negatives. This image is then reversed so that the room is seen as a negative, while the archival portraits—and the individual they recorded—appear as they were originally intended.

Sovereign State (2016) is a single-screen video work by Hew Locke who again interrogates the consumption of official portraiture as an ongoing deference to empire. Recalling his childhood in Guyana, Locke remembers images of Queen Elizabeth II on the covers of his school exercise books long after the country's independence from Britain. He would often be reprimanded for drawing over these images, literally defacing the symbols of monarchy. Reviving this aesthetic of the doodle, Locke's current work challenges the dearth of more complex portraits of the royal family and instead creates images with a very different truth as art. In *Sovereign State*, the monarch's mouth is bound to safeguard her secrets. Sounds of whispers hover in the air as the altered images of the Queen slowly morph and transform. Locke's contemporary take on royal portraiture references medieval and Renaissance imagery of the ruling elite, who were often portrayed with skulls or skeletons as a reminder that 'in the midst of life we are in death' and thus change is always inevitable.[12]

Arrivants was conceived as part of an EU-funded initiative looking at museums, migration and gender. Collectively, the works in the exhibition spanned seven decades, from the pre- to post-independence eras of the Anglophone Caribbean, a period of important change and transformation, and addressed a broad spectrum of issues related to migration in ways that are nuanced, incisive, moving, inspiring, surprising, humorous, thought-provoking and beautiful.

Not only did the diverse works by the participating artists present a broad scope of approaches to the theme, the integration of the works into the museum's existing displays placed these conversations within the broader context of a British colonial history.

A number of the participating artists were exhibiting in Barbados for the first time, and many viewers remarked on the importance of the opportunity to see these works, particularly in relation to one another in the context of the exhibition. Particularly impactful were those works that were commissioned and made for the exhibition and which responded to the context of Barbados and the museum. Several of those artists were able to travel to the island to make and install their work, and the two youngest artists in the exhibition, Kelley-Ann Lindo and Simon Tatum from Jamaica and Cayman Islands, respectively, participated in short-term internships with the Museum, assisting with the installation of the *Arrivants* exhibition. This created significant opportunities for interaction with the local art community and wider public, a social aspect of the exhibition-making process that should not be disregarded.

The intersecting themes of migration and museology provided the potential for a meaningful discussion and analysis about how museums, and particularly small regional museums, can participate in the current interrogations into the meanings and directions of Caribbean art and question the ways in which and where it is presented. While the Barbados Museum as a venue provided opportunities for interventions into existing narratives, the choice was partly in response to the lack of purpose-built spaces in which to exhibit contemporary art. Deficiencies in infrastructure, both physical and professional, were recurring challenges throughout the process. Initial ambitions that the exhibition could travel throughout the region and eventually to the UK were never realised. And the absence of a catalogue documenting the event is a significant missed opportunity.

The challenge for museums and curators to develop innovative strategies, reconciling ambitious objectives with limited resources, to present Caribbean art to Caribbean as well as global audiences is ongoing. As Dornhof et al. observed, the exhibition format, with its potential to assemble and juxtapose diverse objects and perspectives, provides unique opportunities to address issues of migration and diaspora. Migration is often a disorienting process, necessitating strategies of problem-solving and improvisation. Taking on the theme of migration as a critical strategy allows us to rethink the exhibition format, to respond to the specific needs and aspirations of exhibition-making in the region, to speak first to a Caribbean audience, but then also present an inflected, creolised voice to a global audience.

Viable, sustainable strategies will require formulating not only different routes but also potentially, different destinations.

Notes

1 Notable exceptions include the *Global Caribbean* shows, curated by Haitian artist Edouard Duval Carrie, and *En Mas: Carnival and Performance Art of the Caribbean*, curated by Claire Tancons (Guadeloupe) and Krista Thompson (Bahamas). But in both cases, while the curators are Caribbean, these exhibitions originated in North American venues. The *Global Caribbean* exhibition was launched in 2009 by the Haitian Cultural Arts Alliance as part of Miami Art Basel's satellite programming and presented at the Little Haiti Cultural Complex's main gallery. It continued as Global *Caribbean /*

Borderless Caribbean, featuring work of artists from the Caribbean archipelago and its surrounding land mass as well as its diaspora. The organisers have invited guest curators and other academics 'to formulate what a cultural production from the region could consist of [http://duval-carrie.com/global-borderless-caribbean-xii-focus-miami/]. EN MAS: Carnival and Performance Art of the Caribbean opened at the Contemporary Arts Center in New Orleans in March 2015. Another important and early exception is *Carib Art: Contemporary Art of the Caribbean* (1993), which was a travelling exhibition organised by the National Commission for UNESCO of the Netherlands Antilles. The exhibition opened in Curaçao, in 1993, and was exceptional in its efforts to include all countries in the region, regardless of size or artistic production. Reference should also be made to the regional biennials: the Havana Biennial, which began in 1982, the Santo Domingo Biennial, which began in the Dominican Republic in 1992, and the Jamaica Biennial, launched in 2014 as the successor to the Jamaica National Exhibition.

2 The artists were Ewan Atkinson, James Boodhoo, Karl Broodhagen, Ras Ishi Butcher, Eddie Chambers, Paul Dash, Stanley Greaves, Francis Griffith, Caroline Holder, Nadia Huggins, Leasho Johnson, Marianne Keating, Winston Kellman, Kelley-Ann Lindo, Hew Locke, Philip Moore, Kishan Munroe, Lynn Parotti, Keith Piper, Sheena Rose, Veronica Ryan, Simon Tatum, Aubrey Williams, Golde White and Cosmo Whyte.

3 The *Arrivants* exhibition was conceived in collaboration and co-curated with Veerle Poupeye. The concepts, ideas and explanatory texts that have informed this paper were developed jointly with her, and I would like to thank her for her scholarly and collegial contributions. Alissandra Cummins and Karen Brown, as the leaders of the Horizon 2020 EU-LAC Museums and Migration project, provided formative and essential input. Jessica Taylor and Ewan Atkinson also provided valuable insights and assistance, as did Kelley-Ann Lindo and Simon Tatum who assisted with the exhibition install.

4 This blog can be found at https://arrivantsexhibition.wordpress.com (Accessed: 18/10/2022). Images of the artworks discussed in this chapter can also be seen here.

5 There have been artistic interventions at the Barbados Museum previously. Joscelyn Gardner's *White Skin, Black Kin: 'Speaking the Unspeakable'*, curated by Joscelyn Gardner and Denyse Menard Greenidge in 2004, was an intervention into four galleries at the Barbados Museum. In 2008, Sonia Boyce installed the two-screen video *Crop Over* in the Cunard Gallery. And in May 2018, Katherine Kennedy invited five artists (Llanor Alleyne, Annalee Davis, Ada M. Patterson, Adrian Richards and Kraig Yearwood) to join her to create artistic interventions that engaged with the collections.

6 The name comes from the Empire Windrush, a ship that brought an early group from the Caribbean to Britain in 1948.

7 Eddie Chambers's most well-known work is undoubtedly the four-part collage now in the collection of Tate Britain, *Destruction of the National Front* (1979–80). For this four-panel work, the artist tore up an image of the Union Jack and reorganised it to form a red swastika. Reproduced as four screen prints, each successive version is torn until, in the final frame, the image is completely destroyed.

8 See, for example, the digital viewing room created by Hales Gallery entitled 'Mindful Vandalism'. https://halesgallery.com/news/583-hew-locke-mindful-vandalism-online-viewing/

9 Locke refers to his own ambivalent feelings about these public figures. While he admires the technical skill of his fellow sculptors of the past, he wants to draw out the complexities of their readings in a postcolonial context. This is only the second time Locke addressed public monuments in the Caribbean; the first was the statue of Queen Victoria in Georgetown Guyana.

10 Collymore was an important literary figure and publisher of *BIM* magazine and a mentor to both Lamming and Broodhagen as well as other literary figures, such as Derek Walcott and Edward Kamau Brathwaite. Lamming had left Barbados in 1946 to teach in Trinidad; he remained there for four years before emigrating to England, where he

worked as a broadcaster for the BBC Colonial Service. Broodhagen followed him to London two years later when he began his studies at Goldsmiths, University of London.

11 This *Self-Portrait* was recently acquired by Tate and was included in the Tate Britain exhibition *Life Between Islands: Caribbean-British Art 1950s-Now*.

12 Edvard Munch's famous 1893 painting *The Scream* is a significant influence, as are Tudor portraits of Elizabeth I by Marcus Gheeraerts the Younger. These include *The Rainbow Portrait*, in which the Queen wears a dress covered in eyes and ears as the all-seeing ruler, and *The Ditchley Portrait*, where she stands upright on a map of England, storms raging behind her while the sun shines on her.

References

Brathwaite, Edward. (1973). *The Arrivants: A New World Trilogy*. Oxford: Oxford University Press.

Buurman, N. (2018). 'The Blind Spot of Global Art? Hans Ulrich Obrist's Ways of Curating', in Dornhof, S. et al. (eds.) *Situating Global Art: Topologies – Temporalities – Trajectories*. Bielefeld: transcript Verlag, pp. 301–26.

Chambers, Eddie. (2018). *Arrivants Exhibition Artists Statement*. [Online]. Available at: https://arrivantsexhibition.wordpress.com/tag/eddie-chambers/ (Accessed: 05/01/2022).

Dash, Paul. (2018). *Arrivants Exhibition Artists Statement*. [Online]. Available at: https://arrivantsexhibition.wordpress.com/tag/paul-dash/ (Accessed: 05/01/2022).

Dornhof, S. et al. (eds.) (2018). *Situating Global Art: Topologies – Temporalities – Trajectories*. Bielefeld: transcript Verlag.

Gilroy, Paul. (1993). *The Black Atlantic: Modernity and Double Consciousness*. Cambridge: Harvard University Press.

Holder, Caroline. (2018). *Arrivants Exhibition Artist Statement*. [Online]. Available at: https://arrivantsexhibition.wordpress.com/tag/caroline-holder/ (Accessed: 14/01/2022).

Huggins, Nadia. (2018). *Arrivants Exhibition Artist Statement*. [Online].Available at: https://arrivantsexhibition.wordpress.com/tag/nadia-huggins/ (Accessed: 14/01/2022).

Juneja, M. and Kravagna, C. (2013). 'Understanding Transculturalism', in Blimlinger, E. et al. (eds.) *Transcultural Modernisms*. Vienna: Sternberg Press, pp. 23–33.

Lindo, Kelley-Ann. (2018). *Arrivants Exhibition Artist Statement*. [Online]. Available at: https://arrivantsexhibition.wordpress.com/tag/kelleyannlindo/ (Accessed: 14/01/2022).

Locke, Hew. (2018). *Arrivants Exhibition Artist Statement.* Wall text for *Nelson, Bridgetown*. Bridgetown: Barbados Museum and Historical Society.

Morris, M. (1995). 'Overlapping Journeys: The Arrivants', in Brown, S. (ed.) *The Art of Kamau Brathwaite*. Bridgend: Cromwell, pp. 117–31.

Munroe, Kishan. (2018). *Arrivants Exhibition Artist Statement*. [Online]. Available at: https://arrivantsexhibition.wordpress.com/tag/kishan-munroe/ (Accessed: 27/01/2022).

Obrist, H.U. (2014). *Ways of Curating*. New York: Farrar, Straus and Giroux.

Ortiz, F. (1995). *Cuban Counterpoint: Tobacco and Sugar*. Durham, NC and London: Duke University Press.

Poupeye, Veerle. (2022). *Caribbean Art* (second edition). London: Thames and Hudson.

Powell, Richard J. (2010). 'The Systems and Semiotics of Ras Ishi Butcher', in Cummins, Alissandra et al. (eds.) *Ras Ishi: Secret Diaries*. Barbados: Miller Publishing.

Rose, Sheena. (2018). *Arrivants Exhibition Artist Statement*. Wall text for *Flowers and Pearls, Gorgeous*. Bridgetown: Barbados Museum and Historical Society.

Welsch, W. (1999). 'Transculturality—The Puzzling Form of Cultures Today', in Featherstone, M. and Lash, S. (eds.) *Spaces of Culture: City, Nation, World*. London: Sage, pp. 194–213.

15

THE POLITICS OF CHANGE

New pedagogical approaches to Caribbean museology, conservation and curatorship

Alissandra Cummins and Anne Bancroft

Introduction[1]

In 1989, Bridget Brereton cast a critical eye on the work of the Association of Caribbean Historians (ACH) and asked:

> Has the ACH encouraged innovative research methodologies in Caribbean history or helped to open up new subjects or fields of investigation, especially in the context of 'new' social history? On the whole the answer must be, I think, a qualified 'no'. The historiography of the region, especially of the Anglophone Caribbean, which predominates in the output of the ACH, remains somewhat conservative, rather suspicious of the 'new' history and of inter-disciplinary approaches [...] Scholars from disciplines other than history – notably anthropology, but including archaeology, geography, sociology, economics and linguistics – have participated in ACH conferences and presented papers; but the historiography of the region has not been fully receptive to the inter-disciplinary approach, though planning of the UNESCO [United Nations Educational, Scientific and Cultural Organisation] History, to which ACH members have contributed a great deal, is changing this.
>
> *(Brereton, 1992, pp. 5–6)*

A scan of the indices to ACH Conference papers for the first 20 years of the association would support her assessment (Blondel, Knight and Rouse-Jones, 1989, 1990). Neither the words 'heritage' nor 'museum' are referenced in the index before 1989. Barry Higman, as chair of the ACH Conference Programme Committee, encouraged his colleagues, during its meeting in November 1993, to consider expanding the association's themes for the conference, generating a 'list of

DOI: 10.4324/9781003288138-18

possibilities [...] popular history; public history; heritage management/ museums/ heritage interpretation; theory in Caribbean history; [...] pictorial history; film as source' (Juan Gonzalez Mendoza, ACH conference, Puerto Rico, 1994, p. 4).

The museum's role and development in the post-independence Caribbean

Between 1972 and 1993, a series of reports examined the status of museums in the region for the first time since Bather and Sheppard's 1933 report on 'The Museums of the British West Indies' (Bather and Sheppard, 1933, pp. 27–58)[2] some 40 years before (Cummins, 1994, pp. 200–1). These reports were orchestrated through UNESCO, the Organisation of American States (OAS), the United Nations Multi-Country Sustainable Development Framework (UN MSDF) in the Caribbean in association with the Caribbean Conservation Association (CCA) and, afterwards, in collaboration with the Museums Association of the Caribbean (MAC), founded in 1989. While these reports covered a wide range of considerations, including documentation and conservation of Caribbean museum collections, public education requirements and the poor condition of museum (mainly historic) buildings and museum collections, and links between tourism and museum development, they consistently paid critical attention to the chronic need for training of museum staff, given that 'properly qualified museum staff were practically non-existent in the Caribbean' (Lemieux and Schultz, 1973, p. 2). This situation posed a particular challenge for these general or 'community' museums' ability to respond to the expressed needs of their communities. Towle and Tyson, for example, particularly gave priority to the museums' role in defining national and regional identities, especially in their capacity to 'promote a sense of pride, self-esteem and national identity that will help people of developing nations overcome the debilitating sense of cultural inferiority and dependency induced by Colonialism and slavery' (Towle and Tyson, 1979, p. 5).

At a critical juncture in the early 1990s, forty-four museums were surveyed in 13 Caribbean Community (CARICOM) member states, following which a comprehensive programme of training was developed for regional museum personnel including paper preservation, preventive conservation, collections management and cataloguing, travelling exhibitions, museum education, administration and financing. One important finding from this Caribbean Museum Survey was that: '[o]nly 29. 5 percent of all museums surveyed indicated the presence of formally trained personnel on their staff' and 'staff development remained a high priority in the development of Caribbean museums' (Cummins, 1993, p. 61). The report also observes:

> It is regrettable to note that in most instances training opportunities have been limited to short-term workshops and seminars which may give a basic picture, but

> hardly qualified the individual as a professional. The thirst for more at the end of each activity leaves serious museum workers with a frustrating sense of limitation.
>
> *(Cummins, 1993, p. 60)*

These findings remain a reality. A key recommendation was made for '[t]he development of undergraduate and graduate level museum training programmes in conjunction with the University of the West Indies and the Museums Association of the Caribbean' and for fellowship assistance to be provided to CARICOM nationals from St. Lucia, Dominica and Tobago for 'study outside of the region' (Cummins, 1993, p. 69).

The establishment and development of heritage/museum courses in the Anglophone Caribbean

Museology or museum studies originated in the University of the West Indies (UWI) primarily through the efforts of Dr Barry Higman when, in 1992, as the head of the History Department at Mona he introduced the concept of a full-scale master's degree programme in heritage studies. Departmental minutes for the period indicate that despite initial scepticism, approval and comments from Cave Hill and St Augustine regarding the proposed master's diploma in heritage studies supported this proposal and agreed that 'the programme could be offered on a regional basis, moving from campus to campus at appropriate intervals' (UWI Mona, 1992b).

These combined with the development of an Archaeology programme produced "Efforts to obtain a modern Museum/Laboratory...", largely expected to exhibit and preserve the archaeological collections being recovered through the rapidly growing Archaeological Studies programme (UWI Mona 1992a). However, it was clear that "(i) In the longer term, the Department was looking forward to the establishment of a "School of Historical Studies", with its own independent facilities, including audiovisual systems, museum and laboratory." (UWI Mona 1992a). Initial drawings and a prospectus for the UWI History Museum were prepared to launch a fundraising programme. This was done with the funding support of both UNESCO and the Jamaican government who pursued these goals as part of national cultural policy for the island.

Higman's revised programme proposal was later approved, noting 'that a strong Caribbean focus was essential' (UWI Mona, 1992c). The key objectives for the Heritage Studies master's programme were clearly articulated:

> To provide academic training for persons wishing to work in public history and the heritage industry, and to enhance the skills of those already employed in these fields. It offers the opportunity for critical reflection on the practices of public history and for the learning of practical skills.
>
> *(UWI Mona, 1993a)*

Graduates of the programme were expected to be equipped to work in tourism, museums and historic sites and the media, 'interpreting the Caribbean past for a public audience' (UWI Mona, 1993a). Initially it was agreed to be offered part-time in 1993–1994 over a 24-month period (UWI Mona, 1993b), during which it was hoped that funds might be raised to support bursaries and scholarships to enable full-time registration and also to enable students from other parts of the Caribbean to travel to Jamaica for the second year. With the successful launch of the programme in December 1993, by January 1994, the department was able to report that 'Higman, coordinator of the programme [...] reported that attendance had been good and the drop-out rate [...] very low' (UWI Mona, 1994). The programme's continued success was reported at interdepartmental meetings indicating that 'the M.A. Programme had attracted over 80 applicants for... the 30 available spaces' for the 1994–1995 semester'. At the same time, it noted that UWI Cave Hill was considering the possibility of 'a joint venture with the Barbados Museum to establish an Institute of Heritage Studies' (UWI St. Augustine, 1994). The proposal for H67K, the originating heritage studies course, reads as follows:

> Describes and analyses Caribbean heritage, and the attitudes of peoples towards it. It will include the efforts of government and non-governmental organizations to preserve Caribbean heritage in and outside museums. It will examine the politics of heritage management and presentation, and the role and status of public history in the Caribbean. It will investigate the relations between Caribbean history and Caribbean heritage.
>
> *(UWI Mona, 1993c)*

The graduate-level course, which taught that museums have a central role to play in heritage management, preservation and presentation, anticipated that the specialised museums in Port Royal, Kingston and Spanish Town (including the National Gallery and the National Trust) would provide strong institutional support for students and expected that they would be an important resource for the examination of the 'relations between Caribbean history and heritage' (UWI Mona, 1993a). The course 'Artifacts, Museums and Archives', which was eventually transferred from Mona, was the only programme that spoke directly to 'the collection, curation, management and display of artifacts and documents in the Caribbean. Acquisition policies and information systems, Conservation and preparation of exhibitions. Living History Museums. The meaning of artifacts' (UWI Mona, 1993a). It provided, within the limited five weeks allocated, the totality of direct exposure to the study of museums. Typically, the course offered baseline orientation to the precepts of the profession in a series of tutorials covering (1) the Definition and Role of Museums; (2) Curatorship, Collections Management and Conservation; (3) Interpretation and Representation (Exhibitions); (4) Museums and Heritage: The Politics of Culture (Heritage policy, legislation, heritage tourism) and (5) Ethics. However, at Cave Hill, Watson established a direct link with

> the Barbados Museum to teach this component of the heritage studies course which was deliberately included because of the Museum's role as not just the repository of our material culture but also of our natural history and our intangible cultural heritage. I considered the museum to be an integral part of the teaching of heritage. Museum Studies was therefore a natural and logical choice for inclusion in the heritage studies degree programme.
>
> *(K. Watson, personal communication, 2018)*

From the inception of the course, where the emphasis lay on the examination and interpretations of archaeological remains and artefacts and built and natural heritage, lecturers for both the archives and the museum components of the course made clear their concerns about the adequacies of such brief programme offerings to prepare students to enter into their respective fields. While the inclusion of both field trips and a one-month practicum at the end of the course did serve to extend students' exposure to the discipline, these concerns were not really addressed until the artefacts component was removed from the course structure, allowing both museums and archival studies to absorb these essential constructs into their differing contexts and to expand into complementary seven-week programmes.

However, more than two decades were to pass before a full-length museum studies programme at the UWI was developed following its Quality Assurance Review of the Heritage Studies master's programme in 2009. While the programme had enjoyed more than a decade of success at UWI Cave Hill, it had nevertheless existed in its truncated form, despite being the only formal course of study for heritage specialists in Barbados. The Department of History and Philosophy continued its evaluation and commissioned a revised museum studies programme, which finally led to a comprehensive revision of the long-standing programme, updating its course contents to expand

> its course offerings to keep pace with the present growth and future potential of the Caribbean heritage industry. As in the past, the programme will not only continue to create marketable graduates in the field of cultural heritage, but current changes will deepen and diversify students' knowledge base. A new emphasis is placed on innovative types of experiential learning so that graduates can immediately apply their skills across all sectors of the heritage industry, including tourism, museums, archives, government, and university. This new programme has excellent potential to attract a larger and wider (pan-Caribbean/international) student body with its greater interdisciplinary and global approach, AND to foster candidates for higher levels of heritage study and research (MPhil and PhD levels). Furthermore, the revised programme is

> now poised to expand in future new directions, especially into management aspects of the heritage industry.
>
> *(UWI Cave Hill, 2012, p. 61)*

While the Cave Hill Heritage Studies programme began to call for greater synergies between the archaeology and heritage courses, as well as the introduction of Cultural Resource Management in its offerings, the department welcomed opportunities for a broadly multidisciplinary approach and an orientation towards tourism development. It was anticipated that 'the restructuring of the existing/former Museums/Archives course into two separate courses [would provide] students with higher skill sets in both archival work and key museum concepts, including management' (UWI Cave Hill, 2012, p. 2), a development that enabled the 'Artefacts, Museums and Archives' course which evolved from a single elective course into completely separate core programmes.

In 2012, a semester-long course debuted. HIST 6720: Museum Development, Management and Curatorship included provision for

> a new mandatory requirement for practicum work within the research paper [...] The nature of practical work can include any of a number of options, such as oral history documentation, audio-visual work, archaeological fieldwork, documenting museum and archaeological collections, archival retrieval/study, database construction.
>
> *(UWI Cave Hill, 2012, p. 3)*

Exhibit and educational programme planning and design became a particular target early on for students, where students were required to 'demonstrate evidence of independent practical work within a museum or heritage site environment' (UWI Cave Hill, 2012, p. 46). In redesigning the course description and rationale, the university was taking account of an expanding body of new museological writings and specifically Caribbean-initiated research and activity, designed to trace

> the evolution of Caribbean museums from agencies of Empire to symbols of independent national and cultural identities [...] contemporary curatorial practices [...] important for their comprehension of the museum's role in contemporary Caribbean society. [As well as] the foundations of museum/exhibition development and interpretation informed by both museological theory, methodological approaches and technology-based practices required for creative and effective professional practice [...] the establishment of many new, privately operated museum displays, as well as the development or reorientation of publicly operated institutions and sites in response to the demands of local stakeholders [...] illustrated by the new national art galleries and museums, as well as the reinterpretation of national histories.
>
> *(Cummins, 2012, p. 44)*

This course specifically responded to the general growth in the museum and heritage sectors in the Caribbean, which, despite continued economic and societal challenges and the issues of national and cultural representation, both at home and abroad, evidenced a growing need for professional training for deployment in Caribbean museums and other heritage-oriented institutions.

Preventive collection care and conservation practices and training in the region

To date, there is no dedicated course for preservation, collection care and conservation offered in the Anglophone Caribbean. While preservation and collection care topics are incorporated within the museology course, these components have only rarely been delivered by a conservation specialist. More typically these modules have been delivered by curators or museum managers.

Having these critical topics incorporated has provided regional museology and heritage staff with a necessary overview of preventive conservation and collection care. However, the use of the term 'conservation' in the course descriptions is a misnomer, as conservation as a professional discipline is not taught; rather, it is an introduction to the topic.

In practice, collection care is led by curatorial, librarian and archive staff, which can be viewed as an integrated approach to the preservation of objects, as it incorporates the priorities of the different specialists. This approach, however, lacks the input or guidance of a preventive, collection care or conservation specialist, who adds another layer of efficacy to the objects' care. Internationally, heritage courses are typically based on the premise that heritage staff will be working alongside professional preventive and conservation personnel. The reality is that there are many instances in the region where heritage professionals, having completed only short courses or on-the-job training, have been tasked with collection care management. In national institutions in the Global North, this work is executed and led by a specialist, as it is a profession within itself. Realistically, community museums can rarely afford a dedicated preventive conservator on staff nor do they have access to a budget for recruiting preservation and conservation specialists for project-based work. Curatorial and collections staff thus by default are multidisciplinary; working as the 'chief cook and bottle-washer', their resources and capacities are spread thin, with their additional duties in preventive conservation, remedial conservation and as art handlers. The onus has been, and remains widely, on curators and collections managers to inform on and implement preservation and collection care.

Over the decades, only a handful of people have been formally trained in preservation and conservation abroad as opportunities arose. Notably, however, temperate climate strategies are not always appropriate for tropical environments. These conservators afterwards were charged with training others on the job, while facing limited access to resources and to professional development in the region.

Conservation and preventive professional practice in the region remains far less formally established than the curatorial, librarian and archive professions. Similar to the evolution of the profession in Asia and Europe, the specialism within the region has grown out of the development of artists, artisans and craftspeople restoring and repairing others' works, the difference is the lack of a scientific base as its underpinning (Plenderleith, 1998, p. 129). There are historic and contemporary bookbinding and furniture-making apprenticeship practices. However, the nurturing of master craftspeople and skilled professionals was traditionally undermined by a colonial edict that promoted the importation of ready-made objects from the colonial occupier. The conditioning was, and the perception remains to an extent, that the imported is superior, aesthetically and functionally. The artisans who repaired others' objects locally were often constrained by a lack of locally sourced materials and by restricted access to expensive imported materials and equipment as well as by limited expertise or formal training in restoration. It is important to reiterate that temperate materials are not always efficacious in tropical climates. The region is currently reliant on them, however, as the profession and materials have yet to develop indigenously. Until local conservation practices and materials are established, innovative scientifically tested regional options, such as regionally made repair papers (as has been used in Cuba, India, the Philippines and Thailand), will not be available.[3]

'The profession has evolved and concomitantly the skills of those who practice it, thus following the general rule of adaptation for future sustainability' (Margariti, 2019, p. 108). Modern conservation has a scientific foundation, with an emphasis on minimal intervention, reversibility and ethically preserving the physical and intangible integrity of the objects with a high level of motor skills and underpinned by a code of ethics. The International Council of Museums (ICOM) defines conservation as

> all measures and actions aimed at safeguarding tangible cultural heritage while ensuring its accessibility to present and future generations. [It] embraces preventive conservation, remedial conservation and restoration. All measures and actions should respect the significance and the physical properties of the cultural heritage item.
>
> *(https://www.icom-cc.org/)*

> Salvador Viñas describes modern conservation as a complex activity born out of identifying the need for different approaches to care and repair of valued 'heritage' objects as opposed to mending common items.
>
> *(Viñas, 2005, p. 13)*

In addition to the complexity of the variables involved in the act of conserving, 'it may involve many different professionals from many different fields working towards the same goal' (Viñas, 2005, p. 9). There is the confusion of the terms or synonyms used to describe the variations of related 'care' activities by the general public and even within collection repositories. What is clear is that there is a growing need in the region with emerging national identities that as collections and relevant objects are increasing the material culture that reflects them defines identities as significant to the community.

The opportunity to formally study conservation within the region is currently limited to Cuba and Puerto Rico. However, for the Anglophone Caribbean, language remains a considerable barrier. The few conservators that have been formally

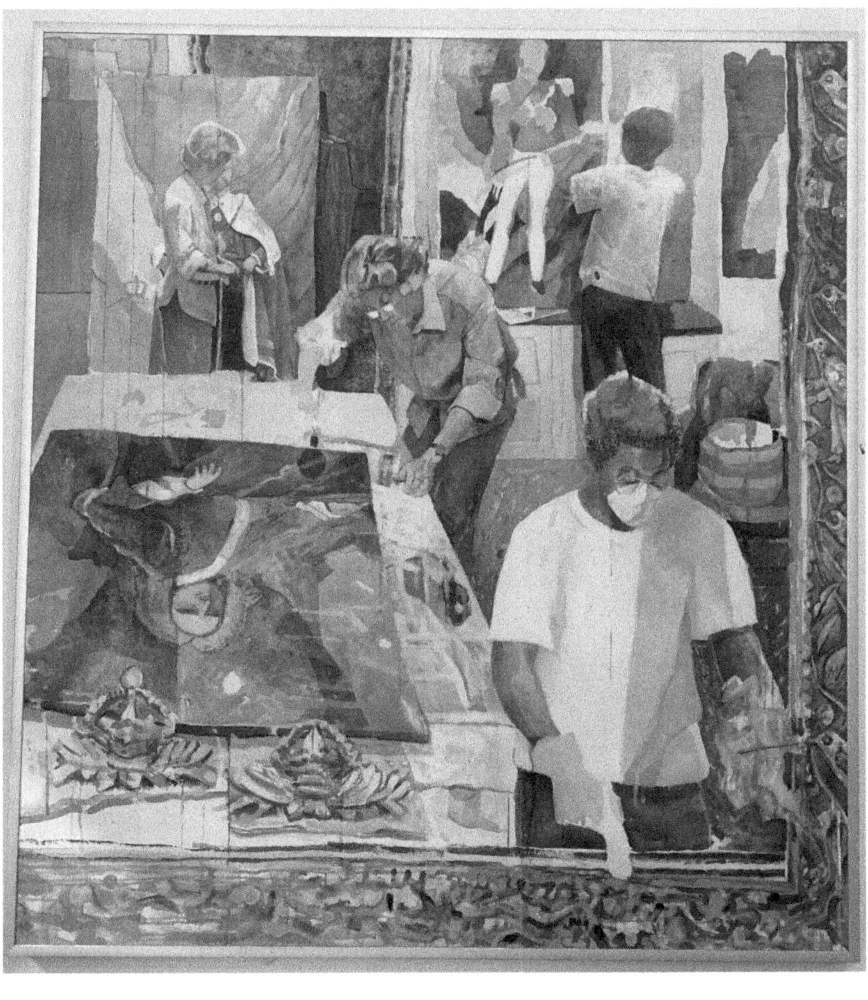

FIGURE 15.1 'The Restoration' 1991 by Allison Chapman-Andrews; depicting the stabilisation of paintings in the National Collection by conservator Patricia Byer, assisted by artists with transferable skills. © Barbados Museum & Historical Society and the artist.

trained in archive, object, painting and paper conservation have done so mainly in Canada, Britain and America. Fewer still have journeyed as far away as Russia, with some of the training including internships with other institutions.

The training delivered by heritage professionals and their guidelines is based on standards for temperate climates, which calls into question why the standards are not formally adapted to the region. There has been a call to have a Caribbean Conservation Code of Ethics mainly to instil accountability and promote awareness (Salkey, 2001, pp. 28–9) but given the limited numbers of conservators the necessity for this is debatable. A temperate climate has major seasonal fluctuations and low humidity compared to a tropical or subtropical one, with a wet and dry season, consistently high humidity and salt air corrosion. The Caribbean requires its own preservation, collection care and conservation standards, practices and procedures that need to be led by regional professional conservators. While the material 'types' can be categorised as the same internationally, there are cultural differences and significance when working with objects of one's own indigenous culture and belief system, and value judgements are collectively made on what to conserve and what can wait.

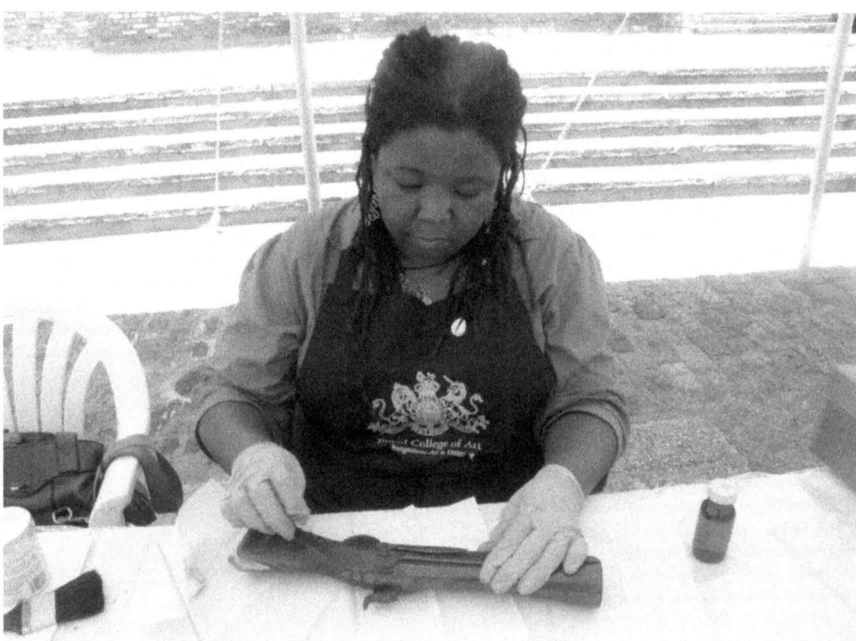

FIGURE 15.2 Louise Parris, the late object conservator who trained in the United Kingdom, demonstrating at a collection care workshop. © Barbados Museum & Historical Society.

Towards the end of the 20th century, paintings and archive conservation laboratories were set up and binderies adapted in Barbados, Guyana, Jamaica and Trinidad and Tobago. In 1977, under the vision of Michael Chandler, the Barbados Archives had a functioning laboratory in which two trained archive conservators shared their skills with other members of staff through on-the-job training, termed 'traditional repair', akin to an apprenticeship.

The challenge with this approach is limited possibilities for both parties to have access to continuing professional development (CPD) as well as reliable succession planning. Feedback from colleagues in these situations includes the constant eroding of their workforce, budget, space, continual training and, quite seriously, the authority to effect change. In the worst cases, the feedback was that procedures, materials used, treatments, techniques and skills had become fossilised. In extreme instances, this had resulted in avoidable damage to objects from interventive and harmful treatments executed with uninformed, though well-intentioned and outdated 'traditional repair' techniques. This scenario continues today.

As conservation expertise largely still lies outside the Caribbean region, the financial implications have historically hindered CPD. In practice, without effective succession planning, internships and formal training conservation can regress to a blanket 'factory' approach to the treatment of unique objects. There remains a need for professionalisation that includes academic rigour, ethical considerations and formal supervised internships with professional conservators.

While there are no current statistics available on the number of conservators and people working as conservators in the region, anecdotal evidence recognises a distinct deficiency. The conservation deficit evidenced in 1993 during the Caribbean Museum Survey referenced earlier remains almost as acute today in 21st-century national institutions. Following an Organization of American States (OAS)-sponsored evaluation workshop, held in Barbados in May 2015 to document identifiable gaps that might exist in regional heritage training opportunities and competencies relevant to cultural heritage and memory work in the Caribbean, the goal was to establish priorities for future educational efforts based on prioritised competencies grouped under: Heritage Management; Preservation and Conservation; Ethics and Professional Practice; Laws, Regulation and Governance; Research and Documentation; Access and Use; General Management; History and Philosophy; and Heritage and Tourism. The first priority, Heritage Management encompassed two areas of cultural heritage work:

> The first is the management of collections held in museums, archives, documentation centres, archaeological storage depots, etc. The second relates to the management of heritage sites. Each of these areas requires its own set of knowledge and neither is being addressed by the educational offerings in the region. Both areas were designated as priorities by the participants in the workshop.
>
> *(OAS, 2016)*

The objectives eventually formulated for this OAS project were to increase the capacity of the Anglophone sub-region to protect and enhance its heritage in a professional and efficient manner. Consultations with heritage professionals made clear that, while capacity building must strengthen 'the knowledge, abilities, skills and attitudes of people with direct or indirect responsibilities for heritage conservation and management', it must also improve

> institutional structures and processes by empowering decision-makers and policy-makers, as well as introducing a more dynamic relationship between heritage and its socio-economic context. It involves an inclusive approach, so that the relevant missions and goals are met in a sustainable way.
>
> *(OAS, 2016)*

Preventive conservation became the baseline of collections care internationally, which was considered both a gap and a priority for the region. The work aimed to slow the rate of deterioration through the control of the environment and the proper exhibition, storage and handling of objects. The UWI Open Campus and the General Secretariat of the Organisation of American States were signatories to a Memorandum of Understanding (MOU) that acknowledged that the development of culture was imperative for socio-economic development. Additionally, the need to recognise, preserve and promote regional customs and ancestral knowledge were underscored as important steps in the effective 'transmission to present and future generations' (UWI Open Campus, 2017, p. 12).

Another component of this MOU, the Caribbean Heritage Network, is an encouraging development. The network has recently started to collate information on heritage education and professional development, heritage legislation and fiscal incentives, inventories and monitoring, socio-economic impacts of heritage, sustainable heritage tourism, traditional crafts and artisans, which includes specialisms, with the intention to create and document the current skills matrix and extrapolate from this for future needs.

Formally trained and skilled conservators in the region have reported not being able to undertake treatments because of limited budgets, particularly to purchase imported specialised materials and equipment. Like the objects they care for, conservators' skills are at risk of deteriorating if not used and they are suffering from the challenges of working in silos. Conservation in the heritage sector is also hindered by a local stigma of working with one's hands. There is lack of input from conservation at the decision-making levels for funding bids, collection care, storage and construction of exhibition spaces, all of which result in preventable physical risks to the longevity of objects. In a recent conversation with formally trained and apprenticed conservation colleagues working in the sector, it was speculated that this practice results from the sector being accustomed to not having conservation historically and a departmental structure that places conservators within a 'service department'.

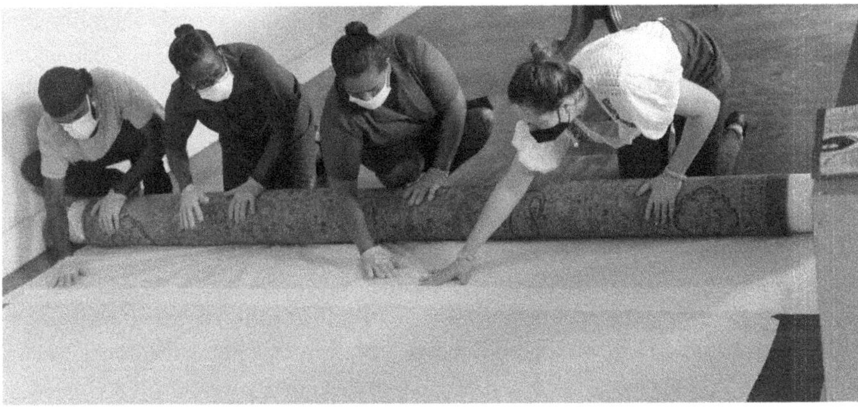

FIGURE 15.3 Training in cleaning carpets and collections storage at the Barbados Museum & Historical Society. © Barbados Museum & Historical Society.

This is opposed to a department that adds value to the understanding of the object—its materiality and the interpretation of the evidence presented within its degradation, its fabrication, its attribution and its authenticity. There is a continued need for advocacy initiatives to explain the role and value of conservation. The suggestion is that the hierarchy and misconception of conservation as a profession by upper management and government has, and is, resulting in the stagnation of advocacy for investment in it. The few conservators that do exist report that they cannot champion effective change to current practice at a strategic level.

> Our archivists and librarians in the region do not always get access to the most ideal resources for professional development as Conservation is not always seen as a priority in fiscal budgets…So, when the opportunity comes around to share with others nationally and regionally and to get some hands-on training from professionals in the Diaspora, it is a resounding 'Yes, please!'.
>
> *(Inniss, 2019)*

In addition to professional development in their own area of specialisation, there is the added pressure in the region to gain skills in conservation. It is felt that online platforms adequately provide theoretical learning opportunities and access to academic papers. The use of online repositories was championed as a resource at the 2018 conference *Itinerant Identities: Museum Communities/Community Museums*. Many of the representatives from the island nations noted that UWI was best suited to host the repository, which would include the documentation of discussions, conferences, workshops and, most importantly, instructional videos. Instructional videos were thought to support community museums without specialists, but to what extent? Can and should archival, curatorial and librarian staff be expected to perform treatments on objects? The truth is that many do undertake remedial treatments and, as this practice is established, would it be better if they had access to instructional videos that include when *not* to intervene and how to have contact with a network of professional specialists, even if only online? There is a wealth of academic and scientific papers available online, many of which can be effectively utilised. The virtual support is approached as a means of supplementing the academic, scientific and practical, hands-on conservation training. The reality is that it is by no means a replacement of it; however, there has been a shift born out of international lockdown restrictions. Instructional videos with experts were deemed a reasonable *interim* measure by universities with conservation courses. Internationally, practical studio time was pared down to a minimum. The online allocation has resulted in hybrid teaching. Its popularity has grown, partly because the efficacy of it is improving with innovations in filming and familiarity as well as accessibility. The logical judgement is that accessibility and a reduced impact on the environment are some of the advantages of online courses, but this also needs to be balanced with the 'in-person' experience, which may be essential in some instances. In the recent British Council-funded project HUNAR: Heritage Unveiled: A National Art Restoration Project (2019–2020) in Afghanistan, professional development shifted to utilise wider dissemination of informational videos via online platforms as well as project awareness and the promotion of gender and social inclusion using social media (R. Mulholland, 2020, personal communication, February).

Much like other hands-on professionals, a conservator's skills are honed experientially. This is optimised when underpinned by the rigour of academic study and scientific and technical analysis, as well as by witnessing and assisting experts in the field. Like the 'Living National Treasures' status of artists and artisans in Japan, where the skills of these artists and artisans are highly valued, the skill of conservation and the succession of it is also necessary to preserve. For complex treatments, there is no question that the hands-on approach and familiarity with the materials are necessary and superior, but while it is challenging to embrace this shift to the virtual, it is claiming its pedagogical place, especially with preventive and remedial treatments such as surface cleaning and object handling.

Conservation courses are expensive, as they necessitate restricted numbers, specialist equipment, multiple lecturers and internships for skills transfer. In recent years, conservation courses such as the Paper and Book Conservation courses at Camberwell College of Arts have been closing because of running costs. In the current financial crisis, most universities are at risk in general, with the arts on the front line of the cuts, and conservation courses are some of the most expensive to run. The options for prospective students interested in conservation specialisation are contracting, but online offerings could be a silver lining for multidisciplinary and preventive conservators.

In the region, the job market for specialised conservators is restricted due to limited funding and lack of advocacy, so that while administrators have been increasingly interested in investing in their capacity to monetise digitised heritage resources few understand the critical importance of conserving the original artefacts as complementary to this process. Notwithstanding a considerable need for such skilled positions and facilities in which conservators could work to stabilise and preserve at-risk heritage, there remains an acute need in the region for collection care, building, metalwork, painting, paper, object, sculpture and textile conservators. The current working practice is to hire specialist conservators, often from both within and beyond the wider region, occasionally with them working on-site or the objects being sent overseas for treatment. The issues with the latter practice are not only the cost, due to the insurance implications, but also that these already fragile objects risk additional damage from regional and international transport. Specialists abroad may also use treatments from a different cultural and environmental perspective. The engagement with the individual object, the collection and the object's use is not as engaged as it is where there are in-house conservators and those of the same cultural background. The absence of national expertise perpetuates the colonial conditioning of dependence, ignorance and indifference.

In recent years, there have been initiatives such as the government policy in Barbados where young 'give back' volunteers who have an interest in the arts and history are engaged with collection-based entities in exchange for state-sponsored education. The programme promotes civic duty as well as raising the profile of careers in this sector. Over the years, there have been project-based possibilities for young people to job shadow visiting professionals. However, viable career opportunities within the heritage sector are rare. For instance, over 20 years ago, one of the authors worked as an assistant with a professional curator and conservators providing practical real-world experience. Most conservators will attest to their internships as the most valuable and utilised learning experience that they draw on in their daily work. This is why we are seeing the resurgence of apprenticeships, internships and fellowships across many practice-based disciplines.

In a survey of the Barbados National Art Gallery Collection in 2020, three young people from a visual arts tertiary institution with previous gallery and museum experience were selected to work as assistants alongside one of the authors, exposing them to the approach of a conservator and preventive conservation. This aspect of the project was championed as key for national succession planning and to provide an opportunity for experience in the heritage sector.

While it was a step towards succession planning, its strength was in the impact the experience had on the assistants. Gallery assistants Malick Storey and Anisha Wood reflected that:

> …these experiences have expanded my interests beyond art production, into other areas in the creative sector such as policy making, curatorial practice, cultural history, community outreach in the arts and, of course, art conservation […]Chances such as these are few and far between in Barbados and would otherwise require traveling to a major metropolitan city for a similar experience.[4]
>
> *(Storey, 2020)*

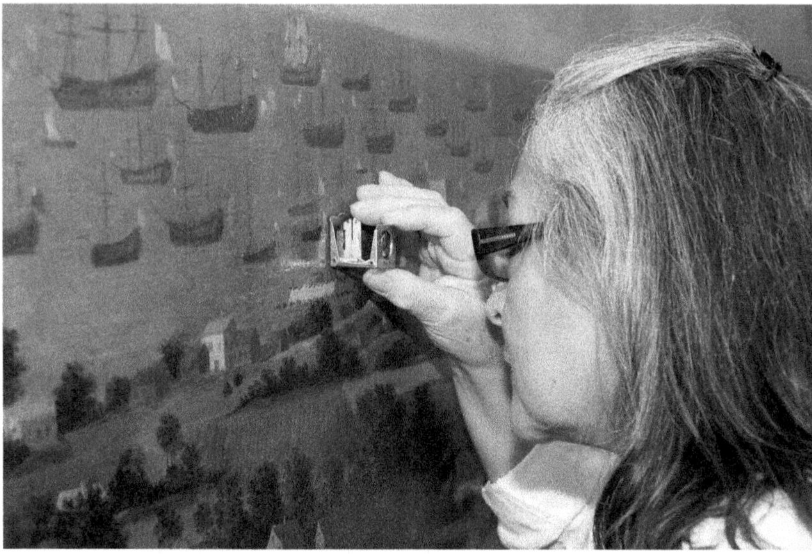

FIGURE 15.4 Lidia Aravena, Puerto Rican Consultant Conservator, assessing a painting under magnification. © Barbados Museum & Historical Society.

> It highlighted for me the linkages between all stakeholders in the sector that aid in the conservation of works for generations to come. This would include the materials chosen by the artist as a substrate, the choices the framer makes when framing a piece, how the art is displayed by the curator and for how long, the best practices used by art handlers when transporting the work and how the work is stored until it's next showing.
>
> *(Wood, 2020)*

These types of projects echo programming for those working in the sector. An interventive treatment of Stipendiary Magistrates Records from St Kitts deliberately became a teaching, networking event and advocacy for current conservation practices (The Blue Road, 2018).

There is a deficit of technical and scientific analysis in the region on objects that would inform about their fabrication and authenticity. Technical history is underpinned by the evidence the object provides and adds value to the understanding of its historic use. Technical art history:

> underwrites everything. It travels in a great sweep from the general to the particular – from global sources of pigment supply to the specifics of extracting dyestuffs in seventeenth-century Holland, from medieval concepts of colour to vivid glimpses in London studios. It is impossible to understand art properly without its insights. It acknowledges – celebrates – the artist at work and the act of making.
>
> *(Bomford, 1998)*

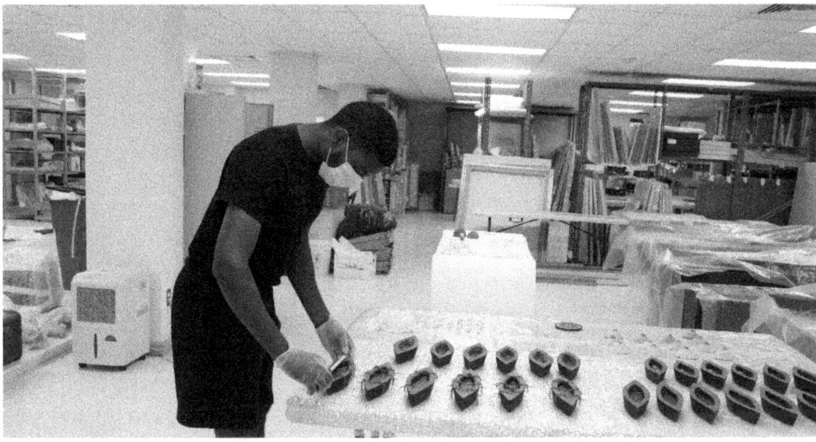

FIGURE 15.5 Barbados Give Back Volunteer assessing a new acquisition at the Barbados National Art Gallery. © Barbados National Art Gallery.

The combination of knowledge of the cultural, historical, fabrication and scientific evidence affords the community with as informed an overview as possible. With these skills, the authority of one's own material cultural heritage would carry more weight internationally.

In recent years, what has organically evolved in the region is the utilisation of professional networks and online platforms to support capacity building for persons charged with preserving and conserving objects. Lectures, discussions and demonstrations have started to be streamed using virtual and social media applications, such as Zoom, WhatsApp, YouTube, Skype and Facebook. MAC has also included online resource hubs on its website providing guidelines on implementing COVID-19 mitigation procedures and papers on conservation.[56] International expertise and regional collaborations are more accessible than ever. At a 2019 workshop on Current Conservation and Preservation Practices in Archives and Libraries held at the Barbados National Archives, round-table discussions included virtual participation from Trinidad, Jamaica, St Kitts and the UK. During the discussions on the workshops, there was a consensus about the need for similar continued practical programming in order to maintain and gain skills as well as networks (Barbados National Archives, 2019).

The feedback was actioned with further online programming such as the three-day Collections Care Workshop for Historic Interiors in a Tropical Environment 2021, which covered

- the agents of deterioration and methods of mitigation,
- an introduction to space assessment,
- an introduction to object handling and object cleaning,
- property maintenance,

- environmental management,
- visitor routes,
- collection care and emergency plans,
- inventories,
- health and safety,
- risk assessments and
- ongoing training needs.

This workshop included videos of assessing exterior and interior spaces, demonstration videos, lectures and exercises. Out of the workshop came the commitment to form a working group in Barbados to take action on shared building maintenance issues, such as pest activity and efflorescence in historic walls.

The training sessions have increased from talks and workshops to formal lecture series, such as the two-week UNESCO programme on Museums and Risk Preparedness in 2021, which comprised a varied series of lectures from different museum specialists, including Caribbean museum directors, international and regional conservators with different specialisms and emergency response specialists. The topics mainly focused on collection care management and risk preparedness but also on fighting against illicit trafficking in cultural property.

An article produced by the Center for Research on Education, Diversity and Excellence (CREDE) at the University of California advances five principles of effective pedagogy: two of these are 'Joint Productive Activity' and 'Contextualization' (CREDE, n.d.). The first indicates a level of cooperation and exchange between the instructor and the student. In line with this process, guidance is given by the course coordinator and the student gains agency in the process of learning. The latter, contextualisation, speaks to the link between instruction and the student's personal, family and community experiences (CREDE, n.d.).

An investment in formal academic training, online resources, practical workshops and apprenticeships within the region in preventive conservation and collection care would go far in safeguarding the objects of national, regional and international importance. Without formal training, procedures and resources for tropical environments, our heritage will continue to avoidably deteriorate: inappropriate treatments will continue to cause irrevocable damage and loss, and the use of damaging chemicals on objects to treat insects and mould will continue. What will be left will be an ever-increasing repository of inaccessible objects requiring interventive conservation. Who will undertake the conservation? Unless the storage facilities, handling and display practices meet appropriate standards, no professional conservator will do it, as ethically conserving an object and placing it back in an environment that will cause damage is unacceptable. Strides will be made when more national and regional positions are created and when persons are professionally trained in different disciplines in conservation; until then, the Caribbean nations will continue to risk and perpetuate dependence on the authority and resources of the Global North. It is also pertinent to realise that the rhetoric

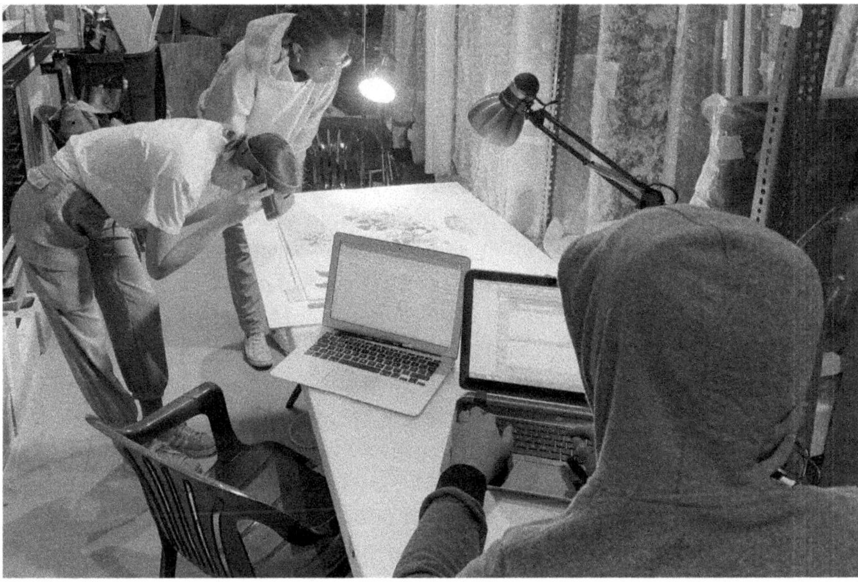

FIGURE 15.6 Work experience at the Barbados National Art Gallery Collection as part of a condition survey 2020. © Barbados National Art Gallery.

used by ex-colonial powers to justify keeping objects that are not indigenous to them is because of a lack of infrastructure and expertise to preserve them. The Tarzan saviour complex is alive and well. While the region does need help, it needs help with implementing sustainable infrastructure.

Until then, the region can continue to be innovative in creating new preventive conservation methodologies and strategies, collaborating with local entities that do have transferable capacities and expertise. For instance, one of the conference papers at the 2011 International Pest Odyssey conference was a Caribbean case study of a collaboration with a local ice cream factory to undertake a low-temperature treatment on a collection which suffered from major insect infestation (Bancroft, Blyth and Watson, 2011).

Proposals for relevant online courses and resources in the region

The increasing traction of e-learning platforms in the region has made clear that there are critical online courses and resources that could serve as the practical foundation of regional collection care.

An introduction to preventive conservation

Participants would be informed about the general issues, such as the 10 agents of deterioration, and the most pertinent issues to the region as well as effective achievable solutions, using local case studies, including environmental maintenance suitable for the tropics.

An introduction to integrated pest management

This course would draw on the established principles of monitoring, mapping, establishing risk zones, housekeeping, incorporating pest management in the design of spaces, pest identification and treatment of infestations. This training would incorporate the creation of a dedicated pest-identification poster for the heritage sector, similar to the one created for English Heritage (English Heritage, n.d.), but which deals with the pests prevalent in the Caribbean such as termites (like the *Coptotermes acinaciformis*), silverfish (*Ctenolepisma longicaudata*) and African powder-post beetle (*Lyctus brunneus*). The establishment of a regional pest working group that incorporates pest control companies' expertise would create a beneficial network for heritage institutions and governments, and it could also establish standards for appropriate treatment options available for heritage objects. The outcomes of such a network would benefit both the heritage sector and the wider community.

Emergency response in the tropics

There is a wealth of information on emergency responses and many comprehensive regional workshops: most recently, the course Building Disaster Resilience in the Caribbean's Culture Sector was offered in 2020 and facilitated by UNESCO and the Caribbean Disaster Emergency Management Agency (CDEMA), for example. However, like the other identified training modules, it would be advantageous to adapt bespoke versions to emergency response in the tropics, in which preparations for hurricanes are at the forefront. The identification of salvage materials, consistent practical training similar to that of the Historic England Salvage and Disaster Recovery Course, and concrete improvement of institutional and national supply stores would afford the region a fighting chance. (Historic England, 2018). The Caribbean has the academic understanding of what is needed from excellent theoretical workshops. Unfortunately, most institutions have not been able to undertake the regular practical training or to implement what they have learned due to lack of resources and access to materials and equipment. This means that the region's already at-risk material cultural heritage is not prepared for the yearly threat of hurricanes, volcanoes and other extreme weather events.

Foundation in object handling

This foundational course would be a quick win and, in the interim, the e-learning module created by the Museum of London could be adapted and utilised (Museum of London, n.d.). Ideally, this training could be improved with video demonstrations, discussions and attendees' physical demonstration of what they have learned.

Object marking guidelines

Producing object marking guidelines would be another easy improvement to current practice, with the proviso that the materials used, such as barrier layers will age appropriately in the local environments.

Condition reporting of objects

While this practice is established in some institutions it should be updated to include input from conservation professionals, to meet the need to be succinct and to utilise virtual reports and imaging as well as to be relevant for loans to other institutions. The region could easily come together to formalise this process.

Surveying object and collection conditions

Institutions need to have appropriate data to strategically prioritise their resources to areas of their collection for preservation and conservation which is critical for securing grants and funding. The worldwide emphasis on digitisation and accessibility of collections can be argued as being more important in the region because of how spread out these nations are, with travel costs remaining challenging. Accessibility via digitisation is advantageous for research and an income-generation avenue for all institutions, but surveying to inform on stabilisation treatments for digitisation is lacking. The collateral issue, however, is that the original object is often damaged in the process to facilitate the 'digital' version, as there is funding for digitisation but not for the resulting conservation required.

Exhibition and display procedures

At the moment, these procedures are mainly based on temperate climate protocols, which, for the most part, are effective, but while adjustments can be made to the environmental parameters compromises are commonplace given restrictive budgets. Many institutions create decent exhibitions and permanent displays with excellent content, but the execution of the physical display can be hampered by dated cases fabricated from inferior materials and ineffective mounts to support the exhibited objects. Again, this would be a quick win with an investment in mount-making professional development and the use, for instance, of bespoke costume mounting along with successful bids for museum-quality display cases.

Gallery and exhibition design in the region does not normally take into account 'dead spaces' in proximity to display cases and objects, which has an increased risk in the Caribbean because of prolific insect activity. Another improvement would be the use of tested inert display materials, within construction of display cases. Training would highlight which objects should be cased, for instance, the barrier distances, appropriate lighting levels as well as air change and circulation is a high priority. A shared and frequently updated list of appropriate materials and suppliers for the region should be generated from this activity.

Gallery maintenance

Many museums need support with establishing, implementing or improving lighting policies, dust deposition management and rotating objects within their permanent displays. Most museums in the Caribbean prioritise a high level of 'housekeeping'[7] in the region, but there are notable improvements, as specified above, which can be illustrated using case studies, including regular repeated training of maintenance staff specifically in bespoke cleaning of heritage objects and spaces. References such as *The National Trust Manual of Housekeeping* could be adapted for this purpose (National Trust, 2011).

Surface cleaning of objects

As many objects remain on open display and others inevitably suffer from dust deposition with high temperatures, humidity and air-borne salt content, surface cleaning of objects in many heritage institutions by non-conservators is necessary to minimise the cementation of dust, the corrosion of objects and biological degradation. With online demonstration videos, the dissemination of these skills, as well as how to identify when an object requires cleaning, could be advantageous in preventive care. The practice of 'dusting' objects exists, but the people doing so should have access to CPD and to specialists to assist them with utilising minimum intervention methods and to avoid preventable damage by 'domestic' cleaning methods and substances.

Offering online resources

All of the highlighted examples could be supported by an online network of specialists and peers, who could meet quarterly to discuss such projects. This can be as simple as the 'Conservation Family' WhatsApp group between professionals working in the sector in the region, which was initiated by participants at the Caribbean Heritage Network's Caribbean Conversations in Conservation conference in March 2020. We anticipate that this informal group will eventually lead to a membership-based and or practice-based entity.

Preventive collection care and conservation modules should be taught by specialists in those areas. There are many institutions internationally for benchmarking online learning platforms; these e-learning opportunities could be similar to the Victoria and Albert Museum's Online Learning Academy or the Courtauld Short

Courses (The Courtauld, n.d.). The repositories of resources could be akin to the Getty's Teaching and Learning Resources, Our Collections Matter Toolkit of the International Centre for the Study of the Preservation and Restoration of Cultural Property (ICCROM) and the Emergency Response Online Workshops hosted by ICCROM (The Getty Conservation Institute, n.d.). What is being proposed is to draw on what already exists, to collaborate where possible without losing autonomy. It would be advisable to incorporate what is relevant for the Caribbean and to use regional case studies and the region's own objects to illustrate the training and to collaborate with people that are on the front line of this region's heritage.

The Caribbean requires formalised bespoke approaches to the care and conservation of its material heritage in a tropical environment. There should be clear and achievable guidelines for museums having to work with external specialists, particularly on how to recruit and commission relevant expertise. Consistent practical exercises in and reviews of emergency response procedures and training are critical, especially with the increase in natural disasters in the region that are unique to it. Regional and national institutions could, for example, procure materials and equipment collaboratively, sharing critical resources and working across departments, institutions and regions.

Curating the Caribbean: new directions, new methodologies and new museologies

The most recent developments in Caribbean museum studies have developed as a result of two fully funded project-based partnerships between the UWI and the European Union and Latin American and Caribbean (EU-LAC) Museums project, taking advantage of the UWI's region-wide scope and capacity to deliver online. In the academic year 2016–2017, a component of the Caribbean Civilization course assessment was designed to further develop potential approaches to a Caribbean museology. The pedagogical approach in this move was unique, as the students took on the role of researcher, analyst and historian. This initiative was birthed out of a global project and built on an MOU created between the UWI and the University of St. Andrews in 2015. The European Union Horizon 2020 Europe, Latin America and Caribbean INT-12–2015 call (European Commission, 2015) approved funding that enabled a unique consortium of universities and museums in both regions, organised through the ICOM, to embark on a community museums project with the overall aim of creating 'a common vision for sustainable, small to medium-sized local and regional museums and their communities, and reinforce mutual understanding and cooperation between regions' (EU-LAC Museums, n.d.).

Indeed, contextualisation was a key element of this course assignment, as students were encouraged to locate their research in their communities. The students were therefore able to attach greater meaning to their research processes.

In July 2016, two new courses were introduced into the UWI Open Campus curriculum through an MOU signed between the UWI and the OAS: Values-based

Heritage Site Management and Museum Conservation Skills. These courses were eventually delivered between February and May 2017, with the rationale for the latter stand-alone elective course explained as follows:

> Museum conservation is an area in need of more trained professionals active in the field globally and particularly within the Caribbean region. Furthermore, as heritage tourism becomes an increasingly important aspect of the tourism sector in the Caribbean, with an increase in community-based museums that attempt to capture the uniqueness of their environment, exposure to critical aspects of museum management and conservation is indispensable.
>
> With seasonal changes and the threat of natural disasters along with man-made threats, it is imperative that museum workers and those engaged in heritage work are knowledgeable about the ways in which they can safeguard their collections for posterity before an incident occurs [...] Students will be better equipped to make appropriate choices in exhibition design and storage facilities for such objects.
>
> *(OAS, 2017)*

The course was developed in response to the critical need for professional training in the care, management and conservation of museum collections for deployment in Caribbean museums and other heritage-oriented institutions. Specifically, it was designed:

> to provide participants with comprehensive exposure to preventive conservation. Participants will be exposed to methods of mitigating deterioration and safeguarding collections through disaster and recovery planning, thus enabling them to make informed decisions about the handling and management of cultural objects. The course will also equip participants with skills in disaster preparedness and management as it relates to these objects. Additionally, the course allows for the exploration of the different areas of collection policy development, record keeping and documentation, giving participants competence in effective museum collection management.
>
> *(UWI Open Campus, 2016 a/b, p. 34)*

It re/introduced participants, many of whom were already working in regional institutions, to current practices relating to museum collection development and essential documentation, critical heritage interpretation and approaches to defining collection and artefact significance, informed by both the methodological and technology-based approaches required for creative and effective professional practice. However, it was found necessary to introduce many of the participants to essential knowledge of the role and function of museums, particularly in the context of emergency evaluation and preparedness, condition reporting and professional conservation practice.

This intensive induction programme left participants with a better knowledge of collections definition and management, as well as a better understanding of how broad their responsibilities were (or would be). The responsibilities would be made manifest as they addressed community needs and supported national efforts to fulfil the requirements for the implementation of the various conventions to which these nation states are signatory. They were also provided with some knowledge of what was needed to secure their collection holdings for future generations. However, it was also clear that they had become highly sensitised to and, deeply aware of, the vulnerabilities of their collections and of their museums and national heritage in general. The questions they left the course with included the following: When next would the course be held so that they could inform other colleagues to join? Where could they go to acquire further assistance to protect and promote their holdings? How soon would the university be providing intermediate-level courses, as they needed to continue? (Cummins, 2017) The needs raised and the potential offerings identified in response are in the process of being actioned with the intention of sourcing funds to manifest the next steps.

Conclusions: potentialities and possible futures for museum, preservation and conservation studies in the Caribbean

As our pedagogical history has demonstrated, the Caribbean region has benefitted at times from investment in museum studies training, distinct from the fields of history and heritage studies. More recently, there is an urgency to equip the sector with skills adapted to market forces and tourism, among other issues facing the region. However, to maintain professional standards in the Caribbean museum sector, an innovative way forward now needs to be designed among multiple actors and put in place. One need only look to the catastrophic events of 2017 to consider the urgency of this situation. The needs of the region share core training requirements with the world, particularly in the Global South, including collections management, preservation, conservation and principles of museum education. At the same time, a future course will need to address distinctive needs as articulated for our islands (and possibly other SIDS), such as appropriate provisions, security, disaster preparedness, recovery and resilience as well as specific conservation disciplines and collection care needs in humid climates. Beyond this, there is a need to have formally trained conservators of different specialisations to undertake treatments and to create and implement regional conservation and preservation policies.

The approach to conservation education and training as proposed by the International Council on Monuments and Sites (ICOMOS) UK may offer the most useful model to address training needs in island nations including the Caribbean, where:

1 Specialist courses should be multidisciplinary and mutual learning should be encouraged. In addition to core subjects available to all participants, optional subjects will extend capacities and/or to fill the gaps in previous education and

training. To complete the education and training of a conservation practitioner, on-the-job and practical training is recommended to give practical experience.

2 Courses aimed at continuing career development should build on the initial education and training of participants. Such short courses play a role in enlarging attitudes, updating the knowledge of specialists and introducing concepts and techniques of conservation and the management of the built and natural heritage.
3 Every country or regional group should be encouraged to develop at least one comprehensively organised institution to provide and/or support conservation education and training and specialist courses. Where such institutions do not exist, this may be achieved though regional exchanges and collaboration, by building new initiatives onto existing programmes and building up training capacity in these institutions.
4 The active exchange of ideas and opinions on new or improved approaches to education and training between national institutes and at international levels should be encouraged, including by national, regional and international level exchange of teachers, experts and students. Collaborative networks of individuals and institutions are central to the success of this exchange. (ICOMOS, 2012, p. 3)

There are synergies and quick wins to be had using the ICCROM Our Collections Matter toolkit and underpinning them to the countries' achievement of the Sustainable Development Goals. It is possible that through future strategic alliances among the UWI, local museums, museums associations, conservation bodies and museum professionals, as well as international partners in policy and academia, a sustainable model for museology, conservation and museum studies can be created for the benefit of a region with sophisticated needs in relation to material culture, history, territory, community and sustainability.

Acknowledgements

Research for this chapter has been supported by the EU-LAC Museums project. This project has received funding from the European Union's Horizon 2020 Research and Innovation programme under grant agreement No. 693669.

Additional research for the chapter was undertaken by Natalie McGuire and Kaye Hall of the Barbados Museum and Historical Society; their work is gratefully acknowledged.

Thanks must also be extended to Professor Sir Woodville Marshall for sharing his archival resources and recollections of the initial developments at UWI Mona and Cave Hill with us, as well as Professor Barry Higman for UWI Mona, Dr Karl Watson for UWI Cave Hill and Dr Veerle Poupeye for Edna Manley College—all of whom willingly lent their archival resources and memories in helping to reconstruct the 'history' of museology and conservation in the Anglophone Caribbean.

Notes

1 An earlier version of this chapter, titled 'The Politics of Change: Pedagogical approaches to Caribbean Museum history and curatorship', was presented by Alissandra Cummins, Karen Brown and Anne-el Bain during the annual conference of the ACH, Bridgetown, 2018.
2 The first analytical study of Caribbean museums and their history, audience, collections, facilities and resources appear in Francis Bather and Thomas Sheppard's 'The Museums of the British West Indies' (including the Bermudas and British Guiana).
3 Banana leaf paper has been used in these countries as a paper repair method. Banana leaf paper has also been made by the Northern Caribbean University, Department of Biology, Chemistry and Environmental Science, Cuba.
4 Storey, M., Wood, A. *Feedback from Barbados National Art Gallery Conservation survey Report 2020.*
5 See the MAC website for their COVID-19 resources: https://caribbeanmuseums.com/covid-19-resources/
6 See the MAC website for their conservation publications in the different volumes https://caribbeanmuseums.com/caribbean-museums-volume-4/
7 Housekeeping techniques and materials are typically used in a domestic setting as opposed to museum environment.

Bibliography

Association of Caribbean Historians (ACH). (1994). 'Announcement – UWI Conference'. *Bulletin of the Association of Caribbean Historians*, 42 (December), p. 41.

Bancroft, A., Blyth, V. and Watson, E. (2011). *Minus 20 Degrees in the Sun.* A Pest Odyssey, 28 October, British Museum, London.

Barbados National Archives. (2019). Current Preservation and Conservation Practices in Archives and Libraries. [Workshop].

Bather, Francis and Sheppard, Thomas. (1933). 'The Museums of the British West Indies' (including the Bermudas and British Guiana). In *Reports on the museums of Ceylon, British Malaysia, The West Indies to the Carnegie Corporation of New York.* London: The Museums Association, pp. 27–58.

Blondel, E., Knight, A. and Rouse-Jones, M.D. (1989). *An Index to the Conference Papers of the Association of Caribbean Historians 1969–1988.* St Augustine, Trinidad: University of the West Indies.

Blondel, E., Knight, A. and Rouse-Jones, M.D. (1990). *Supplement No.1 to Index of the Conference Papers, 1969–1988.* St Augustine, Trinidad: University of the West Indies.

The Blue Road (2018). *Conservation of the Cleghorn Stipendary Magistrates Journals of St. Kitts & Nevis.* [Online]. Available at: https://drive.google.com/file/d/1Yem2sxMZf3T9tr_IY-Ae_Wi52l0vQVco/view (Accessed: 7 May 2021).

Bomford, D. (1998). 'Introduction'. In Hermens, E. (ed.) *Looking Through Paintings: The Study of Painting Techniques and Materials in Support of Art Historical Research.* London: Archetype Publications, pp. 9–12.

Brereton, B. (1992). 'The Role of the Association of Caribbean Historians in Caribbean Historiography over the last 20 years'. *Bulletin of the Association of Caribbean Historians*, 38 (December), pp. [page range of paper in Bulletin].

The Courtauld. (n.d.). *The Courtauld Short Courses 2021.* [Online]. Available at: https://courtauld.ac.uk/learn/short-courses-2021 (Accessed: 9 May 2021).

CREDE (Center for Research on Education, Diversity and Excellence). (n.d.). 'Five Standards of Effective Pedagogy'. University of California. *Learning for Justice.*

[Online]. Available at: https://www.learningforjustice.org/professional-development/five-standards-of-effective-pedagogy (Accessed: 9 May 2021).
Cummins, A. (1993). *Report on the Status of Caribbean Museums and Recommendations for Upgrading Museums to ICOM Standards*. CARICOM/UNDP/UNESCO Caribbean Regional Museum Development Project.
Cummins, A. (1994). 'The "Caribbeanization" of the West Indies: The Museum's Role in the Development of National Identity'. In Kaplan, F.E.S. (ed.) *Museums and the Making of 'Ourselves': The Role of Objects in National Identity*. London and New York: Leicester University Press, pp. 192–220.
Cummins, A. (2012). *Course Description and Rationale for HIST 6720 Museum Development, Management and Curatorship, Post-graduate (Masters) Level* [core course for MA Heritage Studies]. Submitted November 2010.
Cummins, A. (2017). *HIST 6821 Museum Conservation Skills, Evaluation Report*. UWI Open Campus Programme, Semester II, 2016–2017. May. University of the West Indies.
English Heritage. (n.d.). 'Insect Pests found in Historic Houses and Museums' [Online]. Available at: https://www.english-heritage.org.uk/siteassets/home/learn/conservation/science/serpentine/insect-pests-historic-houses-poster.pdf (Accessed: 8 May 2021).
EU-LAC Museums. (n.d.). Home Page. [Online]. Available at: https://eulacmuseums.net/index.php (Accessed: 9 May 2021).
European Commission. (2015). 'The Cultural, Scientific and Social Dimension of EU-LAC Relations'. *Horizon 2020 Framework Programme*. [Online]. Available at: https://ec.europa.eu/info/funding-tenders/opportunities/portal/screen/opportunities/topic-details/int-12-2015 (Accessed: 12 June 2018).
The Getty Conservation Institute. (n.d.). 'Teaching and Learning Resources'. [Online]. Available at: https://www.getty.edu/conservation/publications_resources/teaching/ (Accessed: 9 May 2021).
Gonzalez Mendoza, J. (1994). 'Announcement – UWI Conference'. *Bulletin of the Association of Caribbean Historians*, Puerto Rico # 42 (41).
Historic England (2018). *Salvage & Disaster Recovery Course*. [Online]. Available at: https://www.youtube.com/watch?v=xogyhx05lkI (Accessed: 8 May 2021).
ICCROM. (n.d.). 'Emergency Response'. *International Centre for the Study of the Preservation and Restoration of Cultural Property*. [Online]. Available at: https://www.iccrom.org/themes/disaster-resilient-heritage/first-aid-and-resilience-far/emergency-response (Accessed: 9 May 2021).
ICOM-CC. (2008). Resolution on 'Terminology to characterize the conservation of tangible cultural heritage' adopted New Delhi, https://www.icom-cc.org/en/downloads/icom-cc-resolution-terminology-english
ICOMOS International Training Committee. (2012). Draft 'Guidelines for Education and Training – Practical Instructions for Practitioners'. [Online]. Available at: http://cif.icomos.org/pdf_docs/CIF%20Meetings/Guidelines/ICOMOS_CIF_UK_DRAFT_Guidelines_Practitioners.pdf (Accessed: 3 December 2018).
Inniss, T.A. (2019). 'Conservation in Action: Bajan Archivists Upskilling to Protect our Documentary Heritage'. *Caribbean Heritage Network*. [Online]. Available at: https://www.caribheritage.org/blog/conservation-action-bajan-archivists-upskilling-protect-our-documentary-heritage (Accessed: 7 May 2021).
Lemieux, L. and Schultz, H. (1973). *Report on Caribbean Museums*. Bridgetown Caribbean Conservation Association.
Plenderleith, H. (1998). 'History of Conservation'. *Studies in Conservation*, 43, pp. 129–43.
Margariti, C. (2019). 'New Approaches in Conservation: Has the Essence of the Profession Changed?' In *What is the Essence of Conservation? Materials for a Discussion*. ICOFOM, ICOM-CC, pp. 104–111.

Museum of London. (n.d.). 'Handling Museum Objects'. *Collection care e-learning resources*. Available at: https://www.museumoflondon.org.uk/Resources/e-learning/handling-museum-objects/ (Accessed: 8 May 2021).

National Trust. (2011). *The National Trust Manual of Housekeeping*. London: National Trust.

Organization of American States (OAS). (2016). 'OAS Graduate Course Proposal, Museum Conservation Skills'. *Curricular Enhancement of Professional Heritage Education: Expanding the Socio-Economic Potential of Cultural Heritage in the Caribbean, Part 2: Regional Educational Gaps and Priorities*, OAS, August, p. 14.

Organization of American States (OAS). (2017). *Curricular Enhancement of Professional Heritage Education: Expanding the Socio-Economic Potential of Cultural Heritage in the Caribbean, Part 4*. June

Salkey, Joelle. (2021). *The Need for a Caribbean Code of Ethics for Heritage Conservation Practice.* Caribbean Museums Vol. 4, published online by the Museums Association of the Caribbean. Available at: https://caribbeanmuseums.com/wp-content/uploads/2022/02/Salkey_CaribMuse_2021.pdf

Storey, M., Wood, A. (2020). *Feedback from Barbados National Art Gallery Conservation survey Report* 2020.

Towle, E.L. and Tyson, G.F. (1979). *Towards a Planning Strategy for the Management of Historical/Cultural Resources Critical to Development in the Lesser Antilles. A Concept Paper*. St Thomas, US Virgin Islands: Island Resources Foundation.

UNESCO. (2020). UNESCO-CDEMA Final Report of the Sub-Regional Workshop on Disaster Resilience in the Caribbean's Culture Sector, 11 – 12 March 2020 and Guidelines for the Development of a National Strategy and Plan of Action for Disaster Resilience and Recovery in the Caribbean's Culture Sector, distributed by UNESCO Cluster Office, Kingston Jamaica, 2020.

UWI Cave Hill. (2012). *Programme Summary, New Taught Graduate Programme Proposals – MA Heritage Studies: Programme Revision*. Department of History and Philosophy. University of the West Indies.

UWI Mona. (1992a). 'Minute 59: Archaeology' and. 'Minute 61: Matters from the Departmental Retreat' *Departmental Minutes*, 6 March, p. 4.

UWI Mona. (1992b). 'Minute 71: Postgraduate Matters'. *Departmental Minutes*, 10 April, p. 3.

UWI Mona. (1992c). 'Minute 78: Archaeology'p.1 and 'Minute 84: Postgraduate Matters'. *Departmental Minutes*, 5 June, p. 4.

UWI Mona. (1993a). *Heritage Studies MA Programme*, brochure, University of the West Indies.

UWI Mona. (1993b). 'Minute 46: Postgraduate Matters'. *Departmental Minutes*, 1 February.

UWI Mona. (1993c). *H67K History and Caribbean Heritage – Course Outline*. University of the West Indies.

UWI Mona. (1994). 'Minute 46 i.: Postgraduate – M.A. (Heritage Studies)'. *Departmental Minutes*, 14 January, p. 4.

UWI St. Augustine. (1994). 'Minute 9 i. and ii: M.A. Heritage Studies (Min.23)'. *Interdepartmental Minutes*, 8 July, p. 3.

UWI Open Campus. (2016). 'Heritage Studies Courses'. [Online]. Available at: https://www.open.uwi.edu/programmes/heritage-studies-courses. (Accessed: 9 May 2021).

UWI Open Campus. (2017). Annual Report 2016-2017 - Opening Doors to Life-Changing Learning, p. 12 Available at: https://www.open.uwi.edu/sites/default/files/docs/Open_Campus_Annual-Report-2016-2017.pdf (Accessed: 3 December 2018) .

Victoria and Albert Museum. (n.d.). *V&A Academy*. [Online]. Available at: https://www.vam.ac.uk/whatson/programmes/professional-development (Accessed: 9 May 2021).

Viñas, S.M. (2005). *Contemporary Theory of Conservation*. London: Routledge.

INDEX

Note: *Italic* page numbers refer to figures.

For Product Safety Concerns and Information please contact our EU
representative GPSR@taylorandfrancis.com
Taylor & Francis Verlag GmbH, Kaufingerstraße 24, 80331 München, Germany

www.ingramcontent.com/pod-product-compliance
Ingram Content Group UK Ltd.
Pitfield, Milton Keynes, MK11 3LW, UK
UKHW020155200926
475702UK00016B/2721
9781032288413